Climate Justice and Collective Action

This book develops a theory of climate cooperation designed for concerted action, which emphasises the role and function of collectives in achieving shared climate goals.

In debates on climate change action, research focuses on three major goals: on mitigation, on adaptation, and on transformation. Even though these goals are accepted, concerted action is still difficult to realize. *Climate Justice and Collective Action* provides an analysis of why this is the case and develops a theory of climate cooperation designed to overcome the existing roadblocks. Angela Kallhoff starts with a thorough analysis of failures of collective action in the context of climate change action. Taking inspiration from theories of water cooperation, she then establishes a theory of joint action that reframes climate goals as shared goals and highlights the importance of adhering to principles of fairness. This also includes an exploration of the normative claims working in the background of climate cooperation. Finally, Kallhoff puts forward proposals for a fair allocation of duties to cooperate with respect to climate goals.

This book will be of great interest to students and scholars of climate action, climate justice, environmental sociology and environmental philosophy, and ethics more broadly.

Angela Kallhoff is Professor of Ethics with Special Emphasis of Applied Ethics at the University of Vienna, Austria. She works on ethics, climate ethics and political philosophy. Her books include *Why Democracy Needs Public Goods* (2011), and, as an editor, *Plant Ethics: Concepts and Applications* (2018) and *Nanotechnology: Regulation and Public Discourse* (2019).

Routledge Studies in Climate Justice

Climate justice is a rapidly growing field of critical enquiry which concentrates on the social dimensions of climate change, including the unequal nature of its physical, socio-economic and political impacts and humanity's responses to them. From the stark warnings of the Intergovernmental Panel on Climate Change (IPCC) to the direct action protests of Extinction Rebellion, there is growing interest in the study of the global inequalities of climate change.

Routledge Studies in Climate Justice will comprise monographs and edited collections addressing cutting-edge questions in the growing field of climate justice. The series will include a diverse a range of topics, including climate justice and international development, intersectionality and climate inequality, climate governance and policy, gender and climate change, climate migration and displacement, health and well-being, climate justice activism, pedagogy and participation, and urban climate justice.

Series Editor: Tahseen Jafry is Professor of Climate Justice and Director of The Centre for Climate Justice, Glasgow Caledonian University, UK.

If you are interested in submitting a proposal, please contact Annabelle Harris, Editor for Environment and Sustainability: Annabelle.Harris@tandf. co.uk

Titles in this series include:

Climate Justice and Collective Action
Angela Kallhoff

For more information about this series, please visit: https://www.routledge. com/Routledge-Studies-in-Climate-Justice/book-series/RSCJ

Climate Justice and Collective Action

Angela Kallhoff

LONDON AND NEW YORK

First published 2021
by Routledge
2 Park Square, Milton Park, Abingdon, Oxon OX14 4RN

and by Routledge
52 Vanderbilt Avenue, New York, NY 10017

Routledge is an imprint of the Taylor & Francis Group, an informa business

British Library Cataloguing-in-Publication Data
A catalogue record for this book is available from the British Library

Library of Congress Cataloging-in-Publication Data
Names: Kallhoff, Angela, author.
Title: Climate justice and collective action / Angela Kallhoff.
Description: Abingdon, Oxon ; New York, NY : Routledge, 2021. |
Series: Routledge studies in climate justice | Includes bibliographical
references and index.
Identifiers: LCCN 2020051710 (print) | LCCN 2020051711 (ebook) |
ISBN 9780367753863 (hardback) | ISBN 9781003162322 (ebook)
Subjects: LCSH: Environmental justice. | Climatic changes--Prevention--
Citizen participation.
Classification: LCC GE220 .K35 2021 (print) | LCC GE220 (ebook) |
DDC 363.7/0526--dc23
LC record available at https://lccn.loc.gov/2020051710
LC ebook record available at https://lccn.loc.gov/2020051711

ISBN: 978-0-367-75386-3 (hbk)
ISBN: 978-1-003-16232-2 (ebk)

Typeset in Goudy
by Taylor & Francis Books

Contents

Preface

The international debate on climate change once focused on the effects of emissions of greenhouse gases on our planet and on the dire prospects for future developments. More recently, climate change has been approached from a different angle. Authors in the humanities, in law, in political studies, and in philosophy explore institutional responses to climate change. The shared goal of this endeavour is theories of institutional design in order to contribute to a sharp and significant reduction of greenhouse gas emissions. This book is at the centre of that more recent debate. Simultaneously, it aims at breaking new ground. The book combines insights in theories of climate justice with theoretical approaches to cooperation. This double approach rests on two insights: in order to resolve the climate dilemma, humanity needs to cooperate – ideally on a global scale. But cooperation has no chance without also allocating the burdens of climate action according to principles of justice. Taken together, this book presents a theory of fair climate cooperation. Hopefully, it will spur local, national, and global climate cooperation.

The work on this book has also been inspired by former insights in theories of public goods and the ethics of nature. The atmosphere shares characteristics of a typical public good. Without regimes of appropriation that undermine over-use of the atmosphere as a deposit of greenhouse gas emissions, a solution to the climate dilemma is not a realistic expectation. Yet, the interpretation of climate change as a drama of the commons never fully convinced me. Instead, examples from the cooperative appropriation of water resources provide a better example for the capacities of actors to engage in sharing and using natural resources together. Nevertheless, over-exploitation is a fact. I came closer to a solution to over-consumption when learning from theories of joint action in social philosophy. They provide not only a new model of cooperation; they also explore cooperation on a time axis whose end point is a shared goal. The most important twist in climate change action is a re-interpretation of climate goals as cooperative goals. Overall, the optimism in this book stems from recent insights in shared agency as applied to climate action goals.

Even though focusing on ethics and theories of action, this book is aimed at a broad audience. The models used in this book are not highly theoretical.

Instead, theories of shared agency, as well as the ethics of nature, aim at reconnecting our lives with an appropriate use of shared natural goods.

For realizing this project, many people have been of particular importance. One always receives more support and help than one can acknowledge in developing a book. I am particularly grateful for stimulating debates with authors in climate ethics who shaped my ideas from the very beginning, among them Tracey Skillington, Dale Jamieson, Steve Gardiner, Henry Shue, and Simon Caney. Climate ethicists also joined a lecture series on climate justice at the University of Vienna and contributed to an edited volume on climate justice in a University of Vienna book series. I wish to thank all of them for being so gracious in discussing my proposals on climate cooperation on various occasions. I also wish to thank the faculty of the Climate Centre of the Earth Institute at Columbia University who supported this book with a research scholarship in 2015. Furthermore, my thanks go to the students at the University of Vienna who worked hard on climate justice and on my proposals for climate ethics in the classroom. This book, as well as stimulating conferences on the ethics of nature, was supported by a grant from the Austrian Science Foundation (FWF) on "New Directions in Plant Ethics." Finally, I also wish to thank the people who revised the English text and who improved the manuscript in many ways: Linnea Kralik, Lisa Oberhofer, and Paul. I also wish to thank Routledge for a wonderful collaboration.

Angela Kallhoff
Vienna

Introduction

At its heart, this is a book about cooperation. Recent theoretical insights from social philosophy speak in one voice when addressing joint action. Joint action is a mode of human behaviour in which individuals act together without necessarily melting into a collective actor. The new paradigm of joint action explains how individuals who each have their own plans and ideas nevertheless achieve goals together. In this interpretation, cooperation is about change, about goal-driven interaction, and also about the self-reinforcing and constructive powers people develop together.

When it comes to the question of taking action in response to climate change, theories of joint action address only one side of the coin. The other side is environmental degradation, the threat of losing places for viable life on Planet Earth, and the immense powers climate change threatens to unleash. No one really disagrees with the diagnosis of severe dangers being spurred by the current practices of appropriation of nature by humankind and the growing scarcity of nature's gifts. Nevertheless, a cure to this situation is still not in reach.

This book provides cautious steps into a possible new direction of thought. It proposes not changing the diagnosis, but elaborating a concept of cooperation that theorizes resource-intense interactions of individuals in a new way. It relies on an interpretation of natural resources that has been worked out in theories of NCPRs (natural common-pool resources). According to that interpretation, the foundation of all life on Planet Earth is NCPRs. Rivers, lakes, seas, groundwater reservoirs, land and soil, and the atmosphere comprise the most essential resources for life on earth. Civilizations of all times have made use of these resources. Today this appropriation is overly intense, in total disregard of the "limits to growth"[1] and any concept of a "safe operational zone,"[2] and reckless. Climate change is an immediate consequence of this type of appropriation of nature; it is the effect of an over-exploitation of the atmosphere as a dumping ground for greenhouse gases. It is critical, however, to understand that in the context of the interpretation of misappropriation of nature, climate change is just one particularly severe symptom of a much larger crisis. Humankind is threatened with a potential global "collapse"[3] because NCPRs of every kind are being misappropriated and over-exploited.

Although there is no easy cure for this enormous problem, this book proposes a new path, which might prevent further degradation and help to develop what climate ethicists have claimed for some time now: "a new way of thinking."[4] This proposal moves away from cooperation in a market-style fashion. It also works against a mindset known as "climate isolationism." According to that interpretation, the climate is a global phenomenon that unfolds according to its principles and in isolation from other events on Planet Earth. More specifically, in that interpretation it is causally dependent on emissions sent to the atmosphere, and simultaneously it shapes the living conditions on Planet Earth. Although this study does not at all question the outcomes of climate studies, including the warnings issued by the IPCC[5] and the leading climate institutions, including the Potsdam Institute of Climate Change,[6] it rejects the idea of a causal and linear story. In fact, climate research institutes do the same when explaining self-energizing cycles, tipping points, and interrelations among the development in the seas and in the atmosphere. Moreover, it rejects the idea that the causal impact on the atmosphere via greenhouse gas emissions is the whole story.

To be clear on this point: the most important and urgent action needed to address climate change is rapid and effective mitigation of emissions budgets. Still, this is only one part of the solution. It is important to note that life on earth cannot be divided into pieces and that we have only one atmosphere – that one particularly important NCPR that interacts with water cycles and with the multifaceted and diverse surface of our planet and its vegetation. We have to think about cycles, about self-energizing systems, about detrimental paths in the future and about cures to the imbalance in order to develop the means for the successful adaptation of humankind to a changing planet.

This is, of course, a demanding issue. On the theoretical side of the problem, encouraging recent insights into theories of action and environmental studies have helped develop a new approach. Basically, this approach consists of a combination of theoretical insights from social philosophy on joint action and a re-interpretation of development with respect to NCPRs. Hopeful empirical examples have also been discussed in water ethics. On a theoretical level, a joint appropriation of a water reservoir with a clear vision of the future is comparable to the climate problem. A water reservoir is a common-pool resource. As such, it is endangered by misappropriation, resulting – in my interpretation, not exclusively – from over-exploitation. It also results from uncoordinated use, which threatens the integrity of that resource. A water reservoir offers all sorts of eco-services to humankind. The condition for long-term usability, however, is twofold: the appropriation of eco-services needs to cohere with the conditions under which the resource remains intact. In other words, appropriation needs to be adjusted to the affordances of that resource. Whereas this aspect of joint environmental action has already been thought through in many different ways, including in terms of sustainability, the other side has been obliterated. Appropriation also needs to cohere with a clear-cut interpretation of fairness. The concept of fairness spans a variety of issues, including a fair share of a natural resource, a fair share of burdens in supporting a fair scheme of appropriation, fairness in coping with over-exploitation and environmental hazards, and fairness with respect to people who are vulnerable and cannot afford to contribute to sustain a shared resource. In addition, fairness also includes a fair allocation of second-order responsibilities, which are responsibilities for caring for a system of protection and appropriation to preserve the long-term availability of the resource.

As a progressive species, human beings do not stand still. Human life and the life in collectives, in particular, are always changing. But "development" is also a misused word for the undertakings of ambitious leaders of collectives all over the world: dams are being built, resources are extracted in ever-more efficient ways, industries exploit water resources, and infrastructure expands on a daily basis. This type of development is false in two respects: first, it overlooks earth's limitations and thereby damages Planet Earth; second, it destroys opportunities for building up collectives living in places they have shaped according to the affordances of these particular areas and which are "theirs." It is difficult to explain this type of green collectivity without immediately getting into the trap of conservative thinking and exclusive traditionalism. The vision described by the concept of joint environmental action is actually the complete opposite of some idyllic rural community enjoying the harvest of the fields. The most promising examples of joint environmental action are actually megacities that are starting to reduce their environmental footprint, and that generate community-based life by investing in green spaces – think of Vienna's parks. These are only two examples that work against the prejudice of conservationism.

Here is the baseline of joint environmental action: civilizations of all eras have developed plans for the future; today it is imperative to also take into account the natural affordances in the places where civilizations wish to grow. Development of a progressive species needs to build on available resources in the places in which communities live. The vision of the future needs to include what I have termed "transformation goals." This presumes an idea of how the life of a collective is enhanced by means of a fair and gentle appropriation of *in situ* resources. This view needs to be completed by the overarching perspective of simultaneously reducing emissions and building depositories for greenhouse gases and thereby enhancing the protection of the atmosphere. The atmosphere is one of the global life-sustaining resources of our planet.

The basic claim of a theoretical approach to joint environmental action is not only that cooperation, as outlined in this study, is achievable. Instead, the real win-win scenario means learning the lessons of this theoretical proposal and empowering and connecting joint actors who care for integrated environmental goals. Assuming that individuals connect in order to build up joint actors and that, simultaneously, institutions that already have the capacity to act also work together in favour of environmental goals, then individuals will ultimately be the beneficiaries. They will be protected from environmental hazards; they will enjoy better local environmental conditions. They are also part of a good narrative; and what they do may also be morally right. And at this point, moral theory enters the picture and plays an important role.

Although based on a model of voluntary joint action, this book is also about moral duties. In accordance with proposals about "joint climate duties,"[7] the moral duties discussed in theories of climate action are framed here in a new way. The first level at which ethics enters the picture is the model of joint action itself. Outlined as voluntary and goal-driven cooperation, this approach relies on the insights of Raimo Tuomela, who argues that joint actors share what he calls an "ethos" – a reason-giving resource for acting in a "we-mode" and for reaching goals that individuals want to accomplish together.[8] This is a perfect role model for environmental action: when they identify with these goals, individuals are driven to join others in reaching environmental goals such as "greening the city." The shared ethos has two sides. It provides glue for the people who act together in favour of the realization of a shared goal; it also prevents the problem of people immediately starting to cheat and free-ride. Instead of benefitting from others' engagement, individuals interpret themselves as co-actors. The ethos also gives a first idea of the goal-directedness of joint environmental action. People work together in favour of climate goals not only because they are convinced of those goals; cooperation also depends on those goals being fair.

The application of a model of joint action to joint environmental action has two further implications. Unlike either completely voluntary cooperation or cooperation restricted by claims of sustainability, the goals joint actors develop are goals related to their dependence on a shared natural resource. Joint actors are not free to invent cooperation, nor are they free to detach cooperation from the fact that their activities – in particular, their institutionalized cooperation – is dependent on natural resources and the eco-services they provide. Instead of interpreting this as a severe limitation, the vision of future goals of joint action responds to the need to adjust them to available resources. Joint environmental action is a process of active and cooperative "co-designing with nature."[9] But investment in active cooperation is not voluntary when it comes to fair goals. The theoretical proposal includes the view that a fair contribution to climate goals is obligatory, even morally obligatory.

Yet the explication of climate duties is not straightforward. In order to establish climate duties, it is first necessary to explain the application of the theory of joint environmental action to climate change. Although the atmosphere is a global entity, and although mitigation is the most important goal that needs to be accomplished by the world population, there is more in the discussion of climate change and collective action that needs to be re-examined. Imagine a scenario in which local environmental actors achieve their climate goals: they develop green infrastructure, stop emitting vast amounts of greenhouse gases through new methods of agriculture, and develop climate-friendly housing. And imagine that this not only happens in big cities like Chicago or Toronto, but all over the world. And then imagine that, at the same time, farmers all over the world shift their production to climate-neutral production. Governments start to impose massively high taxes on high-emissions infrastructure. This is not happening today. One problem in need of resolution is the question of moral obligations. A theory of joint climate action thus seeks to answer the question of whether people have a duty, or duties, to cooperate in favour of climate goals – and under which conditions they hold. If duties can be reasoned, there will also be ways to defend institutions that support compliance with climate duties. This book develops a positive answer to the issue of moral duties. And it discusses the question of who the addressees of duties actually are.

The last part of the book addresses these questions. It still works on the foundation of the theoretical model of joint climate action. It is not difficult to adapt joint climate action to various actors, including organizations and representative institutions. The way is prepared by theories that investigate the conditions under which collectives – both unstructured and structured – count as true actors. The more difficult problem is that climate duties appear to be useless unless they are framed in cooperative schemes that give some guarantee of cooperation from other actors. The question "what if I am the only cooperator?" needs to be taken seriously. In addition, theories of moral obligation have primarily been discussed against the background of mutual obligations among individuals. Climate duties not only address obligations in a new – that is, global and long-term – perspective. In my view, they also need to be framed as obligations to fulfil distinct environmental goals, not exclusively as obligations to rescue climate victims or to not over-exploit life-sustaining resources.

The proposal for climate duties in this book works on the premise that the only way to achieve climate goals is cooperation. In addition, it takes seriously the claim that unless mitigation goals and adaptation goals are fulfilled, the prospects are dire. It interprets these prospects to bluntly mean that "non-cooperation is suicide." But instead of addressing climate ethics as an emergency ethics, the book works on a new balance between joint environmental action as valuable enterprise and as morally obligatory. At the end of the book I also propose methods of enhancing cooperation, as well as a second-order duty of political institutions to remove barriers of cooperation and to empower joint environmental actors by all possible means. I additionally recommend a new international institutional setting, in which climate sinners actually receive just penalties. Institutions and people who jeopardize the most basic conditions for a decent life on earth must be held liable for killing and for mass species extinction. In the narrative of this book, they also need to be interpreted as *defectors* who sabotage the expectations and goals of the cooperators.

By arguing climate duties from scratch, this approach also tackles certain doctrines that have become common ground: the transition to green technologies is costly and needs to be supported by subsidies; economics will solve the problem by its own means; environmental protection is undesirable and hampers development; tragedies of the commons cannot be resolved, or can only be resolved by means of either perfect markets or enclosure. My main goal, however, has not been to offer criticism of those standard problems. Instead, the book provides a narrative that works against fatalism and in favour of action. As part of a massive panorama of different joint actors, every single joint effort makes a difference. But joint actors need to be empowered; the most important empowerment is, of course, to end non-cooperative and egoistic politics and the high-emitting industries. In the context of this study, they are not only defectors, they are also sabotaging the efforts of cooperators. What they do is morally wrong. Hopefully, they will become ever more marginalized by a united group of environmental cooperators.

Plan of the book

This introduction has introduced the argument and offered a clear idea of the broad context of this exploration. The remainder of the book is dedicated to arguing the claims thoroughly. It is composed of ten chapters, which I now introduce.

Chapter 1 explains the theoretical frame that the book's exploration of joint climate action presupposes. It starts with a sketch of the existing debates on climate justice and introduces the concept of joint agency. It also explains the ways a theory of joint climate action rejects some common reservations against a close tie between climate change and action. Finally, it highlights the normative implications of the proposed interpretation of climate change action.

The subject of Chapter 2 is the "tragedy of the commons." One of the most striking descriptions of climate change is that it is just another "tragedy of the commons," one which is seemingly irresolvable. Climate ethicists allude to the observation of Garrett Hardin that goods that are common-pool resources, and are therefore not equipped with natural access systems, are prone to overuse, and draw parallels and differences with respect to the over-exploitation of the atmosphere. This chapter claims that although it is correct to interpret the atmosphere as a common-pool resource, a solution to that problem remains unproposed. More specifically, it connects the "causal tragedy" of exhaustion and over-exploitation with another tragedy, which is the human incapacity to cooperate in favour of systems that protect the atmosphere from over-exploitation.

Chapter 3 continues the preceding chapter's discussion of the "tragedy of the commons," but highlights another twist. In addition to causal depletion, climate change actions suffer from an additional drawback. A range of collective action problems are attached to cooperation once it gets started, including abusive behaviour and free-riding. This chapter delves deeper into the observation that social dilemmas create a real problem in climate cooperation, but argues that they result to some degree from a false interpretation of cooperation. Instead of focusing on voluntary exchange processes, climate theories can work with theories of collective action that follow a different theoretical path, in that they are goal-oriented, rest on joint contributions in favour of a shared goal, and help to overcome dilemmas of collective action.

Chapter 4 explores theories of "joint agency," which has been a focus of research in social philosophy for some decades. More specifically, ethicists in the current debate (Tuomela, Bratman, Gilbert) have developed an account of working together" and "acting together" that breaks with the premise of theoretical individualism, stating that action is primarily an undertaking of individuals. "Acting together" is not only an everyday phenomenon, but can also be analysed with respect to the key notions in theories of action. The outcome is a joint action approach that explains how groups of people succeed in achieving a goal that they want to achieve together. This chapter also enquires into the conditions under which that proposal can be applied to "joint environmental action." To explain this application, it draws from examples of water cooperation.

Chapter 5 discusses the concept of an ethos in joint environmental action. Although the details are still being debated, many researchers share the claim that joint action includes a normative element. People who act in favour of accomplishing a shared goal are not only willing to do so; in some particular situations, they also feel obliged not to cheat or to undermine the achievement of a shared goal. I use this initial idea to support the claim that environmental goals are conceived as urgent goals, and as goals that need to be brought about by a joint effort. However, this alone does not suffice. In addition, Chapter 5 gives elements of an ethos that includes normative propositions that people who act in favour of environmental goals are supposed to share.

Chapter 6 applies these observations to climate goals. One of the claims of this book is that if climate goals are to serve as joint action goals, they need to be reframed. Quantitatively framed goals, such as reduction proposals, are still important, but they are only part of a bigger picture. Climate goals include "green goals," which not only support the protection of the climate but also have positive side effects on local communities, on the water supply, and so on. In sum, this chapter includes a debate on the overarching goals of mitigation and adaptation, but also maintains the theoretical idea that climate goals are best defined as goals that people really want to bring about together.

Up to this point, the theoretical proposal of "joint climate action" has been restricted to the activities of groups of people who tend to invest in climate goals and environmental goals more generally. The remainder of the book discusses two more questions: are individuals or groups of people, possibly even organizations and other collectives, morally obliged to contribute to climate goals? And under which conditions is that the case?

Chapter 7 defines the first step. It provides the central argument for individuals having climate duties. It defends the claim that in a situation in which non-cooperation threatens to prevent the achievement of goals that are not only fair, but are also critical, the only alternative is either action in favour of those goals or inaction (if not defection). In that scenario, cooperation is obligatory, yet the conditions under which "climate duties" are valid still need to be rendered more precise. Presenting those elements is the task of the remaining chapters.

Chapter 8 takes up a discussion of climate duties as "joint action duties." It explores the view that climate duties are justified, but that they need to be interpreted as duties of individuals to collectivize in order to really summon the capacities needed to bring about environmental change. This argument is tricky, and ultimately flawed. It does not solve the problem that individuals now appear to have two duties: they first have to collectivize, and then to perform duty-based actions as a joint actor. Rather than take that deviation – which is in my view also maintained because of a capacity condition regarding joint actors – it is reasonable to discuss the obligations of groups and collectives straight away.

Chapter 9 discusses conditions under which the general claim that individuals are obliged to contribute to climate goals is also applicable to groups and collectives. The main argument is that there is no general answer to the question of whether obligations can be established. Instead, it is necessary to discuss "fairness in accumulative goals." In order to achieve climate goals, responsibilities and goal-related burdens need to be allocated to actors according to principles of fairness. The first step is an argument about the conditions under which collective actors can be held responsible at all. The second step is an exploration of principles of fairness in allocating burdens to various types of actors. As an outcome, the insufficient distinction between political actors (and governments in particular) and private actors breaks down. What counts instead is a distinction along the lines of capacity to act, ability to pay, and liability due to former activities.

Chapter 10 takes the lessons from the insights into this study and puts them to work. More specifically, it puts forth conclusions on the institutional and political level. The most important insight is that in addition to institutions that translate climate duties into law, and in addition to claims for leadership, coalitions of the willing, and international institutions, another exigency also surfaces. We – the world community – have to build institutions on several levels of agency that favour the type of cooperation outlined here. Above all, governments have to take responsibility for institutions that enable the citizenry and a variety of collective actors to contribute to climate goals according to the prerequisites defined in this book.

Before getting started, another specific trait of the conversations and debates in this book has to be explained. The debate into which this book jumps is rather complex. Climate ethics is not just another ethical theory. Instead, it needs to address human action, in particular collective action; it needs to take into account research on climate change; and it has to address particularly difficult issues of responsibility, of duties to act and not to act, of the rights of individuals to profit from externalities that are detrimental to other individuals' lives. In sum, the theoretical approaches in climate ethics are far more complex than moral philosophy – that does not delve into a mix of moral claims, methods of a new theoretical framing, and a reflection on the meaning of natural facts.

Simultaneously, all the problems have an immediate impact on our common life and on everyday life. This book is an intellectual enterprise. But it is also a debate about a very pressing issue of our day. It is important to get clarity about possible modes of action with respect to climate change, about ways to act together, about moral claims and even duties.

In order to disentangle the complexity, I have invented two characters, Cindy and Bert, who wish to understand the complexity of the issues raised in each chapter. They discuss the issues at the beginning of each chapter. They explain the research questions in an accessible way. By doing so, they also summarize the most important outcomes of the previous chapter. In particular, they also help to highlight the red thread that runs through this book.

Before starting the chapters, Cindy and Bert, who might be regarded as students who are interested in the climate ethics debate, discuss the issues that will be raised in the chapters. It is astonishing to see that this method really works. Obviously, it is not possible to trace each single step of the debate in the chapters. But it is possible to summarize the leading questions, to highlight the most important theoretical pathways into new directions, and also to rediscover the red thread chapter by chapter. I hope this narrative helps the reader to detect the most important insights. And it can be helpful for selecting chapters of interest.

They start their conversation with questions (Q) and answers (A):

Q: What is this book about?

A: This book presents a proposal to solve a very complex problem regarding climate action. In climate science, debates have focused on three major goals: mitigation (the reduction of greenhouse gas emissions), adaptation (using social instruments to cope with changes in the natural environment), and transformation (the shift from carbon-intensive production and energy supplies to green technologies). Even though these goals are broadly accepted, concerted action is not being taken. This book gives an analysis of why this is the case, and it develops a theory of climate cooperation in order to overcome this standstill.

Q: What are the main challenges?

A: In climate philosophy, some authors argue that the biggest challenges are collective action problems. Even though some actors are now willing to contribute to the goals of mitigation, transformation, and adaptation in response to the climate crisis, these efforts to date have been largely ineffective. One reason for this is that the good effects from the efforts of some are offset by the actions of "free-riders." For example, on a local level, we see some people using their bicycles for transportation to reduce carbon emissions while at the same time others undermine these efforts by continuing to drive their SUVs.

Q: Is there any solution to the problem of non-compliance?

A: Frankly, there is no simple solution. Immediate and decisive enforcement of environmentally friendly behaviour is neither possible nor desirable. As this book argues, the behaviour of free-riders is only one aspect of the problem. Essentially, the problem has two different aspects. The first is *physical*: the atmosphere has been overused as a waste dump for greenhouse gases, and this situation leads to climate change and its problematic causal effects. This side of the problem can only be mitigated – not resolved – but mitigation is necessary to prevent worst-case scenarios. The physical effects of climate change have been conceptualized as a typical "tragedy of the commons." This terminology, disseminated by Hardin in the late 1960s, describes the behaviour of individuals acting in their own self-interest to deplete shared resources with no entrance barriers, such as the atmosphere. The other layer of the problem is *social*. Addressing the climate crisis requires joint action, but cooperation is often thwarted in favour of competing interests and self-serving behaviours.

Q: Does this mean that a theory of joint action could resolve the problems associated with climate change?

A: This book argues that while it is desirable to solve the problems that prevent individuals from contributing to collective action, more is needed. As a minimum first step, addressing the crisis will require global and multi-faceted efforts towards joint climate action. This book delivers a theory of joint climate action that would entail the cooperation of governments, collective actors, as well as individuals to achieve shared climate goals.

Q: What are the arguments for this theoretical model?

A: The most fundamental argument in favour of this model of cooperation is that we are living in a collective age. Whereas many theorists take this idea for granted, it has never been directly applied to theories of climate change action. This book addresses this gap in the research. But any model of action must unite the physical and social layers of the climate change problem to (a) achieve physical goals and (b) explain how cooperation with respect to these goals works. Research on joint action and social action theory, such as Bratman's work on shared cooperative activity, makes it clear that joint action – that is both concerted and goal-driven and that plans towards common goals that can only be achieved jointly – does not require identical motivations or interests on the part of the individuals taking part in the common action. In the book, I try to demonstrate that climate goals can be conceived as "joint action goals," and I investigate the conditions under which we can treat them as such. These conditions are necessary but not sufficient for explaining joint climate action.

Q: What else is needed?

A: Two more theoretical insights are important. The first is that goal-driven action calls not only for a desirable shared goal, but also for decisions about fairly allocating the burdens that come with goal-driven action. Attaining climate goals is expensive, and even when individuals accept the urgency of the problems they seek to solve, they may not be willing to act unless a fair scheme for allocating burdens is in place. This appears to be one of the lessons of the long-standing debates over climate justice. Simon Caney has argued convincingly that a hybrid approach is needed to allocate burdens of climate change action according to principles of justice. The book includes a debate on climate justice, but restricts it to the allocation of burdens for climate change action. I do not share the view that we need to look back and define today's burdens according to pathways in the past. Even authors (e.g., Henry Shue) who used to emphasize historical justice – and for good reason – argue today that looking backwards can interfere with developing solutions today. While it is important to take historic burdens into account, other factors play a more significant role in thinking through joint actions today. These factors include the capacity principle of allocating burdens according to the abilities of actors and the principles of role ethics. Some actors, such as nation-states, have a duty to shoulder more of the burden for the sake of serving the common good.

Q: This means that the model of social action includes elements of a theory of justice?

A: Yes. But since the problem has both social and physical aspects, we also have to make sure that the physical side is taken into account. On this point, the book turns to lessons from water ethics. At first glance, the atmosphere appears to be a common-pool resource comparable to water: both types of resources suffer from the lack of natural entrance barriers, and both can be overused and misused as waste dumps. But water resources have also played another role. As theories of "water cooperation" have shown, rivers and lakes have also served as sites of cooperation. In fact, common-pool resources – even when non-renewable – play an important role in human activity and offer ecosystemic value. In some cases, though, the integrity of that resource is threatened. Authors such as Elinor Ostrom have argued that societies are sometimes able to invent regulatory frameworks to help restore the integrity of a resource while also achieving good outcomes in profit-seeking behaviour. However, the comparison between water resources and the atmosphere is not exact: it is almost impossible to restore stable and pre-industrial levels of greenhouse gases in the atmosphere.

In the realm of so-called integrated water management systems, some researchers have developed a model of cooperation that does not assume the restoration of a previous state but rather uses the shared resource as a basis for "active co-designing with nature." Civilizations have always impacted natural resources and exploited water and other necessary resources. But a wise planner does not destroy these resources, but rather "actively co-designs with nature." This requires the formulation of shared future goals. Climate stability is itself not a reasonable goal because it has already been forfeited, but the goals that have been set by the IPCC are reasonable and can be achieved through concerted efforts.

Q: Does this complete the model of climate cooperation?

A: It is almost complete. The most important point is to frame climate action as concerted and goal-driven action. The goals need to be conceived as social-cum-physical changes that actors are willing to envision in the context of fair burden sharing.

Q: What does this theoretical proposal mean with respect to governments, industrial firms, and individuals?

A: This is an important question that can only be answered by introducing two more aspects of the theory. One aspect draws on the relationship between concrete group actors and specific climate goals; the other is part of a normative theory of collective responsibility. Let me begin with the first aspect relating to specific groups actors.

The model of joint action leaves plenty of space to apply it to different types of groups and actors. Theories of joint agency are not restricted to specific groups or to small groups only. Collectives engaged in joint action do not even need to be structured in any particular way because they have institutions for decision-making. The theoretical model is open to all sorts of collective actors with a shared goal. Another presupposition of theories of collective action is that collective actors can be conceived as true actors and each group and collective actor is capable of realizing a joint action in favour of a shared goal. I argue that in order to initiate climate action, it is reasonable to think about redefining climate goals with respect to specific capacities of distinct groups. And since the most desirable form of collective action is often voluntary, I also argue in favour of letting groups of all sorts define their local and specific climate goals – as long as they cohere with scientific outcomes. Greening a city might be a good project for the town's citizenry, whereas regulating traffic is an appropriate project for the city government. Switching to green energy sources is a good way for corporations to contribute to a climate goal, whereas storing solar energy for a quarter of a city's population might be a fitting project for a community of homeowners.

Q: This first part of the model for joint climate action helps to illustrate that climate cooperation really is comprehensive: it is suitable for different types of group actors and allows them to stick to their goals. But what is the second aspect?

A: So far, the arguments have all been about voluntary climate cooperation. The reason for this is that voluntary self-organization is the most rapid kind of change that can be realized. Institutions are slow, and governments are even slower. Yet this does not mean that climate cooperation needs to be conceived as purely voluntary action on a theoretical level. Quite the contrary – climate cooperation is obligatory. This book includes two chapters on climate duties and all of the difficulties that come with collective duties, whose outcomes are only convincing when taken as a whole. Philosophers have already gone a long way towards studying "duty gaps," "duties of collectives" as opposed to "duties of collectivization," duties of individuals that are not necessarily unilateral duties, and so forth. This literature must be taken into account in order to understand duty as it relates to the question of who is obliged to act on climate goals.

Q: What are these outcomes, and who has duties when it comes to climate goals?

A: I think I can demonstrate convincingly that climate goals are collective goals and that it is reasonable and desirable to act in accordance with what I call moderate eco-centrism. This does not mean that valuing nature should serve as the utmost rationale for action, but it supports the view that reasonable goals are consistent with "active co-designing with nature" and not with utilizing nature without limits. Duties come into play when the achievement of collective social-cum-physical goals rests on coordinated efforts that can be destroyed by non-compliance. Defectors and actors who do not contribute to the goals place the joint efforts to address the climate crisis at risk. And this is unfair and morally wrong.

In arguing that individuals are subject to climate duties, I deviate from the positions of authors who regard the foremost duty as not harming those they refer to as climate victims. But the precise nature of these obligations as they apply both collectively and individually requires a fuller account. John Broome is of course right to claim that the duties of governments in responding to the climate crisis are particularly important, but there is no reason to exempt individuals and other actors from their responsibility to cooperate in favour of helping to achieve climate goals. One condition for any proposed course of action is that for it to be appropriate, it should be realistic. This practical framework takes into account the potential significance of a single action in service of the shared goals of a joint action. The actions of a single person who has given up eating meat might not make a difference in reducing greenhouse gases, and as long as the individual's action is ineffective, it is also difficult to argue that a duty exists to eat less meat. But once there is a movement of people who abstain from eating meat, a framework of collective action has been created that clarifies the duty of the rest of humanity not to undermine the efforts of the many who try to realize a desirable and overall reasonable goal.

Q: Does this mean that Bill Wringe is right to argue that the world community has a duty to act now?

A: Yes, of course he is right. But his argument is far too general, as other authors have already pointed out. Moreover, arguments in favour of climate duties do not have to rest on the assumption that they stem from the duty to avoid engaging in or contributing to collective harm. Instead, we can proceed from the insight that the efforts of joint actors are undermined both by those who do not contribute to existing cooperative schemes and by those who undermine these schemes. Note that this argument does not work for any goal, but it is particularly relevant in the case of reasonable and shared climate goals. Note also that this argument does not rely on framing climate goals as emergency goals. The reason for this is that it is important to integrate climate goals into broader agendas of collective actors of all sorts. According to role ethics, some actors have definitive and special responsibilities – not only to comply with shared goals, but also to initiate concerted action and to provide normative guidelines for improving cooperation. Governments as well as supranational political institutions are examples of actors that bear these duties. Yet, role ethics also prescribes certain kinds of behaviours for private corporations and individual citizens: regarding the capacity principle, wealthy countries and wealthy individuals carry heavier burdens than poor countries. Overall, it makes sense to avoid predetermined fair allocations of duties and re-examine our assumptions about collective and individual duties according to various principles of justice. This approach would help address the notorious problems of shifting burdens and scapegoating.

Q: One final question: is climate cooperation, even in this detailed picture, ever a realistic assumption?

A: I think it is. We are currently experiencing a moment of great momentum among all sorts of actors who are now engaging with this issue. What still needs to be done on a practical level is twofold: first, the time has come to begin "moralizing" on climate issues, albeit in a productive and thoughtful manner. Climate action should be considered obligatory, and efforts to undermine (e.g., through thoughtless consumption and wilful waste) the actions of those actively seeking to mitigate the climate crisis should be called out. Decades ago, Henry Shue argued that there was a fine line between luxury emissions and subsistence emissions. And even though other authors such as Simon Caney have added further distinctions to Shue's – for example by arguing that emissions are part of a portfolio of behaviours that are not easily changed and that individuals have no obligation to atone for being born in a rich country – I would like to use this distinction to further develop my argument concerning climate duties. There is no right to spoil the efforts of climate-actors by action that is luxury behaviour. We need ideas of where to draw this line; and I think at some point this line also needs to be translated into law. Reasoning this argument through thoroughly would require another book with much to say about institutions and control mechanisms in economic systems. The outcome would likely be a picture as complex as climate cooperation itself.

Second, at the end of the book I propose a scheme to single out various types of "defectors." I think this is particularly important. Noncompliance with climate goals has several causes, including disinformation and a lack of alternatives. However, in some cases noncompliance with climate policy may relate to deep-seated beliefs or interests, as in the case of the coal industry. Here again, moralizing may not be an appropriate response. "Active co-designing with nature" is meant to replace certain practices that may be harmful or outdated, but in general those who continue to produce carbon-intensive energy are not setting out to destroy nature. Rather, they have not yet found another means of achieving their legitimate goals of generating profit and a high GDP. Instead of condemning those who fail to respond to the crisis, a better first step for philosophers is to investigate the reasons and motives behind their decisions. Only then can the problems of noncompliance or free-riding instead of engaging in climate cooperation be addressed in ways that have the potential to become enforceable.

Q: Is there reason for optimism?

A: Yes, certainly. And we should not forget that there is still time to act, change current practices, introduce new technologies, lead more climate-friendly lives, and make institutional changes.

Notes

1 This term was coined in the Brundtland report. See World Commission on Environment and Development, ed., *Our Common Future* (Oxford/New York: Oxford University Press, 1987).

2 Rockström et al. defend the view that there are "safe operational zones" for basic natural life cycles, including the climate. They propose to keep human action within this space in order not to endanger the integrity of natural resources. See Johan Rockström et al., "Planetary Boundaries: Exploring the Safe Operating Space for Humanity," *Ecology and Society* 14, no. 2 (2009).

3 See Jared M. Diamond and Christopher Murney, *Collapse: How Societies Choose to Fail or Succeed* (New York: Penguin Audio, 2004).

4 Gernot Wagner and Martin L. Weitzman, *Climate Shock. The Economic Consequences of a Hotter Planet* (Princeton, NJ/Oxford: Princeton University Press, 2015), 17.

5 IPCC, "AR6 Synthesis Report," accessed April 3, 2019, www.ipcc.ch/report/ sixth-assessment-report-cycle/.

6 The Potsdam Institute of Climate Impact Research, accessed April 3, 2019, www. pik-potsdam.de/en.

7 Concerning the advanced debate on climate duties, see the arguments in Chapters 7–9 of this book.

8 See Raimo Tuomela, *Social Ontology: Collective Intentionality and Group Agents* (New York: Oxford University Press, 2013); Raimo Tuomela, *The Philosophy of Sociality: The Shared Point of View* (Oxford/New York: Oxford University Press, 2007).

9 This term is derived from arguments in integrated water management. See Jerome Delli Priscoli and Aaron T. Wolf, *Managing and Transforming Water Conflicts* (Cambridge/New York: Cambridge University Press, 2009), 121.

1 Theoretical frame

Bert: Hi Cindy. Let's talk about climate change today. I've heard that climate ethics is now on the rise. Can you explain to me what climate ethics does?

Cindy: Sure, I'd like to do that. Climate change is a fact, but it's also man-made. Philosophers think that this insight has to bear on the actions we now have to take in order to prevent an unfettered raising of temperatures on Planet Earth. But justice is difficult to argue, because justice means at its heart that everyone gets the same. What should this be in times of climate change?

Bert: That's interesting. But do you think there is an alternative to thinking about justice? I think that it's unfair that some people cause the high emissions of climate gases that cause climate change and that others just go along as usual.

Cindy: You're right. Many think that justice is an issue. But it has limits. In particular, justice cannot be realized by single actors. Instead, we need collective action. But in order to understand where collective action is useful and possibly even obligatory, it's also necessary to think once again about the spaces for collective action. And this requires an interpretation of climate change.

Bert: Sure, I understand this. And I know that there is a big debate about how to frame climate change. Some argue that climate change needs to be regarded as the effect of a certain amount of greenhouse gases; each rise in greenhouse gases also causes a rise in temperatures. But other scientists have argued that conceiving of climate change as an isolated phenomenon is wrong.

Cindy: Right. And there are more so-called "dogmas" that need to be addressed. It's also necessary to think about the possibility of quantifying climate change. Some authors have recently argued that it's not so much about the quantities, but about qualities of nature that are affected. Moreover, almost everybody thinks that individuals cannot change anything. The problem is huge, and individual actions won't make a difference.

Once these problems have been clarified, it's obviously the case that people should work together in order to achieve better outcomes regarding climate change action. But the arguments aren't straightforward. Instead, moral philosophers also have to explain how to address the problem by means of moral arguments. Let's hear what philosophers have to say!

Climate change is clearly an ethical challenge. This vast transformation is undermining the potential for the fair distribution of life-sustaining goods, jeopardizing the prospects of current living generations, imposing severe risks on groups of people who have not contributed to the causes of climate change, and provoking deep disagreement about lifestyles involving high emissions of greenhouse gases. Although pioneering work towards developing an ethics of climate change dates back to the 1980s, a collective effort has gained momentum only recently.[1] The fundamental and recurrent theme of this debate is climate justice.[2] This book carries the debate forward by offering a new theoretical perspective. It argues that the best possible way to achieve the urgent and shared goals of climate change mitigation is to adopt a mode of action that has been thought through in theories of joint action – action undertaken by many different actors to achieve a shared goal.[3] The theoretical prerequisites for carrying out this type of cooperation have been studied in detail. The aim of this study is to build on this scholarly work to construct a theory of joint agency for climate goals.

At its core, this study presents a mode of cooperation that coheres with schemes for the appropriation of a shared natural resource, which (a) keeps the resource intact or restores its integrity and (b) accomplishes what I term "fair goals" by means of joint action. Fair goals in appropriating a shared natural resource are subject to two groups of principles. The first group defines a "fair share" of a resource in terms of the appropriation of the services that the resource provides. Here, the debate on climate justice, and environmental justice more broadly, is particularly helpful. The second group requires a fair amount of proactive responsibility from various actors for bringing about the shared goal. This second group is about second-order responsibility.[4] To argue for second-order responsibility, it is not enough to discuss the existing responsibilities of actors. Several intricate philosophical problems must also be resolved. To allocate accountability for climate goals according to moral principles, it is necessary to discuss intricate philosophical problems anew. This includes research on the criteria for addressing collectives as responsible collective actors,[5] on whether duties of collectives can be established in the case of climate change,[6] and on the justification of climate duties in the first place.[7] In order to clarify the meaning of the terms, I use the term *collective actor* to designate entities having the full potential to act and simultaneously composed of many subjects. Unless a more specific meaning is introduced, the concept of *duty* is used as a synonym for any justified moral obligation.

When introducing the theoretical framework for climate duties, it is important to address a common misunderstanding. Although this approach is based on a theory of joint action, defined as the voluntary action of individuals who are accomplishing a shared goal together, the theory does not stop there. Instead, after having outlined why this model of cooperation is particularly well suited to overcome the often-debated *tragedies of the commons*[8] and its particularly severe exemplification in the case of climate change,[9] this discussion is primarily needed for exploring the theoretical prerequisites for collective action as related to environmental goals, and climate goals in particular. The second part of the study is dedicated to the transformation of this model of cooperation into a valid normative *climate duties* approach.

This chapter starts with a brief sketch of the climate justice debate. Section 1.2 explains how "joint action," as a theoretical model of cooperation, is helpful in the context of climate goals and environmental action. Section 1.3 argues against *climate dogmas* as a recurrent theme in debates on climate change. I use the term *dogma* to explain that certain doctrines are not only widely shared assumptions, they are specifically *theoretical* assumptions that – although broadly accepted – need to be criticized for theoretical reasons. Section 1.4 explains that this approach to climate ethics raises new and distinct moral questions. It also defines new spaces in which ethical reasoning is obligatory. Section 1.5 explains the intricate philosophical problems that need to be resolved in the course of this book.

1.1 Climate justice and its limits

The concept of "justice" and "theories of justice" abounds in works of philosophy that take theoretical approaches to climate change.[10] In that context, justice is primarily framed as *distributive* justice. From the beginning of the debate on climate justice, three different distributive scenarios have gained particular attention. First, distributive fairness relates to the distribution of the right to access a scarce resource – namely, the remaining space in the atmosphere for greenhouse gas emissions. This approach has also been reduced to a debate on the fairness of allocating emission rights to various actors.[11] Second, the intergenerational implications of the distributions of burdens and of benefits that correlate with the use of the atmosphere as a depository for greenhouse gases have been discussed. This debate includes approaches to historical responsibility for the causation of damage resulting from high emission levels in the past.[12] It also includes the discussion of responsibility regarding the needs of future generations.[13] Third, the debate on climate justice also addresses duties of assistance, especially the necessary and global assistance for climate victims in processes of adaptation and the assistance in developing green technologies.[14]

In sum, the debate on climate justice has reached an advanced stage. The concept of justice has been applied successfully to a scenario in which the over-exploitation of a shared natural resource, the atmosphere, needs to be addressed through a fair appropriation of what remains and through a fair allocation of burdens resulting from the consequences of climate change. But this theoretical approach still faces severe limitations. Justice among individuals in most spheres of life differs significantly from justice among individuals in relation to natural resources. In the second case, individuals not only benefit from natural resources in a variety of ways, but also need to share them. More specifically, the atmosphere as a shared natural resource needs to be interpreted as a natural good that is not fully renewable. In sum, natural common goods have been regarded not only as posing a particular set of justice challenges, but also as goods that need a different framework for justice than other goods.

Overall, efforts to address climate justice face three important challenges. First, justice is a key concept in every normative approach to the climate problem. Nevertheless, it needs to be reshaped in order to address justice as related to a natural common-pool resource (NCPR).[15] Second, the limitations of a justice approach result from the fact that too much emphasis has been placed on the fair distribution of the remaining space in the atmosphere as a dumping ground for greenhouse gases. Since this dumping ground has been overused to the point that there is no space left, the idea that climate justice is about the distribution of the remaining space in that dumping ground is flawed. Instead, as Gernot Wagner and Martin L. Weitzman state: "Climate change requires an entirely new way of thinking."[16] Although the prospects vary in relation to the theoretical assessment of climate change, emissions need to be reduced quickly and enormously to prevent a climate catastrophe. Third, although justice is a crucial element in each approach to climate ethics, the area this concept applies to needs to be discussed anew. Justice is a sword and can divide people. But when used in a forward-looking way and when addressing not only fair rules for the appropriation of a shared resource but also fair burdens in shaping future agendas for the *sharing* of a shared resource, it is particularly helpful. In the remainder of this section, I focus on this last aspect of the question. But first, I address the problem of why cooperation with natural goods is likely to fail.

1.2 Joint action and the dramas of collective action

Discussions of climate ethics have recognized that climate change is causing deep normative challenges. The acknowledgement of problems of collective action is also a recurrent theme. In *Reason in a Dark Time*, Dale Jamieson notes that "climate change can be seen as presenting us with the largest collective action problem that humanity has ever faced, one that has both intra- and intergenerational dimensions."[17] He also states that it is a distinguishing feature of the problem of climate change "that it is the world's largest and most complex collective action problem."[18] The problems of collective action are not only aggravated by an extension in time, but also by irregular causal patterns. According to Jamieson, the complexity results from "the high degree of connectivity in the climate system, the non-linear nature of many of the relationships, threshold effects, and buffers that exist in the system."[19] This diagnosis is in line with similar arguments from Stephen M. Gardiner, who adds that the complexity of collective action problems as related to the appropriation of the atmosphere also result from an intergenerational problem. Gardiner argues that individuals do not take into account the fact that future generations actually need to be counted as beneficiaries. To do justice to future generations, they would have to be involved in collective action problems and their solutions.[20] In addition, Wagner and Weitzman state that secondary collective action problems are also severe. The climate is not the only thing the failure to resolve collective action problems puts at risk. But once cooperative schemes are in place, *free-drivers* and *free-riders* spoil successful outcomes.[21]

Collective action problems arise when actors cooperate in a way that prevents the achievement of a social goal, although this goal is highly desirable and favourable to all participants. More specifically, actors forfeit a social goal, even when they act on premises that are rational. The parallel to climate change is straightforward: a stable climate system – at least one that does not contribute to catastrophic consequences for humankind – is desirable for every single person living on Planet Earth. But instead of achieving the goal of climate stability, the world population jeopardizes it. The result is not irrational in terms of individually stupid behaviour. Instead, each self-interested actor is right in not investing in climate protection, because climate protection is not only costly, but a wasted investment as long as it is undermined by other self-interested actors. As a consequence, climate stability cannot be achieved.

In order to explore this interpretation of failures in achieving climate goals, it is necessary to take a closer look at the meaning of "cooperation." In the explanation of collective action problems, *cooperation* already has three different meanings: first, cooperation in the context of climate means to contribute effectively to policies and strategies that mitigate greenhouse gases. An actor who does not choose to deceive, but instead to cooperate, invests in actions that support shared climate goals. More specifically, she tries to adjust cooperative behaviour in such a way that overarching goals such as mitigation are not forfeited.

Another meaning of cooperation in this context is compliance with already existing strategies that have been proven helpful in mitigating the causes of climate change. Non-cooperators not only fail to invest in climate-friendly activities but also proactively choose to invest in climate-intense goods and perform acts that undermine strategies that would be successful if supported by many actors. This second interpretation assumes knowledge not only about climate change, but also about a successful scheme of cooperation. In this interpretation, cooperation is different from a market-like exchange of assets, but rather means compliance with norms that support climate goals.[22]

Both of these meanings of cooperation need to be distinguished from a third type of cooperation: a coordinated effort among various individuals to achieve a shared goal. In this interpretation, and as related to climate change, cooperation is investment in strategies and actions that contribute to joint goals that can only be accomplished by joint efforts. This last type of cooperation assumes that the actors are aware of a goal and are willing to invest in that goal. Goal-driven cooperation is best explained in a model of "joint action." More specifically, this third mode of cooperation can imply the second type of cooperation, when the mode of co-agency is rendered explicit and the individual share of each actor is settled.

Although these initial insights still need to be translated into the various modes of cooperation involved in environmental action, it is important to understand, right from the beginning, the logics of cooperation – and how they change when it comes to joint action. In particular – although this is related to the diagnosis of collective action problems – the solution to the problems differs from former solutions. In order to protect NCPRs, it has been proposed to engage in practices that are friendlier to the environment, that are accompanied by a reduction of negative externalities, and so on. This also includes "respect for planetary boundaries."[23] But this is not far-reaching enough regarding demanding climate goals. Instead, a new mode of cooperation is needed – one that places the goals of cooperation front and centre.

Nor is this interpretation restricted to remedies for failures of cooperation, in terms of non-compliance with possibly successful schemes of cooperation. Nor, as well, can cooperation be reduced to a climate contract or a set of mutual promises.[24] A joint action approach instead argues that a different type of action is promising for addressing climate goals successfully. What is needed is a joint effort from a variety of actors – especially, but not exclusively, powerful actors – to achieve difficult goals.

Unlike in contract theory, the best possible strategies cannot be framed as mutual promises to comply with rules that everyone accepts. Instead, a commitment to future goals is needed. Therefore, the goals need to be qualified so that cooperation in favour of these goals includes an appropriate reaction to the fact of accelerated climate change. It is also assumed that the goals will best be achieved by a variety of actors and by different means – including creative solutions of single actors, strong commitments of effective joint actors, innovative technologies, and effective politics.[25] The underlying rationale of this distinct type of cooperation has been explored in social philosophy in terms of "joint action." Joint action is an action that is carried out by individuals; at the same time, it is composed of structures that resemble the elements of action undertaken by individuals. Joint action differs, in particular, from the actions of actors who are already constituted as collective actors – companies, schools, or nation-states, for example.[26] Joint actions, in contrast, include actions such as "dancing together" or "walking together" – actions that are undertaken by individuals together. The debate on "joint action" as a specific and collective type of agency is not only advanced, in that authors have developed a detailed theory of action as applied to various groups of people,[27] it also helps us understand the strength of groups of actors who act jointly to reach shared goals.

A common misunderstanding also needs to be addressed. Reframing climate action as joint action does not necessarily mean that climate action needs to be interpreted as *voluntary* action. Although it starts with a model of joint action based on the commitments of individuals to a shared goal, this theoretical proposal can be used to argue for an obligatory model of cooperation. After having outlined a voluntary joint environmental action approach, I argue that joint action with respect to climate goals qualifies as obligatory. More specifically, a case can be made for a duty to support climate goals – but only under conditions of goals that follow principles of fairness and an allocation of the weight of duties that also relies on moral insights. Overall, the joint action approach will be used to explain environmental cooperation in a new theoretical light.

1.3 Against standard reservations

So far, I have explained how the approach developed in this book relates to normative theories of climate change and to theoretical approaches to cooperation. In order to portray the theoretical premises, it is also important to discuss the description of climate change in theories that interpret what is happening with our warming planet.

Climate change is a fact. "Climate change," however, is an umbrella term that covers a range of events. According to the Intergovernmental Panel on Climate Change (IPCC), what we call "climate change" is caused by a multitude of events in the past, including high emission rates of greenhouse gases that remain in the atmosphere and disturb its cooling effect.[28] In addition to the interpretation of the fact of climate change and its consequences, a theoretical framing also includes ideas about how humankind relates to these facts. In elaborating on climate ethics, it is necessary to explore these interpretations thoroughly. To start with, it is a well-known insight into theories of action that each action is only given "under a certain description."[29] This means that action is not an event in the world like other causal events. Instead, action is caused by actors who try to achieve specific goals. The interpretation of this insight is developed by choosing descriptions that accord with these presumptions.

Accordingly, a theory of joint action needs to explain how climate change and the action of individuals as well as of collectives correspond to each other. Although the power of individuals and actors to change the current situation is limited, since the future effects of climate change are already to a high degree predetermined by past emissions, future actions, and collective actions in particular, will have an impact on the fate of natural common goods. Moreover, in climate ethics, the interpretation of climate change is not detached from a prognosis of what humankind can achieve and *how* things can be changed.

In order to explore options for action in response to climate change, it is necessary to first explore standard interpretations of it. The debate is overshadowed by what I call "climate dogmas." A dogma is an interpretation of facts that has a normative backdrop and that many people share, without necessarily being aware of the underlying facts. It is not my goal here to provide arguments against the alleged dogmas; instead, I intend to name them in order to debunk them as dogmas. For the rest of this book, they can then be set aside.

Before exploring climate dogmas, two more provisos need to be made. I address only those that are relevant to the thesis of this book. There are many other interpretations that either neglect the human impact on climate change altogether, that propose sheer fatalism, or that reject the thesis of anthropogenic change. Although I think it is important to address all these issues, this is not the place. Nor is it the place to juxtapose climate dogmas with empirical facts. Detailed reviews of the science would be necessary – and even then, it would not be easy to expose the underlying rationale of the presumptions. The dogmas that will be discussed in this section cover ideas about climate change in isolation from other environmental problems, from the presupposition that climate change can be fully quantified, and from the presupposition that actors cannot make a difference with respect to future developments.

1.3.1 The dogma of isolationism

Climate change has been interpreted as an exceptional event. Its sheer magnitude, long-term effects, and potentially disastrous outcome are indeed exceptional. But exceptionalism has also been conflated with another thesis: *Climate change needs to be addressed in isolation from other environmental problems.* This is the *dogma of isolationism*.

Isolationism is different from exceptionalism in that it argues that climate change needs to be defined as a stand-alone problem. This approach to climate change is reasonable in one respect. In order to explore a problem, it needs to be distinguished from other problems; each environmental problem is unique. Still, the problem with understanding climate change is not only that specialists are needed to interpret it thoroughly. Rather, it is assumed that climate change needs to be addressed in isolation from other environmental goals.

When climate change is regarded as part of a much larger problem, it is also clear that cuts in emissions – though desperately needed – will not suffice. Instead, high emissions are the result of a way of life that damages the natural environment in many different ways. Looking at the problem in a more comprehensive context is necessary to get a clear grasp of the transformation needed to protect the environment from further irreversible damage. Against this background, climate change should not be regarded as an isolated event, but as part of a problem that results from an unsustainable mode of existence – and from institutions that protect and support that mode. In short, the problem of climate change needs to be addressed not in isolation, but rather in the context of a broad, inclusive picture.

1.3.2 The dogma of quantification

Climate scientists argue that the amount of greenhouse gas emissions is correlated directly with the warming of the atmosphere. The IPCC warns that the average increase in temperature must be kept within the limits of two degrees Celsius,[30] otherwise the effects could possibly become uncontrollable. Obviously, the two-degree limit depends on estimates, not on exact data. It has also been put forward to push for an effective reduction of greenhouse gas emissions. The two-degree limit can be translated into an amount of additional emissions (or "carbon budget") that the atmosphere is capable of absorbing without additional warming beyond two degrees. This argumentation has been challenged in various ways. Again, it is not my goal to question the seriousness of the underlying research that prompts this estimate. Instead, I intend to highlight another climate dogma: the *dogma of quantification*.

An exclusively quantitative approach not only underestimates the effects of cooperation that do not result from adding one amount to another. It also runs into trouble when it comes to generating motivation for supporting mitigation goals. Quantitative goals are abstract, making it difficult to motivate people to subscribe to them. Moreover, since quantifiable climate emissions are the effect of industrial production and transportation, emission budgets are correlated with economic performance and economic growth – at least as long as economic growth and economic performance result from the use of carbon-intensive energy. In this framing, there is no place for the effects of cooperation other than in terms of addition and subtraction of emission budgets. Although it is important to reduce greenhouse gas emissions, it is doubtful whether the logic of climate cooperation is one of a linear process in terms of either adding to the overall emissions or reducing the amount of emissions. This interpretation overlooks another logic that might unfold in terms of cooperation. This is not the logic of adding and subtracting total amounts, but rather the logic of group behaviour.

1.3.3 The dogma of individual shiftlessness

Climate change confronts humankind with particularly threatening possibilities. The dramatic development is also fed not only by a rapidly growing amount of greenhouse gases in the atmosphere, but also by a range of systemic reactions that complicate the already unfolding dynamics. Many changes in nature are having a deep impact not only on the shape of our planet, but also on the climate system. Accelerating factors in climate change include the warming of the seas, the melting of ice, the loss of permafrost zones, the disappearance of wetlands, the sealing of massive areas of land surface, and the loss of forests and primeval forests. These developments are sometimes cited in support of the claim that individuals are helpless and cannot bring about change. When introduced as a necessary element in the assessment of climate change, this becomes the *dogma of individual shiftlessness*.

Because climate-change mitigation goals are so big, individual action is not only inefficient, since there is far too little that individuals can do to efficiently reduce emissions, individual action is also considered useless in terms of causal correlations. When framed as a purely quantitative problem in isola-tion from all other problems, individual contributions are indeed hardly worthy of attention. Yet it is far from clear that climate change needs to be addressed in exclusively quantitative terms. In addition, this approach underestimates both the possible dynamics of joint action and the impact of groups of actors on climate change and on possible remedies. As a consequence, the actions of individuals as well as of collective actors need to be reintegrated into a framework that explains ways to accomplish joint goals, including mitigation, but not limited to it. In sum, the theorems that have been addressed as "dogmas" contribute to a standardized interpretation of cli-mate change. Philosophers, however, do not have to accept this interpretation.

1.4 New spaces for moral arguments

Philosophers and researchers in climate ethics today are fortunate in that our understanding of climate ethics has reached an advanced stage. It has become part of the debate about mechanisms for either reducing emissions or adapt-ing to a changing environment. As a discipline that engages in outlining and exploring normative content, ethics is in a position to contribute important insights about human agency. It provides guidance for exploring values that play a role in addressing the environment as a resource for individuals to lead a good life. It also explores constraints on actions in the context of coopera-tion and societies working together. Ethics can be especially helpful in that many regard climate change as closely related to a particular form of injustice.

In debating environmental justice, Gordon Walker elaborates on a variety of different types of environmental justice. From a normative perspective, Walker argues that "climate change makes the most persuasive case for a justice framing."[31] He elaborates on this claim and states:

> With climate change we are confronted with evidence of patterns of inequality and claims of environmental injustice that span the globe, that permeate daily life and which pose threats to the current and future health and well-being of some of the poorest and most vulnerable people around the world. Climate change demands more than ever that we think relationally, about how things interconnect, about who benefits at the expense of others and about the spatially and temporally distant impacts of patterns of consumption and production. The consequence is that, for many already economically, politically and environmentally marginalized people, climate change presents compounding forms of injustice.[32]

Since climate change produces unfair burdens, it is appropriate for ethics to address it.

Overall, this study breaks new ground also by transcending claims of distributive justice. More specifically, the new role ethics can play is not self-evident. Instead, the new spaces for ethical concerns – for example, duties to cooperate, normative presumptions regarding value commitments, and moral obligations with respect to climate victims – need to be argued for cautiously. An ethics approach to joint climate action includes normative arguments, which have three facets.

The most intriguing point to argue is that individuals as well as institutions *should cooperate* in favour of mitigating climate change. Although I first outline a model of joint action that is based on the voluntary cooperation of individuals, this study also includes arguments for a change to a normative model. More specifically, this presumes a thorough discussion of the role collectives play in joint climate action and whether various types of collectives have obligations and duties towards climate goals.[33] It also presupposes a debate on the types of actors involved, including the moral role of individuals.[34]

Another argument I make is that climate cooperation is just another realization of a type of cooperative behaviour that is not atypical in human behaviour. People – as social philosophy explains – tend to "work together." And people do so quite successfully. Concepts of joint action, however, have not yet been applied to action on climate change. The first remedy for collective action problems that resonate with environmental problems is a theoretical approach to joint environmental action informed by social philosophy. More specifically, goals need to be formulated with caution. This approach to climate action also includes a revision of climate goals. With regard to environmental action, and climate action in particular, I also argue that in order to motivate individuals to work together in favour of those goals, an environmental *ethos* is necessary.[35]

A third point I make in this book is that cooperation, in terms of joint action, also relies on normative arguments. I argue that principles of justice are a necessary element in framing environmental goals and climate goals. They help to assign obligations regarding environmental goals to various types of actors. Principles of fairness, however, are not restricted to distributive principles. They also include special obligations of particular types of actors and principles that relate to the capacities of given actors. And a new discussion of responsibility also includes claims regarding institutional design.

Although a debate on the fair distribution of the burdens and benefits of climate change is part of the analysis here, it is not central. It is necessary to go one step further, to argue for a climate justice approach that entails, as a core principle, the obligation to cooperate and to shoulder a fair share of the burdens of climate change. More specifically, regarding natural goods such as the atmosphere, individuals as well as collective actors are not free to choose whether to benefit from that resource. Once a fruitful scheme of cooperation has been proposed, the freedom is limited to either cooperating or defecting. In these specific contexts, I contend, complicity with that scheme is obligatory.[36]

Both for defending a concept of joint environmental action and for defending the normative principles that should guide cooperation, ethics provides the theoretical tools. It is especially helpful for outlining distinctively normative claims and their underlying foundations. The broader question ethics deals with is the question of how individuals, groups of individuals, and humankind should rebuild and remodel institutions and ways of life in order to attempt to keep nature intact. It is not too late for that.

Optimism is justified for two reasons. On the one hand, engaging in ethics would be irrational if people were not able to listen to arguments and to model their attitudes and values accordingly. As individuals we may be helpless, but as groups effectively engaging in collective action, it is possible for us to change as we need to in order to achieve important climate goals. On the other hand, theories of joint action are appropriate in climate scenarios because they are non-paternalistic and non-deterministic. They take the ingenuity of individuals and of groups seriously in leaving the formation of goals to the actors and to the acting collectives.

1.5 Theoretical challenges

An approach to climate ethics that works with a model of joint action faces particular theoretical challenges. To avoid underestimating them, I also outline them at the beginning of this study. In order to give a concise description of theoretical challenges, it is important to first identify the starting point. Much work has already been done to introduce ethics as an element in theoretical approaches that deal with climate change.[37]

All the same, the approach developed in this study needs to address several intricate philosophical problems. First, it is far from clear that theories of joint action are suited to addressing practices of appropriation of a shared natural resource. Although examples of environmental action are mentioned in theories of joint action, they are used as negative examples that exemplify failures of joint action.[38] In my view, these shortcomings do not have to be taken as the last word. To some degree, they rest on a misunderstanding of environmental cooperation as just another version of market-style exchange processes. More specifically, environmental cooperation needs to be reframed as a goal-directed activity. Nevertheless, the interpretative process regarding joint environmental action is still demanding. It starts with a reassessment of two distinct models of joint action as argued by Bratman[39] and by Tuomela.[40] It then uses both models to argue for a theoretical approach to joint environmental action, which is a joint practice of the appropriation of a natural resource under premises of a shared ethos and shared goals accordingly.

A theory of joint environmental action also needs to explain how "the environment" figures into theories of action at all. Usually, and specifically in climate debates, what is recognized as problematic is not action itself, but is instead the negative externalities resulting from action and from collective action. In this theoretical framing, the atmosphere and other natural resources are not destroyed by wilful action, but by side effects of actions that are part of our daily life and of industries that support our lives. To develop an alternative model, it is necessary to take a fresh view on "the environment" and how it relates to action of collectives. One way to do so resonates with approaches in water ethics. In theories of integrative water management, it has been argued that processes in which collectives regulate shared resources need to be interpreted as "active co-designing with nature."[41] This not only presumes that collectives determine goals that they share with respect to the future of a natural resource, it also involves an expression of the values that support certain goals. In accordance with insights into water ethics,[42] the normative perspective in defining joint climate goals also includes insights into "planetary boundaries"[43] and in remaining spaces for agency. In short, environmental goals and climate goals need to be revised in order to outline a theory of joint environmental action.

A third challenge, the focus of the second half of the book, results from the attempt to transform a well-reasoned model of joint environmental action into a model of obligatory environmental action and specifically of climate action. Assuming that joint climate action has been explicated convincingly, the question of whether and why individuals and collectives should invest in this model still remains unanswered. This is a normative question, one that can be abbreviated as the question of whether "climate duties" are justified.

Several problems stand in the way of arguing for climate duties in the context of a model of joint action. One can best be called the *placement problem* of duties. It concerns the issue of identifying the right addressee of climate duties. Assuming that joint climate action were to prove a successful model for confronting climate change in such a way that individuals as well as groups achieve climate goals successfully, who then holds the duties? Is it the individual, is it a group of actors, or are already-existing collectives responsible for bringing about these climate goals? This problem resonates with the problems of allocating duties in collective action scenarios, which philosophers have already explored.[44] Those problems, moreover, entail further complications, resulting from the fact that even if collectives were successful at joint action in favour of climate goals, their duties might derive from the duty to collectivize in the first instance.[45] In addition, it cannot be presupposed that every kind of actor must fulfil the very same duties. Instead, it appears necessary to introduce fairness once again into a scenario of climate obligations. Principles of fairness need to support a just allocation of burdens according to the capacities of various actors and according to their already-existing moral obligations. Whereas this used to be taken as an argument that highlights the obligations of governments,[46] according to a theory of joint climate duty, a much more detailed picture emerges. The responsibility to cooperate in pursuit of climate goals belongs to individuals as well as to collectives. More specifically, the arguments for following through on climate goals depend on principles that link a fair share of responsibility to various types of actors and of second-order responsibilities.[47] Whereas responsibilities apply to a variety of actors, political institutions are also responsible for building and supporting institutions to enable joint climate action.

In addition to the problems of determining the correct allocation of duties to duty-holders, it is also important to address defection. Here again, two different perspectives can be distinguished. On the one hand, it is important not to underestimate the voluntary model of joint environmental action. Instead of starting with duties and moral claims, it is helpful to first argue for methods of removing obstacles in the way of environmental cooperation. This requires a scientific analysis of why people defect and why environmental cooperation is hindered.[48] In addition, it is helpful to distinguish various types of defection in relation to the reasons why actors defect. This, however, is only half of the picture. It is also necessary to develop not only active support for some types of joint climate action, but also to discuss ways to institutionalize joint action and to support it by means of law and by means of institutions that can set the right incentives. Although it falls outside my goals here to offer a study of the best institutional solutions, at the end of this book I offer some ideas on the subject.

Notes

1 Simon Caney provides an overview of the debate on climate justice in "Cosmopolitan Justice, Responsibility, and Global Climate Change," in *Climate Ethics. Essential Readings*, eds. Stephen M. Gardiner et al. (Oxford/New York: Oxford University Press, 2010).

2 The literature on climate justice is broad and advanced. Publications that are particularly important in the context of this study and which provide important new insights include Derek Bell, "How Should We Think About Climate Justice?," *Environmental Ethics* 35, no. 2 (2013): 189–208; John Broome, *Climate Matters. Ethics in a Warming World* (New York/London: W.W. Norton, 2012); Simon Caney, "Cosmopolitan Justice, Responsibility, and Global Climate Change," *Leiden Journal of International Law* 18 (2005): 747–775; Stephen M. Gardiner, *A Perfect Moral Storm. The Ethical Tragedy of Climate Change* (Oxford/New York: Oxford University Press, 2011); Margaret Gilbert, "Rationality in Collective Action," *Philosophy of the Social Sciences* 36, no. 1 (March 2006): 3–17; Dale Jamieson, *Reason in a Dark Time. Why the Struggle Against Climate Change Failed – and What It Means for Our Future* (Oxford/New York: Oxford University Press, 2014); Eric A. Posner and Cass R. Sunstein, "Climate Change Justice," *Georgetown Law Journal* 96 (2008): 1565–1612; Tracey Skillington, *Climate Justice and Human Rights* (New York: Palgrave MacMillan, 2017); Steven Vanderheiden, *Atmospheric Justice: A Political Theory of Climate Change* (Oxford: Oxford University Press, 2008); Gernot Wagner and Martin L. Weitzman, *Climate Shock. The Economic Consequences of a Hotter Planet* (Princeton, NJ/Oxford: Princeton University Press, 2015).

3 For this study, the most important publications include Michael E. Bratman, *Shared Agency. A Planning Theory of Acting Together* (New York: Oxford University Press, 2014); Michael E. Bratman, "Shared Intention," *Ethics* 104, no. 1 (1993): 97–113; Philip Pettit and David Schweikard, "Joint Actions and Group Agents," *Philosophy of the Social Sciences* 36, no. 1 (March 2006): 18–39, https://doi.org/10.1177/0048393105284169; John Searle, *The Construction of Social Reality* (New York: The Free Press, 1995).

4 Simon Caney, "Two Kinds of Climate Justice: Avoiding Harm and Sharing Burdens," *Journal of Political Philosophy* 22, no. 2 (2014): 125–149.

5 For discussions of criteria for addressing collectives as responsible agents, see Peter A. French, *Collective and Corporate Responsibility* (New York: Columbia University Press, 1984); Christopher Kutz, *Complicity: Ethics and Law for a Collective Age* (Cambridge: Cambridge University Press, 2000); Philip Pettit, "Collective Persons and Powers," *Legal Theory* 8 (2002): 443–470; Raimo Tuomela, *Social Ontology: Collective Intentionality and Group Agents* (New York: Oxford University Press, 2013).

6 Stephanie Collins, "Collectives' Duties and Collectivization Duties," *Australasian Journal of Philosophy* 91, no. 2 (2013): 231–248, https://doi.org/10.1080/00048402.2012.717533; Anne Schwenkenbecher, "Joint Duties and Global Moral Obligations," *Ratio* 26, no. 3 (2013): 310–328, https://doi.org/10.1111/rati.12010.

7 Elizabeth Cripps, *Climate Change and the Moral Agent: Individual Duties in an Interdependent World* (Oxford: Oxford University Press, 2013); Baylor Johnson, "Ethical Obligations in a Tragedy of the Commons," *Environmental Values* 12, no. 3 (2003): 271–287.

8 This term was coined by Garrett Hardin in "The Tragedy of the Commons," *Science* 162, no. 3859 (1968): 1243–1248, https://doi.org/10.1126/science.162.3859.1243.

9 Stephen M. Gardiner, "The Real Tragedy of the Commons," *Philosophy and Public Affairs* 30, no. 4 (2001): 387–416.

10 For a representative collection of the most basic discussions in this debate, see Stephen M. Gardiner et al., eds., *Climate Ethics. Essential Readings* (Oxford/New York: Oxford University Press, 2010).

11 For a defence of these critical claims, see Dominic Roser, *Ethical Perspectives on Climate Policy and Climate Economics* (Zurich: University of Zurich, Faculty of Economics, Zurich Open Repository and Archive, 2010).

12 On the central claims in the discussion on intergenerational justice, see Lukas H. Meyer and Axel Gosseries, eds., *Intergenerational Justice* (Oxford/New York: Oxford University Press, 2009).

13 See Richard P. Hiskes, *The Human Right to a Green Future: Environmental Rights and Intergenerational Justice* (Cambridge: Cambridge University Press, 2008), https://doi.org/10.1017/CBO9780511575396; Edward Page, *Climate Change, Justice and Future Generations* (Cheltenham, UK/Northampton, MA: Edward Elgar, 2007).

14 On claims of assistance, see Margit Ammer, "Klimawandel und Migration/Flucht: Welche Rechte für die Betroffenen in Europa?," in *Klimagerechtigkeit und Klimaethik*, ed. Angela Kallhoff (Berlin/Boston: De Gruyter, 2015), 81–103; Henry Shue, "Climate Hope: Implementing the Exit Strategy," *Chicago Journal of International Law* 13, no. 2 (2013): 381–402.

15 Some important steps in this direction have already been made. See David Schlosberg, *Defining Environmental Justice: Theories, Movements, and Nature* (Oxford: Oxford University Press, 2007), https://doi.org/10.1093/acprof:oso/9780199286294.001.0001; Gordon P. Walker, *Environmental Justice. Concepts, Evidence, and Politics* (London/New York: Routledge, 2012).

16 Wagner and Weitzman, 17.

17 Jamieson, *Reason in a Dark Time*, 61.

18 Ibid., 162.

19 Ibid.

20 Gardiner, *A Perfect Moral Storm*, 165–174.

21 Wagner and Weitzman, 38–40.

22 In the remainder of this study, I use "climate goals" as an umbrella term for goals that have been proposed either to mitigate the effects of climate change or to adapt to the consequences generated by conditions of climate change. A thorough exploration of "climate goals" is offered in Chapter 6.

23 I derive this term from Rockström's model . See Johan Rockström et al., "Planetary Boundaries: Exploring the Safe Operating Space for Humanity," *Ecology and Society* 14, no. 2 (2009).

24 For a proposal of a climate contract following the lines of Rawls's proposals in *A Theory of Justice* (Cambridge, MA: Belknap Press, 2005), yet building on a win-win scenario, see Eric A. Posner and David A. Weisbach, *Climate Change Justice* (Princeton, NJ: Princeton University Press, 2010).

25 A more detailed exploration of environmental goals is given in Chapter 6.

26 Searle, *The Construction of Social Reality*; Pettit, "Collective Persons and Powers."

27 For reasons that will be outlined in Chapters 4 and 5, the main theoretical references for a model of joint action are Bratman, *Shared Agency*; Bratman, "Shared Intention"; Tuomela, *Social Ontology*; Raimo Tuomela, *The Philosophy of Sociality: The Shared Point of View* (Oxford/New York: Oxford University Press, 2007).

28 For recent proposals to adjust institutional designs, see John S. Dryzek, Richard B. Norgaard, and David Schlosberg, eds., *Oxford Handbook of Climate Change and Society* (Oxford: Oxford University Press, 2013).

29 This insight is arrived at in the context of a theory of action in G. E. M. Anscombe, *Intention* (Cambridge, MA: Harvard University Press, 2000).

30 IPCC, "AR6 Synthesis Report," accessed April 3, 2019, www.ipcc.ch/report/sixth-assessment-report-cycle/; United Nations, "The Paris Agreement," accessed October 4, 2018, https://unfccc.int/sites/default/files/english_paris_agreement.pdf.

31 Walker, *Environmental Justice*, 179.

32 Rebecca L. Walker, "The Good Life for Non-Human Animals: What Virtue Requires of Humans," in *Working Virtue: Virtue Ethics and Contemporary Moral Problems*, eds. Rebecca L. Walker and Philip J. Ivanhoe (New York: Oxford University Press, 2007), 179.

33 For central insights into this discussion, see Collins, "Collectives' Duties and Collectivization Duties"; Schwenkenbecher, "Joint Duties and Global Moral Obligations."

34 See Cripps.

35 For the interpretation of the notion of "ethos" in joint action, see Raimo Tuomela, *The Philosophy of Social Practices: A Collective Acceptance View* (Cambridge: Cambridge University Press, 2010), 39.

36 To support this claim, I rely on arguments presented in Kutz and on arguments in Robert H. Myers, "Cooperating to Promote the Good," *Analyse & Kritik* 33, no. 1 (2011): 123–139.

37 For a reassessment of the arguments for and against climate ethics with an emphasis on climate justice, see Stephen M. Gardiner and David A. Weisbach, *Debating Climate Ethics* (New York: Oxford University Press, 2016).

38 Authors sometimes mention environmental examples, yet usually only in order to exempt these cases from a standard theory of joint action. See Bratman, *Shared Agency*, 144; Tuomela, *The Philosophy of Sociality*, 112–113.

39 Bratman, *Shared Agency*.

40 Tuomela, *Social Ontology*.

41 Jerome Delli Priscoli and Aaron T. Wolf, *Managing and Transforming Water Conflicts* (Cambridge/New York: Cambridge University Press, 2009), 121.

42 For a model of water cooperation that rests on this assumption, see Angela Kallhoff, "Water Ethics: Toward Ecological Cooperation," in *The Oxford Handbook of Environmental Ethics*, eds. Stephen M. Gardiner and Allen Thompson (Oxford/New York: Oxford University Press, 2017); Angela Kallhoff, "Transcending Water Conflicts: An Ethics of Water Cooperation," in *Global Water Ethics. Towards a Global Ethics Charter*, eds. Rafael Ziegler and David Groenfeldt (London/New York: Routledge, 2017), 91–106; Angela Kallhoff, "Water Justice:

A Multilayer Term and Its Role in Cooperation," *Analyse & Kritik* 36, no. 2 (2014): 367–382.

43 Rockström et al.

44 For groundbreaking insights into the complexity of duties that belong to collectives, see Kutz.

45 For arguments about duties to collectivize, see Collins, "Collectives' Duties and Collectivization Duties"; Stephanie Collins and Holly Lawford-Smith, "Collectives' and Individuals' Obligations: A Parity Argument," *Canadian Journal of Philosophy* 46, no. 1 (January 2, 2016): 38–58, https://doi.org/10.1080/00455091.2015.1116350.

46 Broome, 50–51.

47 Caney, "Two Kinds of Climate Justice."

48 An example of that sort of investigation is offered in Shazeen U. Attari, David H. Krantz, and Elke U. Weber, "Reasons for Cooperation and Defection in Real-World Social Dilemmas," *Judgment and Decision Making* 9, no. 4 (2014): 316–334; Elke U. Weber and Paul C. Stern, "Public Understanding of Climate Change in the United States," *American Psychologist* 66, no. 4 (2011): 315–328, https://doi.org/10.1037/a0023253.

2   The tragedy of the commons

Bert: Don't you think that it's good that philosophers have started to think about moral arguments regarding climate change?

Cindy: Sure. But this might all be in vain. The reason is obvious: some philosophers have already made the convincing claim that it's too late – both for action and also for moral argument. They say that, regarding climate change, humankind is currently facing a non-reversible catastrophic event. They call it a "real tragedy of the commons." "Commons" are natural collective goods and they think that the atmosphere is such a good. It has been filled up with greenhouse gases; and since there are no limits to using the atmosphere, humankind has overexploited the atmosphere as a waste-dump for greenhouse gases. But this event can't be cured, because there is no way to remove greenhouse gases once they've been deposited.

Bert: But haven't you heard that this view of the events has recently been challenged? People who explore another commons – water resources – challenge the story of the tragedy of the commons. They say that the concept of "tragedy" is neither particularly helpful, nor is it right. It mixes too many different problems; and it leaves no space for a solution. Have you heard about that?

Cindy: Yes, I have. Let's see what scientists say. I think they try to disentangle a variety of problems that are all unduly called "tragedy."

Bert: Yes, they do. And they use water systems as a comparison. Isn't that interesting?

Cindy: Of course it's interesting. And let's have a look at how they relate this problem to climate change. I think they now speak of "two sides" of the drama. And they try to distinguish the side of man-made tragedy not as a natural outcome, but as a failure of cooperative behaviour. Let's have a look at the debate!

Public goods are a specific group of entities with two distinctive properties. No potential beneficiary can be excluded from them, and they have no competitors for consumption. An example is the lighthouse in a harbour. Whenever a boat enters the harbour, it can make use of the light to navigate, and no additional boat reduces the benefit of the lighthouse for any other pilot. Yet pure public goods are rare. As a subclass of public goods, common-pool resources can be called "impure public goods." They have only one feature in common with pure public goods: non-excludability. Though non-excludable, common-pool resources are still highly competitive goods.[1] More specifically, common-pool resources can be exhausted by overconsumption. Common-pool resources have already been of particular interest in environmental theory, in relation, for instance, to water reservoirs and the atmosphere. In those cases, natural common-pool resources (NCPRs) do not share only the characteristic of a lack of entrance barriers. Instead, the debate on NCPRs, often referred to as the "natural commons," includes a normative perspective.[2] Common natural resources need to be shared among various stakeholders. As a consequence, claims of providing fair patterns of distribution have been raised.

To begin, it is necessary to explain the theoretical framing of resources such as water reservoirs and the atmosphere as natural commons. This interpretation provides the background against which conflicts relating to the characterization of natural commons can be addressed. More specifically, shortcomings that result from chaotic patterns of exploitation and environmental hazards are at the centre of this debate.[3] This chapter explains that because of irregular entrance barriers and because of a growing demand for these goods – both as resources and due to the lack of space for waste disposal – they have become increasingly at risk. Yet instead of explaining this "tragedy of the commons"[4] either as the result of a scenario of collective choice that allows for no constructive alternative or as the result of self-interest, this chapter contributes to a slightly different interpretation: common-pool resources suffer from uncoordinated behaviour and from patterns of exploitation that are destructive for several reasons. These common-pool resources are not under threat solely because of tragic choices or due to the egoism and self-interest of beneficiaries, but because of a variety of factors that contribute to invisibility and self-energizing effects. If there is a tragedy, it is the invisibility of vicious circles and mutually destructive behaviour. More specifically, this chapter focuses on the material side of the tragedy of the commons: the over-exploitation of shared resources. It does not address collective action problems, which have often been connected with the discussion of the tragedy. Instead, collective action problems are approached when discussing future-oriented politics and "joint action."

Section 2.1 explains the theoretical background of the drama of the commons. It also explains the specific characteristics of natural commons in the light of this interpretation. Section 2.2 demonstrates that the drama of the commons is far more complex than failures-of-collective-action and over-exploitation approaches insinuate. Indeed, the drama results from a variety of conflicts. A linear and one-sided interpretation is therefore not particularly helpful. In order to illustrate this complexity, some conflicts on shared water resources are described. Section 2.3 explores what I call the "material side" of the drama, which is a structure of causal effects that together drive over-exploitation. These effects and their causes need to be distinguished from collective action problems. Section 2.4 explains the connection to the climate debate. It looks at proposals for interpreting "one atmosphere"[5] as a common-pool resource, yet it also transcends this interpretation. It argues that the atmosphere is not only problematic as a "waste dump," but that there is also the problem of geophysical stability of the system that has been termed "climate." Climate instability results from a host of effects, among which the high emissions of greenhouse gases are one factor, but not the only one. Section 2.5 summarizes the outcomes by listing shortcomings that need to be addressed as part of the material side of the drama of the commons. Section 2.6 draws the conclusion that the drama of the commons, as resulting from uncoordinated patterns of exploitation of shared resources, needs to be kept apart from the discussion of shortcomings of collective action. Collective action can only be diagnosed against the background of desirable social goals. These goals lie between the limited capacities of public resources to offer social services and the risk that those who use these resources will exploit them. Overall, this chapter serves as a preparation for further discussion, for two reasons. First, I want to defend the view that the analysis of shortcomings in addressing natural resources is important for preventing a scenario that has been described in debates on water resources: unless governmental systems of conflict solutions have been established successfully, common-pool resources will function as conflict catalysts. This means that already-existing conflicts among various parties or individuals are intensified through conflicts provoked by the usage of one shared resource. This interpretation has been ignored in the climate debate thus far, although it has been prominent in the water debate. Here, the shared river or the shared water basin has already been regarded as a factor for polarizing communities. Either contracts are elaborated and peaceful solutions to conflicts are established – and the range of examples is impressive here – or conflicts are dramatized to such a degree that even "water wars" are likely to occur.[6] Yet instead of subscribing to a theory that stops here, I will side with analysts of research in water conflicts. Recent research in water ethics elaborates on various solutions to these types of conflicts. This includes a decisively ethical standpoint and a vision of shared future goals.[7]

   Second, I intend to reject two interpretations of the "tragedy of the commons" that have been of major importance in the climate debate as well. The first interpretation holds that the most difficult problems were solved when entrance barriers were erected that led to an "enclosure" of natural commons. This, however, is not a promising strategy to protect NCPRs from over-exploitation and environmental hazards. More specifically, it falls short of acknowledging that the natural commons are continuously needed as the life source of the world population. It also overlooks that it is not only the amount of services that are gained from a resource, but the qualitatively distinct ways of appropriation that matter. To put it bluntly, breathing the air does not damage the atmosphere, but a huge amount of greenhouse gases will. Rather than systems of enclosure to protect NCPRs from over-exploitation, what is needed are institutions that channel human action and side-products of civilization in such a way that it becomes possible to continue to jointly benefit from that resource through successful cooperation. Because of climate change, this needs to include a sharp reduction in greenhouse gas emissions.[8] The second interpretation of the "tragedy of the commons" I reject is the view that regards collective action problems as causally responsible for climate change. In the view outlined in this study, collective action problems need to be interpreted as the failure to achieve socially desirable goals. As long as these goals have not been declared and the relationship to single actors has not been outlined, it is false to regard failures in collective action as *causes* of climate change. What is right, though, is that cooperation and joint action need to succeed if they are to successfully mitigate climate change and to achieve shared future goals.

2.1 The drama of the natural commons

As a subclass of public goods, common-pool resources are – by nature – non-excludable, yet competitive goods.[9] Public goods have been the focus of research in financial studies and in political economy. In the context of welfare economics, public goods have revealed their problematic sides. As goods that are strictly antithetic to private property – they have been considered undividable goods and as goods that belong to groups of people, not to singular owners – public goods cause major market failures. Moreover, they will neither be produced nor upheld by a market economy, because investments in them are not beneficial.

This narrative, already negative, has been aggravated by theories of collective action. According to this line of thought, public goods provoke dilemmas and failures that have been described as dilemmas of collective action. They prompt a behaviour that has been characterized as "free-riding." Moreover, they resemble a scenario that is well known as the "Prisoners' Dilemma." As non-excludable goods, each potential beneficiary is free to derive benefits from a public good. She will not have to pass an entrance barrier. She is therefore basically invited to take a "free ride." The assumption in explaining this type of behaviour is the following: when people are free to benefit from a collective good without paying for it, they will do so. Moreover, people who are not in a position to know that other people will display cooperative behaviour will act as if each person seeks their own individual benefit. Regarding public goods, the result is disastrous. No one is willing to invest voluntarily into a good, even if it would ultimately be beneficial for everyone involved, without first knowing whether the other people involved will cooperate.[10]

As a consequence, authors in the field of political economy have argued that public goods will not be sustained unless political institutions take control of them.[11] Yet this produces major shortcomings, too. Instead of creating goods that are at least in some respect similar to market goods, artificially supported public goods will be inefficient, tending to become "natural monopolies" and grow beyond reasonable limits. Moreover, they will generate "cultures of dependence" instead of strengthening a sense of responsibility and engagement in their beneficiaries.[12] Many of these narratives are right – at least in some respects. Think of public health care systems and public transport. However, these dramas not only represent just one side of public goods, they also do not fit common-pool resources. This latter aspect will be explained shortly, but first it is necessary to reject the general idea of the necessary destruction of public goods.

Many of the mechanisms that have been described – the inefficiency of the system, cultures of dependence, and so on – are not caused automatically, but rather are brought about by bad management structures. For decades, authors in the field of political economy have contributed to debunking some of the myths that still surround public goods. More specifically, there are options to work together successfully, even in the context of global public goods.[13] Public goods have also primarily been debated against the background of political economy. Yet this obscures another side of public goods, which, in my view, is as important as the economic view of public goods. Public goods perform services for society that can be summarized as contributions to upholding democratic values. Public goods support social cohesion, they contribute to structures of social justice, and they provide the background against which people are free to engage in higher-ranking societal activities. In short, public goods serve the basic values of a democratic and fair society in a number of important respects.

Since pure public goods are rare, attention has shifted to impure public goods. This group includes two types of goods, distinguished by the prevalence of one of the two characteristic traits of a public good. Common-pool resources are presented as highly non-excludable, but rivalrous goods. Club goods, on the other hand, are excludable, but non-rivalrous to the fixed group of users.[14] Because of the fact that potential beneficiaries need to pass an entrance barrier in advance, these goods are also called toll goods. In fact, many artificially produced public goods belong to the second category. Institutions for recreation and cultural institutions, public transport, free media, public health-care systems, as well as public education, are all toll goods. Public institutions have, of course, an interest in erecting entrance barriers to refund these goods to some degree. If intended as goods that support social justice, public institutions have an equal interest in keeping the entrance barriers low. Tolls should thus be affordable for all citizens.

Unlike toll goods, common-pool resources do not possess entrance barriers, or do so only imperfectly. This is the most important subcategory of a public good for addressing NCPRs. More specifically, the physical characteristics of some types of natural goods contribute their most interesting characteristics. Examples of NCPRs are sunlight, water in reservoirs, rivers, the seas, the Antarctic, biodiversity and – most importantly for this study – the atmosphere. These goods are not provided artificially by groups of people. Instead, they are simply present, available. This is not to say that entrance barriers do not exist at all or that institutions have not taken care of these goods. Quite the opposite is true. Since the rise of civilizations, NCPRs have been transformed into resources that support civic life. In doing so, they have undergone continuous change; access systems have basically been shaped by civilizations. Water pipelines have been built, irrigation systems have been erected, and landscapes have been transformed by cultivation and agriculture. The phenomenon of culture-related transformation is not new.[15] Overall, it is important to note that the interpretation of natural goods as common-pool resources does not entail the claim that entrance barriers have not been erected. Instead, it claims that they are – in various ways – imperfect.

One approach that has been proposed for protecting NCPRs from further harm is "enclosure." Enclosure means that artificial and secure entrance barriers are established to achieve complete control over the access of a good.[16] One of the underlying rationales of processes of enclosure is related to the observation that economic activities, as well as the societal use of natural goods, need to respond to "planetary boundaries." In 1987, when the "limits to growth" had already been put forward by the Brundtland report, one important concept in protecting NCPRs was the proposal to respect limits.[17]

Instead of setting limits on economic growth, from the perspective of environmental protection, erecting access barriers to NCPRs is a much more convincing idea. Yet this proposal has serious flaws. I intend to discuss three of these shortcomings to show why one should reject the idea that enclosure is a solution for environmental problems, including enclosure by means of privatization. Such a solution can only be part of "joint environmental action," as I explain later. Here are three failings of enclosure:

First of all, the open borders of natural goods such as water basins, rivers, open landscapes, and the atmosphere are too huge for complete surveillance. It may be helpful to have the technical means of enclosing the atmosphere to establish fair distribution of access rights, but for the time being, no device for monitoring these things exists.[18] Even regarding smaller entities, such as groundwater reservoirs, lakes, and open landscapes, there are too many possible loopholes. Complete surveillance or complete fencing by artificial means is simply not feasible.

Second, the transformation of NCPRs by means of enclosure will not necessarily transform them into toll goods, but more likely into private goods.[19] Yet – to use a metaphor – toll goods are fenced goods. And fences may disrupt natural processes that are important for natural systems. Disruptions might in turn contribute to undermining natural processes of repair.

Third, a further reason for supporting the view of the incompleteness of the transformation of common-pool resources into toll goods is normative. Natural common-pool resources are different from private goods in that they belong to the public. This means that appropriation also needs to resonate with common interests.

In a situation in which scarcity of natural resources is a rapidly growing problem, negotiations about best practices for harvesting natural goods must be continued. Overall, fencing these goods is not a reliable solution. Instead, to get closer to possible solutions, the conflicts over NCPRs need to be studied in more detail.

2.2 Multi-layered conflicts: Examples from water reservoirs

Natural common-pool resources are at the centre of serious conflicts. Studies on environmental justice demonstrate that unequal distribution of environmental goods is not the only cause of conflict. Conflicts result from a host of different issues, including unequal distribution of environmental goods, but also unfair political procedures, a lack of recognition for the claims and needs of groups of people, as well as pre-existing social tensions that are worsened by environmental conflict.[20] Moreover, the legal situation regarding NCPRs is still being worked out. As Radkau explains, even when conflicts have been settled in terms of water laws and institutions that monitor water laws, they are enshrined in customary law that remains local law, and thus solutions remain dependent on political agendas.[21] As a consequence, natural water reservoirs as well as rivers are at the centre of severe conflicts.[22] Some experts even think that "water wars" need to be considered real possibilities.[23]

A review of the literature on this subject shows that the following issues are entangled in water conflicts:[24]

- Unsettled property rights;
- Exclusive claims on a water reservoir, which excludes groups of beneficiaries – resulting from either legal titles to that good or from the conviction of being the "real" owner;
- A growing scarcity of a resource, possibly caused by climate change, which leads to an escalation of already existing conflicts;
- A particular vulnerability of groups of beneficiaries who cannot protect themselves against a diversion of water to other sources;
- Practices of benefit-seeking behaviour that undermine other practices – as for instance industrial use spoiling groundwater reservoirs;
- Lingering processes of degradation that at some point become visible and felt;
- Agendas and visions for best possible future uses that result from particular perspectives (as for instance dam projects from governments).[25]

Water resources provide a good example of common-pool resources because the common usage of a shared resource can be a source of severe conflict. In addition to conflicts that occur because of the scarcity of a resource, conflicts may also result from adaptation processes to overcome natural resource scarcity.[26]

Yet conflict over natural resources is only part of the picture. Examples also demonstrate that in some scenarios peaceful alternatives are possible. As for water conflicts, authors have also investigated peaceful solutions to conflicts in the context of water management systems. Jerome Delli Priscoli and Aaron T. Wolf conclude: "The history of social organization around river basins and watersheds is humanity's richest records of our dialogue with nature. It is amongst the most fertile areas of learning about how political and technical realms interact."[27] They warn, moreover, against overstating the threat of coming "water wars":

> In the modern times, only minor skirmishes have been waged over international waters – invariably other interrelated issues also factor in … War over water is not strategically rational, hydrographically effective, or economically viable. Shared interests along a waterway seem to overwhelm water's conflict-inducing characteristics and, once water management institutions are in place, they tend to be tremendously resilient.[28]

Although limited to certain cases, these authors' studies do show that peaceful resolution of conflicts is possible.

On the one hand, NCPRs give rise to conflicts that are multi-layered. As a consequence, they cannot simply be set aside or forgotten. On the other hand, however, they press humans to find a solution. Because common-pool resources are necessary for the survival of societies, access conditions to these common-pool resources need to meet the exigencies of societies and people. The need for common-pool resources simultaneously requires that societies come up with a peaceful solution for addressing emergencies like the water crisis.[29] The outcome of these conflicts is either an agreement between parties or an accelerated conflict with no winners. More specifically, authors in the field of water studies emphasize that conflict is neither inevitable nor the standard case. Instead, a broad range of cooperative arrangements has been reported and documented.[30]

In addition, the concern that current practices are not sufficient to keep water reservoirs intact is not reserved to our present century. As an example of how far back this problem goes, I want to cite a document from 1909 printed in the *Annals of the American Academy* that demonstrates that more than 100 years ago scientists were already finding it necessary to be cautious with shared and limited water resources. McGee states:

> Our stock of water is like other resources in that its quantity is fixed. It differs from such mineral resources as coal and iron, which once used are gone forever, in that supply is perpetual; and it differs from such resources as soils and forests, which are capable of renewal or increase, provided the supply of water suffices, in that its quantity cannot be augmented. It differs also in that its relative quantity is too small to

permit full development of other resources and of the population and industries depending on them. Like all other resources, it may be better utilized. It must be better utilized in order to derive full benefit from lands and forests and mines.[31]

The theory that I present here resonates with these insights into the particular quality of NCPRs. It contributes to a normative framing of potentially successful cooperation as an approach to the natural commons. Yet before elaborating on that approach, another zone of serious conflict needs to be considered. Natural common-pool resources appear to be particularly endangered by a conflict that has been summarized as the "tragedy of the commons."

2.3 The material tragedy of the commons

So far, I have described NCPRs as subjects of multifaceted conflicts. A survey of the literature on water conflicts portrays water resources as conflict catalysts. This means that these types of conflicts are particularly severe, yet they can lead to peaceful solutions, as well. Some authors claim that ideal strategies for solving these conflicts include means of negotiation and water contracts.

The metaphor of the "tragedy of the commons" not only posits that overly exploitative practices need to be stopped to prevent an irreversible exhaustion of natural resources, it also states that some effects of actions, either of individuals or of collectives, are particularly influential, even "tragic." A tragedy is not something that simply happens and causes a bad event. Instead, the diagnosis of a tragedy juxtaposes a particularly dramatic outcome and the goodwill of actors. Steve Gardiner has provided the most vivid picture of such a tragedy when exploring climate change as the "real tragedy of the commons."[32] He argues that the climate tragedy is particularly difficult to resolve not only because of its vast global scale, but also because of its intergenerational dimension.[33]

Tragedies have unintended, yet particularly severe effects that result from the actions either of individuals or of collectives. I reinterpret what is often called the "tragedy of the commons" to explain that tragedies also entail unintended misfortunes, and that actions have causal effects.

In order to highlight this side of the tragedy of the commons, I describe a whole array of tragedies: the tragedy of exhaustion, the tragedy of chaotic exploitation, the tragedy of gradual depletion, the tragedy of invisibility, and the tragedy of reinforcement through channelling. All of these tragedies result from causal effects on the resource that together are particularly harmful. The effect is that the common-pool resource is severely endangered.

2.3.1 The tragedy of unintended exhaustion

As long as people and institutions are not forced to coordinate their activities when benefitting from natural resources, unplanned depletion is a likely outcome. In optimizing private gains, many parties together contribute to overconsumption. This first tragedy of the commons is *the tragedy of unintended exhaustion.* [34] It results from a situation in which over-exploitation takes place, although long-term sustenance would be the best for at least many of the beneficiaries of a shared resource. Access barriers might even exist and be incorporated in rules that communities choose as a means of preserving their life-sustaining natural resources. [35] Yet such efforts might not lead to success, because they may not be all-encompassing. Often, potential beneficiaries cannot literally be hindered from using a shared resource; sometimes, institutions that should guard sustainability fail to do so. As long as an effective means of stopping the over-exploitation of an endangered resource is missing, depletion is still a real danger. That being said, exhaustion is not something that beneficiaries intend.

2.3.2 The tragedy of chaotic exploitation

The second material problem is the *tragedy of chaotic exploitation.* This tragedy results from an altogether different cause than uncontrolled access to a common resource. Since exploitation of natural resources does not presuppose any fixed rules or procedures, patterns of exploitation do not have to result from cooperative schemes that have been negotiated among various beneficiaries. Even when some regulation is in place – which is usually the case – this does not necessarily prevent depletion from resulting from chaotic patterns of exploitation. Imagine a situation in which a person tries to benefit from a lake as a freshwater resource and another person wants to benefit from it as a waste dump for sewage water. Even if their actions were guided by a desire to optimize their benefits from a resource, there are no fixed rules that could contribute to optimizing those benefits. Instead, water can be used as freshwater, as fishing grounds, as a means for industrial production, and as a waste dump. The same concept holds true for the atmosphere. [36]

Chaotic exploitation does not necessarily contribute to the depletion of common resources, but it might contribute to a situation in which some types of usage are undermined. The desire to benefit from a river or a water basin for recreation cannot be fulfilled when another actor spoils the water or has another desire, as in the instance of using water for drinking. Although each person just wants to benefit for her own sake, one type of usage will rule out another type of benefit-seeking behaviour.

2.3.3 The tragedy of gradual destruction

Yet there are further tragic events regarding NCPRs. The third tragedy is the *tragedy of gradual destruction.* Air, water, and soil are not renewable resources – at least not in the literal sense of being resources that reproduce themselves on a regular basis. Instead, these goods need to be protected from pollution to keep them intact. Yet much environmental pollution does not come as one catastrophic incident, but rather as a gradual depletion. More specifically, common-pool resources have "memories." Pollution of the sea or the air does not wash out. Instead, some pollutants that do not degrade naturally remain in the sea and in the air. Plastic in the seas is one example. Unless these pollutants have been extracted by technical means, they will remain in the system. Moreover, the dispersion of pollutants in the sea or in the air is not controllable. Instead, there might be places where pollutants accumulate. Given this, air, water, and soil are particularly vulnerable to heavy incidents of pollution. Such incidents will produce traces that remain in the system for a long time. Moreover, they will shape the properties of the system – sometimes for a long time.[37] One of the main challenges in coping with climate change resonates with the tragedy of destruction by accumulation: pollutants have accumulated in the atmosphere. So far, there is no way to wash CO_2 out again; instead, there are repercussions. The warming of the earth is an effect of an accumulation of pollutants that have been deposited in the atmosphere for at least several decades. Moreover, tipping points may contribute to rapid and major changes of the whole system.

This type of tragedy has an additional aspect when debated in the context of the psychology of climate change. Elke Weber argues that the perception of climate change is not one of serious and immediate threats, but rather of a risk that rests on gradual change.[38] Yet psychological reactions result from one of two systems: an affective and an analytical system. Reactions to risk usually relate to the first of the two systems, yet gradual change does not elicit a reaction that provides the resources with a comprehensive answer.

2.3.4 The tragedy of invisibility

Another type of climate-change tragedy is *the tragedy of invisibility*. The air is literally invisible. The same applies to greenhouse gases, including methane and carbon dioxide. Gases that are deposited into the atmosphere cannot be seen, they can only sometimes be smelled or felt. That incidents of pollution are invisible is not only an aggravating factor but a real dramatic aspect that contributes to the failure to address the problem of climate change effectively.[39]

Yet the tragedy of "invisibility" not only refers to something that in the literal sense cannot be seen with our eyes. Instead, for many reasons, the availability of water, air, and the soil has been taken for granted. It might happen that no one actually cares about these goods unless a situation of shortages or – in the case of water – of dramatic overflow happens. The reasons for this situation are also theoretical. In a political economy, pollution of the air and water have been rationalized as "negative externalities." The analysis focuses on economic activities; it then analyses negative side effects. Yet they are only considered to be negative to the extent that they do harm to people; they put our well-being and health at risk. The air and the water are treated as mere mediums, not as entities deserving attention. Their qualities and needs are simply invisible.

2.3.5 *The tragedy of reinforcement through channelling*

Finally, there is the *tragedy of reinforcement through channelling*. Civilizations need natural resources; usually, they exploit natural resources by means of institutions that channel benefit-seeking behaviour. Yet the channels civilizations erect to facilitate the appropriation of natural goods are not focused on sustainability, but on optimal gains. Moreover, the channels that bring water and air to the population as well as direct waste into the environment follow the laws of the availability of resources as well as the needs of populations. The outcomes are again often not planned, but instead chaotic. Sometimes, the effects of erecting institutions that provide channels for ecosystemic goods go into directions that cannot be foreseen, causing events that cannot be stopped once set into motion. Although an action may be carried out with the best intention, the subsequent chains of effects are channelled by factors that drive them in a harmful direction. A particular tragedy arises when institutions that were erected with the best of intentions nonetheless provide channels that support the over-exploitation of a good. Imagine a wonderful beach and a road that channels traffic directly to that beach. Channels of greenhouse gases similarly contribute to the rapid depletion of the atmosphere.

Burning coal and fossil fuels was a precondition for the type of industrial development that has contributed to enormous wealth first in Europe and the New World. Yet no one could have anticipated that this type of wealth accumulation would be as all-encompassing as it appears to be today. Selective mechanisms were at work; some of them are still responsible for channelling events in one direction or another. The first channelling factor is that people choose commodities that make their lives easier. Production of such items has therefore risen enormously. A second factor is that the market system contributes to channelling production in a certain way. Coal and fossil fuels continue to be the cheapest sources of energy. Therefore, the techniques for exploiting these energies have spread all over the globe. Yet exploiting carbon-intensive energy also channels the waste of that energy use directly into the atmosphere; climate change is to a large degree the result of that channelling.

In sum, the over-exploitation and harmful exploitation of natural commons can be seen as the result of a range of "tragedies." Yet instead of tying them exclusively to either collective-action failures or to over-exploitation, I have argued that they result from patterns of appropriation of a resource that fail to result in sustainable practices. The effects have not been intended; perhaps they could not have been avoided at that time. They must nonetheless be acknowledged and their causes explored. Overall, a theoretical approach to future success in managing the natural commons needs to include tools to overcome these failures.

2.4 Climate change and the drama of the commons

Climate change can be framed perfectly as just another tragedy of a natural commons. In "One Atmosphere," Singer explains that the way the atmosphere is being used is comparable to a "giant waste dump." It has been overused and now the space for additional waste has become scarce. Henry Shue has written of Ostrom's helpful insights on the question:

> This makes planetary absorptive capacity for carbon (and other GHGs) a classic example of what Elinor Ostrom has called "common-pool" resources, with each use of the capacity being "subtractable" from the remaining total and "rival" to all other possible uses, most notably use by our descendants ... The "budget" of absorptive capacity is being consumed, depriving the people of the future of options.[40]

As a global common-pool resource, the atmosphere has been overused as a waste dump. This interpretation coheres with what I have called the "material side of the tragedy of the commons." More specifically, some of the causal tragedies can also be applied to former practices that have contributed to over-exploitation. Applying these distinctive interpretations already has some explanatory power.

Overall, the exhaustion of the atmosphere as a collective waste dump was certainly not intended. This statement differs from another in the climate debate. Instead of assuming that people could not and did not know that greenhouse gases have the effects we are facing today,[41] this first tragedy simply states that the effect of over-exploitation was unintended. And this is true – there has been no purposefully destructive actor intending to destroy the climate system, not even in the sense of a collective actor. Moreover, chaotic exploitation is also a disturbing fact we must take into consideration.

This interpretation of climate change as a causal tragedy also resonates with the interpretation of climate change as a crystal-clear case of the over-exploitation of the atmosphere as a deposit of greenhouse gases. As Wagner and Weitzman put it:

> Think of the atmosphere as a giant bathtub. There's a faucet – emissions from human activity – and a drain – the planet's ability to absorb that pollution. For most of human civilization and hundreds of thousands of years before, the inflow and the outflow were in relative balance. Then humans started burning coal and turned on the faucet far beyond what the drain could handle. The levels of carbon in the atmosphere began to rise to levels last seen in the Pliocene, over three million years ago.[42]

This interpretation of climate change can easily be completed by also discussing the reasons humankind was so stupid as to turn on the faucet that much. The reasons given in the former section are plausible. The faucet was not turned on that much intentionally and by a single actor. Instead, the effects of many different activities were invisible; industrial development also meant "channelling" side effects in terms of ever-higher emissions rates; and the atmosphere's "memory" now contributes to the irreversibility of what happened. Overall, this story is much more plausible than the narrative of collective action failures, which posits that too many too self-interested beneficiaries do not care for a social goal, but instead over-exploit a limited resource.

As for further elaboration on the commons in this book, it is also important to notice that "the climate" is a rather abstract category. Natural common-pool resources are instantiated as entities that have a systemic and material side, and that perform eco-services when kept intact. Eco-services are causally related to common-pool resources; they have a beneficial effect, either on persons as beneficiaries or on the quality of the environment. This type of resource is best framed as somehow always having a local dimension, as well. Moreover, it is important to discuss common-pool resources as embedded in more complex natural surroundings, including a variety of different natural commons. Local climate cannot be discussed without also considering the local natural water resources and water cycles. And the opposite direction is equally important, given that water cycles and water resources are not only part of a microclimate, they are also shaped by processes that are part of the larger event "climate change."[43]

Overall, it is important not to forget the material side and the complexities of the problem of climate change. Although experts all agree that a sharp reduction of greenhouse gas emissions is a necessary element of successful climate politics, the restriction of flawed practices also needs to be regarded as addressing only part of the problem. The climate suffers from high emission rates; yet the degradation of natural commons is a problem with many facets, in particular also including the mutual effects of degradation. Even when efficacious and timely mitigation is obligatory, this is only part of a more comprehensive approach to addressing the manifold dramas of the commons.

Another issue in need of scrutiny is not only the conflation of the material side with collective-action problems, but that climate stability and even climate politics have to be regarded as "commons." As for the interpretation of public goods as non-rival and non-excludable goods, this resonates with the fact that both goods are to some degree and in some specific respects non-excludable and non-rival. For the purposes of this study, I restrict the concept of "natural commons" to real and material natural goods such as microclimates, lakes, and rivers. It is helpful to distinguish between artificial public goods and natural public goods to also explain their specific compositions and rationales.[44]

2.5 A list of failures

Natural common-pool resources, including the atmosphere and climate, suffer from a variety of "dramas." As scarce resources, they serve as catalysts for social conflict. They provide groups of beneficiaries with all sorts of normative challenges, including unsettled property rights and clashes over fairness in terms of appropriation and distribution. These facts will not be neglected. Yet for the material side of the drama of the commons, the narrative is far simpler. It posits that with resources that lack clear-cut entrance barriers, whose appropriation is not regulated in advance, uncoordinated exploitation and invisible and mutual reinforcing effects will contribute to the degradation and destruction of that resource. It is not only tragic that these resources are needed for the survival of humankind. Tragedy can also be seen in the fact that over-exploitation was not intended, but actually a hidden and unintentional result.

In addressing conflicts over water and air, a range of different types of problems need to be faced. Neither the often-discussed typology of the "tragedy of the commons," nor the interpretation in terms of "collective action problems" suffices to address the range of various sources of conflict. Instead, the material side is portrayed by a list that includes the outcomes of this chapter.[45] The list can be split into two categories: environmental hazards that are typical of NCPRs, and destructive patterns of exploitation that portray tragic aspects.

2.5.1 Environmental hazards typical of NCPRs

Exhaustion. The lack of clear-cut entrance barriers contributes to problems of overuse and depletion – of groundwater, water basins, or fishing grounds, for example.

In-pool pollution. Containment of environmental hazards is particularly difficult with respect to undividable substances and particularly hazardous materials (water and air, in particular).

Long-term accumulative effects. Due to the "memory" of collective, cohesive goods and due to the fact that NCPRs are systems and therefore not completely renewable, impacts on natural goods may be long-lasting.

2.5.2 Destructive patterns of exploitation

Chaotic or conflicting usage. Types of benefit-seeking behaviour conflict with each other. In an extreme case, the same NCPR is used simultaneously as both a source and a dumping ground; in that case, none of the interests can be accomplished satisfactorily. Patterns of exploitation do not rest on cooperation, nor on an assessment of long-term effects of channelling.

Invisibility of destructive effects. Destruction of a resource passes unseen, because the side effects of patterns of exploitation are not visible, but nevertheless are detrimental to the resource.

Channelling of negative effects. When a natural resource is used in a one-sided way, one that is simultaneously reinforced by institutional framings, side effects and negative consequences can be particularly destructive because of that channelling.

2.6 Two sides of the drama

So far, I have explored the material side of the tragedies of the natural commons. It was part of my analysis to demonstrate that reducing it to either problems of injustice or to institutional failings does not suffice to give a reliable assessment of conflicts surrounding common-pool resources. The main aim of this chapter was not to give evidence of the assessment of climate change, neither was it to portray other processes of depletion of natural resources as particularly dramatic. Reports in that spirit already abound. Instead, the analysis focuses on the application of the category of a "common-pool resource" as a specific subclass of public goods.

Overall, scarcity and the threat of the depletion of NCPRs provoke deep conflicts in society. Yet the analysis of the multi-layered conflicts in this chapter serves another goal. A closer look at conflicts surrounding water reservoirs also explains why societies have – sometimes – succeeded in addressing scarce and shared resources. Conflicts are not irreparable. Yet it is the material side of the tragedy of over-exploitation and destruction that is the real problem. More specifically, without coordination of exploitative behaviour, common-pool resources are put at risk. In order both to develop future scenarios of good practices of benefit-seeking behaviour and of peaceful exploitation, it is critical to not only develop goals that serve as yardsticks for working together, but also to avoid the failings resulting from tragic patterns of over-exploitation. Future goals of cooperation need to avoid these failings; in particular, they need to render bad side effects visible and avoid mutually destructive exploitation of a shared resource.

As for tragedies, I have argued that the metaphor of a "tragedy" covers a range of aspects that are comparable to tragic events. Yet – unlike the diagnosis of the tragedy of the commons as resulting from collective action problems – the tragedy of the commons needs to be seen as a material problem. More specifically, it has diverse causes, though chaotic and uncontrolled patterns of exploitation of shared resources are particularly salient. A tragedy of the commons is best explained as the outcome of patterns of exploitation that – because of the characteristics of these patterns – contribute to the depletion of a resource.

Notes

1 For an analysis of this concept of public goods, its current critique, and an alternative approach in political philosophy, see Angela Kallhoff, *Why Democracy Needs Public Goods* (Lanham, MD: Lexington Books, 2011).
2 For global common-pool resources, see Susan J. Buck, *The Global Commons. An Introduction* (Washington, DC/Covelo, CA: Island Press, 1998); Angela Kallhoff, "Water Justice: A Multilayer Term and its Role in Cooperation," *Analyse & Kritik* 36, no. 2 (2014): 367–382; Peter Singer, "One Atmosphere," in *One World. The Ethics of Globalization* (New Haven, CT: Yale University Press, 2002), 14–50; Steven Vanderheiden, *Atmospheric Justice: A Political Theory of Climate Change* (Oxford: Oxford University Press, 2008).

3 This is part of the interpretation of failures of collective action underlying climate change as resulting from reckless behaviour, which Stephen M. Gardiner supports. More specifically, Gardiner takes the problem of moral corruption and of the temptation of what he calls "buck-passing" – handing over problems to the next generation while benefitting from them in the present – seriously. For this, see Stephen M. Gardiner, *A Perfect Moral Storm. The Ethical Tragedy of Climate Change* (Oxford/New York: Oxford University Press, 2011), 302.

4 This term was coined by Garrett Hardin, "The Tragedy of the Commons," *Science* 162, no. 3859 (1968): 1243–1248, https://doi.org/10.1126/science.162.3859.1243. Since Hardin's article, it has also been applied to climate change. For the latter, see Stephen M. Gardiner, "The Real Tragedy of the Commons," *Philosophy and Public Affairs* 30, no. 4 (2001): 387–416.

5 This is the title of Singer's contribution to climate ethics in Peter Singer, *One World: The Ethics of Globalization* (New Haven, CT: Yale University Press, 2002).

6 For a discussion of various solutions to situations of water scarcity that refer explicitly to ethics, see Peter G. Brown and Jeremy J. Schmidt, eds., *Water Ethics. Foundational Readings for Students and Professionals* (Washington, DC: Island Press, 2010); Magdy A. Hefny, "Water Management Ethics in the Framework of Environmental and General Ethics: The Case of Islamic Water Ethics," in *Water Ethics. Marcelino Botin Water Forum 2007*, eds. Manuel Ramón Llamas, Luis Martinez-Cortina, and Aditi Mukherji (Boca Raton: CRC Press, 2009), 25–44; Kallhoff, "Water Justice"; Manuel Ramón Llamas and Luis Martinez-Cortina, "Specific Aspects of Groundwater Use in Water Ethics," in *Water Ethics. Marcelino Botin Water Forum 2007*, eds. Manuel Ramón Llamas, Luis Martinez-Cortina, and Aditi Mukherji (Boca Raton: CRC Press, 2009), 187–204. For case studies that report options for peaceful solutions, yet also highlight potential for conflicts, even wars, see Hussein A. Amery and Aaron T. Wolf, *Water in the Middle East. A Geography of Peace* (Austin: University of Texas Press, 2000); Karen Bakker, ed., *Eau Canada. The Future of Canada's Water* (Vancouver/Toronto: UBC Press, 2007); Peter Beaumont, "Conflict, Coexistence, and Cooperation: A Study of Water Use in the Jordan Basin," in *Water in the Middle East. A Geography of Peace*, eds. Hussein A. Amery and Aaron T. Wolf (Austin: University of Texas Press, 2000), 19–44. For a systematic approach to water conflicts and to water management practices, see Jerome Delli Priscoli and Aaron T. Wolf, *Managing and Transforming Water Conflicts* (Cambridge/New York: Cambridge University Press, 2009).

7 For this, see Eran Feitelson, "A Hierarchy of Water Needs and their Implications for Allocation Mechanisms," in *Global Water Ethics. Towards a Global Ethics Charter*, eds. Rafael Ziegler and David Groenfeldt (London/New York: Routledge, 2017), 149–166; Franklin M. Fischer, "Water Value, Water Management, and Water Conflict: A Systematic Approach," in *Mountains: Sources of Water, Sources of Knowledge*, ed. Ellen Wiegandt (Dordrecht: Springer, 2008), 123–148; Hefny.

8 This is consensual in climate science, as represented by the Intergovernmental Panel on Climate Change. See the fifth report: IPCC, "5th Report," accessed October 25, 2020, www.ipcc.ch/report/ar5/.

9 In this section I summarize the outcomes of my study of public goods in Kallhoff, *Why Democracy Needs Public Goods*; Angela Kallhoff, "Why Societies Need Public Goods," *Critical Review of International Social and Political Philosophy* 17, no. 6 (2014): 635–651. I do not defend this interpretation of public goods here.

10 For a forceful critique of the overgeneralization of that sort of dilemma, see Elinor Ostrom, *Governing the Commons. The Evolution of Institutions for Collective Action* (Cambridge/New York: Cambridge University Press, 1990).

11 For political-economic treatments of public goods, see Joseph E. Stiglitz, *Economics of the Public Sector* (New York/London: W.W. Norton, 1988); Joseph E. Stiglitz, *Whither Socialism?* (Cambridge, MA: MIT Press, 1994).

12 For a short summary of these insights, see Kallhoff, *Why Democracy Needs Public Goods*.
13 As for global public goods, see e.g., Scott Barrett, *Why Cooperate? The Incentive to Supply Global Public Goods* (Oxford/New York: Oxford University Press, 2007); Inge Kaul, Isabelle Grunberg, and Marc Stern, eds., *Global Public Goods: International Cooperation in the 21st Century* (New York/Oxford: Oxford University Press, 1999).
14 Stiglitz, *Economics of the Public Sector*, 131–140.
15 What might be classified as only a recent development is actually the immense impact that growing economies and a growing world population have on natural common-pool resources (NCPRs). Moreover, waste has become a serious problem. More specifically, the atmosphere now serves as a "giant sink" for the greenhouse gases that civilian life produces. For this interpretation, see Singer, "One Atmosphere."
16 For a similar, yet particularly critical interpretation of the term "enclosure" as related to processes of transformation of common goods into private goods, see David Bollier, *Silent Theft: The Private Plunder of Our Common Wealth* (London: Routledge, 2002), 1.
17 World Commission on Environment and Development, ed., *Our Common Future* (Oxford/New York: Oxford University Press, 1987).
18 One might argue that treaties that build on caps provide tools for enclosure: they set limits to effective use of the atmosphere as a waste dump for greenhouse gases. Yet even if a world-wide system that controls greenhouse gases could be established, the effect would not be an enclosure of the atmosphere; instead, the protection of sinks of greenhouse gases and the care for precipitation cycles is still important; exchange of gases is only a part of the inputs and outputs of climate systems.
19 For an example of a privatization of water that provoked violent conflict in Bolivia, see Irma van der Molen and Antoinette Hildering, "Water: Cause for Conflict or Co-Operation?," *Journal on Science and World Affairs* 1, no. 2 (2005): 138.
20 For an analysis of these structures that rests on a thorough exploration of data, see Gordon P. Walker, *Environmental Justice. Concepts, Evidence, and Politics* (London/New York: Routledge, 2012). The central conclusions of his study demonstrate the fact that environmental conflicts over NCPRs are multi-dimensional; see Walker, *Environmental Justice*, 214–221.
21 Joachim Radkau, *Natur und Macht. Eine Weltgeschichte der Umwelt* (München: Beck, 2002). [*Nature and Power. A World History of the Environment* (Cambridge: Cambridge University Press, 2008)].
22 Radkau presents examples from practices of colonialization (Radkau, 2002, 183–225) and discusses processes of globalization that affect natural resources as responsible for the loss of natural goods (Radkau, 2002, 284ff).
23 For examples of present-day water wars in India and an eco-feminist analysis, see Vandana Shiva, *Globalization's New Wars. Seed, Water & Life Forms* (New Delhi: Women Unlimited, 2005), 66–84.
24 For examples of empirical studies and theoretical investigations of the consequences of and solutions to water conflicts, see Emmanuel M. Akpabio, "Cultural Notions of Water and the Dilemma of Modern Management: Evidence from Nigeria," in *Water Management Options in a Globalised World. Proceedings of an International Scientific Workshop (20–23 June 2011, Bad Schönbrunn)*, ed. Martin Kowarsch (Institute for Social and Developmental Studies (IGP) at the Munich School of Philosophy, 2011), 156–171; Amery and Wolf; Kristin M. Anderson and Lisa J. Gaines, "International Water Pricing: An Overview and Historic and Modern Case Studies," in *Managing and Transforming Water Conflicts*, eds. Jerome Delli Priscoli and Aaron T. Wolf (Cambridge: Cambridge University Press, 2009), 249–265; Bakker; Beaumont; Dan W. Brock, "Forgoing Life-Sustaining Food and Water: Is It Killing?," in *By No Extraordinary Means. The Choice to Forgo Life-*

Sustaining Food and Water, ed. Joanne Lynn (Bloomington: Indiana University Press, 1989), 117–131; Bryan Randolph Bruns and Ruth S. Meinzen-Dick, "Negotiating Water Rights: Implications for Research and Action," in *Negotiating Water Rights*, eds. Bryan Randolph Bruns and Ruth S. Meinzen-Dick (London: ITDG Publishing, 2000), 353–380; Delli Priscoli and Wolf; Dipak Gyawali, "Water and Conflict: Whose Ethics is to Prevail?," in *Water Ethics. Marcelino Botin Water Forum 2007*, eds. Manuel Ramón Llamas, Luis Martinez-Cortina, and Aditi Mukherji (Boca Raton: CRC Press, 2009), 13–24; Hefny; Peter Lawrence, Jeremy Meigh, and Caroline Sullivan, *The Water Poverty Index: An International Comparison* (Newcastle-under-Lyme, UK: Keele University, 2002); Llamas and Martinez-Cortina.

25 This list is far from exhaustive – many more types of conflicts over water have been reported, including particularly violent conflicts, ethnic clashes that are accelerated by scarce resources, and even international wars. See van der Molen and Hildering, 135.

26 Irna van der Molen and Antoinette Hildering distinguish between "first-order conflicts," which result from resource scarcity, and "second-order conflicts" resulting from strategies of societies to overcome a situation of scarcity; see van der Molen and Hildering, 134. Conflicts might also be caused by using NCPRs for geopolitical goals. For the global commons in geopolitics, see Zbigniew Brzezinski, *Strategic Vision: America and the Crisis of Global Power* (New York: Basic Books, 2012), 110–120.

27 Delli Priscoli and Wolf, 32.

28 Ibid. This positive assessment is reiterated by van der Molen and Hildering, who regard the cooperative scenario not only as an alternative to a conflict scenario, but as one that is prevalent. See van der Molen and Hildering, 29–30.

29 The opposite, the "collapse" of societies because of failures of management of natural goods, has been explored by Diamond and Murney in their impressive study: Jared M. Diamond and Christopher Murney, *Collapse: How Societies Choose to Fail or Succeed* (New York: Penguin Audio, 2004).

30 A field of research in which this option has been particularly well documented is studies of treaties among nations that share a river. Transboundary regulation among upstream and downstream countries provides examples for peaceful solutions among rival parties. For a summary, see van der Molen and Hildering, 139–140.

31 W. J. McGee, "Water as a Resource," *Annals of the American Academy of Political and Social Science* 33, no. 3 (May 1909): 532.

32 See Gardiner, "The Real Tragedy of the Commons."

33 I put off discussing Gardiner's interpretation of the "tragedy of the commons" to Chapter 3, where I explore collective action problems, which are central to his interpretation.

34 Much of what has been argued by Garrett Hardin in "The Tragedy of the Commons" can be reduced to the fact that goods whose access systems have loopholes and that are over-claimed will necessarily be exhausted at some point.

35 This is a very abbreviated description of Ostrom's proposal to address local common-pool resources in terms of institutional settings that regulate access. See Ostrom, *Governing the Commons*.

36 Dale Jamieson argues that this is not only a standard case that has been reasoned in economics, but is also exemplary for the problem of climate change. See Dale Jamieson, *Ethics and the Environment. An Introduction* (Cambridge: Cambridge University Press, 2008), 14.

37 The long-term effects of disasters have been studied particularly well for their social effects, as in studies exploring the long-term effects of radiation. See Deborah Oughton, Ingrid Bay-Larsen, and Gabriele Voigt, "Social, Ethical,

Environmental and Economic Aspects of Remediation," *Radioactivity in the Environment* 14 (2009): 427–451; Deborah Oughton, "Social and Ethical Issues in Environmental Risk Management," *Integrated Environmental Assessment and Management* 7, no. 3 (2011): 404–405.

38 For this, see Elke U. Weber, "Experience-Based and Description-Based Perceptions of Long-Term Risk: Why Global Warming Does Not Scare Us (Yet)," *Climatic Change* 77 (2006): 103–120.

39 For similar arguments, see Dale Jamieson, *Reason in a Dark Time. Why the Struggle Against Climate Change Failed – and What it Means for our Future* (Oxford/New York: Oxford University Press, 2014), 101–103.

40 Henry Shue, "Climate Hope: Implementing the Exit Strategy," *Chicago Journal of International Law* 13, no. 2 (2013): 395.

41 Lukas H. Meyer and Axel Gosseries, eds., *Intergenerational Justice* (Oxford/New York: Oxford University Press, 2009).

42 Gernot Wagner and Martin L. Weitzman, *Climate Shock. The Economic Consequences of a Hotter Planet* (Princeton, NJ/Oxford: Princeton University Press, 2015), 15.

43 For an analysis of this context, see David Lewis Feldman, *Water Policy for Sustainable Development* (Baltimore, MD: Johns Hopkins University Press, 2007).

44 For this argument, see Kallhoff, "Why Societies Need Public Goods."

45 I have deliberately listed these in an order that deviates from the way I ordered the sections of this chapter. This helps to render my proposal for a minimum shared goal and a practice of cooperation more plausible. It has no effect on the outcomes of this chapter.

3 Collective action problems

Cindy: Oh man, this is difficult. But it's helpful to understand that there is such a nest of problems. It's not a single problem, but many things cling together with other things. I have understood that philosophers do not wish to solve all problems at once. Instead, they try to disentangle complex problems in order to determine the best possible way to address a problem.

Bert: Yes, I think that's how they work. But what's the next step?

Cindy: I think the most important outcome of the last chapter is insights into the social dimension of climate change. I already knew that humankind has had a huge impact on climate change. What I understand now is something different: even the best possible way to address climate change appears to depend on concerted action. If it's right that climate change is to some degree the outcome of shortcomings in collective action, it's logical that mitigating climate change needs some form of collective action. Am I right?

Bert: Hm, I think there is some error in your conclusion. Causing a problem is different from resolving it. But what appears to be right is that concerted action is needed in order to mitigate climate change at this point. Am I right?

Cindy: I think there is something more in it. But I'm not exactly sure what it is. I've heard that philosophers now talk a lot about individual fault and false behaviour regarding climate change. I think some of them even think that climate change is caused by false behaviour. But this sounds strange. On the other hand, it also sounds strange to think that nobody is responsible. What do you think?

Bert: I think both extreme views cannot be right. Actually, I think the debate possibly goes in another direction. Think about the water examples that we have already heard about. With regard to water resources as common resources, there have been successful types of cooperation. Possibly climate philosophers want to apply these insights to behaviour with respect to the atmosphere. I think they shouldn't even continue to debate all the shortcomings, the tragedies etc. This doesn't help. They should look into the future and think about ways to bring people together to act for future goals.

Cindy: You're right. But philosophers do not simply jump into the future. They need time. And they use this time for getting things right. They analyse problems before presenting solutions. But I think you're right that they have started to discuss a completely new issue. They don't want to discuss any longer the failures that have been made or the tragedies that have occurred. They want to develop proposals for how future goals can best be achieved. Let's look at how they think and what they want to address.

The atmosphere shares the characteristics of a common-pool resource. This subclass of a public good is non-exclusionary towards potential beneficiaries; it does not have access-barriers that allow the exclusion of potential beneficiaries. Yet unlike pure public goods, a common-pool resource is competitive. More specifically, rising demand spurs competition over a scarce resource. The problem called "climate change" has been interpreted as intimately related to these properties of the atmosphere as a common-pool resource. In "One Atmosphere," Peter Singer famously argues for an interpretation of the atmosphere as a global common good, contending that it serves as a sink for greenhouse gases, and when performing this service, part of it is appropriated as private property. According to Singer, this appropriation needs to be advanced so that other potential beneficiaries will have the same chance to get their share. "If we begin by asking, 'Why should anyone have a greater claim to part of the global atmospheric sink than any other?' then the first, and simplest response is: 'No reason at all.'"[1] Singer then goes on to explore principles of distributive justice that allow a more detailed answer. Since then, the interpretation of the atmosphere as a common-pool resource has been immediately related to a debate on distributive justice.

The atmosphere is not just one more common-pool resource among others. Steve Vanderheiden is right in claiming that the atmosphere is a special good, at least in the following respect:

> The planet's atmosphere is a common good that provides vital climate services to all the world's persons, with its absorptive capacity allowing for a finite quantity of GHG emissions before heat-trapping gases begin to accumulate in the atmosphere, destabilizing those climatic services and causing harm to persons and peoples. When viewed in this way, several problems for cosmopolitan justice are revealed, and a powerful claim for recognizing the terms of justice applied among the world's nations and persons becomes apparent.[2]

Vanderheiden also spends much effort outlining these problems and possible solutions. Other authors, however, though they also start with an analysis of the atmosphere as a global common-pool resource, propose another way of unfolding the problems surrounding a natural commons. They view common-pool resources as particularly problematic in that they cause collective action problems. Since the atmosphere is a truly global good, the resulting problems of collective action are particularly severe. Dale Jamieson, for example, argues that "climate change can be seen as presenting us with the largest collective action problem that humanity has ever faced, one that has both intra- and inter-generational dimensions."[3] And Steve Gardiner regards climate change as the "real tragedy of the commons."[4] In addition, Gardiner regards non-cooperation among climate beneficiaries and potential climate victims as one of the most challenging problems standing in the way of our ability to address the normative side of climate change.[5] That people are trapped in a "perfect moral storm" instead of moving forward to good climate solutions also results from the incapacity to resolve problems of cooperation.[6]

In studying this line of thought, the first thing that has to be acknowledged is that its proponents regard "collective action problems" as intricately related to the over-exploitation of the atmosphere as a shared natural resource. Yet this is only one interpretation, and it is part of a much more complex picture. Although many authors hold that climate change, the characteristics of a common-pool resource, and failures of collective action are closely related to each other, there are many ways to explain this tripartite relationship. More specifically, the inadequacies of collective action do not necessarily apply to practices of appropriation of the atmosphere as a shared sink for greenhouse gases. Instead, some think that collective action problems supervene on climate politics,[7] whereas others believe that collective action problems are part of the causal history that has produced the physical conditions of climate change.[8]

This chapter is dedicated to exploring the many facets of collective action problems and social dilemmas that come with the interpretation of the atmosphere as a common-pool resource. Although relating collective action problems to the atmosphere as a global common-pool resource has some initial plausibility, some claims need to be kept apart. Authors who investigate collective action problems identify them as causes of the current climate crisis, but they have completely different background assumptions in mind. Some of them say that collective action problems account for the failure of climate politics. As already noted, both Jamieson and Gardiner share this assessment of collective action problems. Others think that the description in terms of collective action problems and – more broadly conceived – in terms of social dilemmas is indispensable for understanding the failures of climate politics so far.

In analysing the facets of collective action problems, the chapter defends two claims: first, various interpretations of collective action problems need to be kept apart. More specifically, the debate on collective action problems as part of the causal history of climate change differs from the interpretation of the inadequacies of collective action of another type. The latter will be discussed as social dilemmas. Social dilemmas have also been diagnosed as thwarting successful climate politics, which is primarily a collective endeavour towards reducing emission rates. Second, social dilemmas are interpreted as parasites upon schemes of cooperation that would serve climate beneficiaries much better. When social dilemmas occur, schemes of cooperation that would otherwise be successful in realizing a socially desired goal are undermined. This parasitic character becomes apparent when social dilemmas are not primarily interpreted as part of the "tragedy of the commons," but are regarded as failings of cooperation. This second step provides a useful first step in addressing cooperation towards the aim of achieving climate goals anew. It therefore opens a perspective that will be outlined and explained in the remainder of this book.

This chapter proceeds in five steps. I first portray the interpretations of climate change as resulting from severe collective action problems as tied to a common-pool resource. Authors argue that there is some truth in acknowledging those problems. Yet they also argue that climate change differs significantly, at least in terms of scale. Section 3.1 is meant, not to give a survey of that literature, but to portray a widely shared interpretation of collective action problems as tied to the atmosphere as a common-pool resource. Section 3.2 explores another obstacle to getting a clear view of collective action problems. This is the interpretation of collective action problems as natural and tragic events. In the literature on environmental ethics, this has been framed as a "green paradox." I discuss this proposal and offer a critical assessment. Section 3.3 explores a set of "social dilemmas" that are important in the climate debate, yet adopt a rationale different from the logics of dilemmas as presented in game theory. Free-riding and other social dilemmas are distinct in that they are parasitic on pre-existing cooperation. This coheres with an interpretation of natural common-pool resources (NCPRs) that is given in Section 3.4. This interpretation says that NCPRs are necessarily part of a web of interaction and exchange processes. Yet instead of supporting the view that these webs are either simply there or necessarily chaotic, I explain why I think these chaotic webs can and should – with some adjustments – be supplemented by distinct patterns of cooperation. Section 3.5 also introduces the idea of conceptualizing cooperation in this particular case as "joint agency."

3.1 Failures of collective action as causes of climate change

Common-pool resources are not only at the centre of societal conflict, they also provoke a range of collective action problems. Yet the diagnosis of a collective action problem presents more than a trait of the more comprehensive problem of climate change. Indeed, it has been regarded as a key causal factor in the degradation of climate stability. In diagnosing climate change as ultimately related to collective action problems, authors are aware of the irregularity of this set of problems in the case of climate change.

Gardiner, who explores the forfeiture of forward-looking climate politics as resulting in part from amoral behaviour of our contemporaries, regards the analysis of it as a collective action problem as sound. Climate change is a drama of the commons. It is aggravated by three factors: by the fact that climate change is an intergenerational problem, by the fact that it is global, and by theoretical challenges that remain unresolved.[9] One particularly aggravating factor is a moral failing, summarized as the "tyranny of the contemporary," enshrined in the collective behaviour of "buck-passing."[10] Instead of internalizing the environmental costs of its consumption, the current generation does not pay for them, but instead passes them to subsequent generations. Gardiner thinks that the complexity of the moral storm might even be "convenient for us" – in that it hides the need to act on behalf of it.[11]

Yet this is only one side of Gardiner's interpretation of the unfettered path of climate change. He has also spent much effort on explaining how collective action problems are enmeshed in the theoretical challenges already reported. One particularly important aspect relates to the claim that when related to common-pool resources, the interpretation of dilemmas of collective action is probably slightly different from the more general approach taken by game theory. In *A Perfect Moral Storm*, Gardiner reconstructs the Prisoners' Dilemma and puts altruistic people in the place of the prisoners. The structure remains identical, but he explores a setting in which both people involved in the prisoners' scenario try to get the highest possible donations, one for a project in Sudan, the other in Cambodia.[12] The best strategies for both to win the money is an aggressive approach, but if both do so, they will alienate the donor. This setting works precisely the way the original Prisoners' Dilemma works, but it neither includes nor assumes self-interest.[13] Yet the even more interesting conclusion is that for the motivational setting, it suffices to assume that the motivations of people who cause the degradation of the atmosphere are just "here-and-now" decisions that do not take long-term effects into account. Gardiner ties this together with the behaviour of consumers:

> Suppose we take the consumption decisions of individuals as a leading cause of emission growth. Then, assume that such decisions are overwhelmingly driven by the judgments that individuals make about the short- to medium-term consequences for themselves, their families and friends, and (perhaps) their local communities … [W]e see the prominence of the kind of self-referential and time-indexed motivation that generate tragic structures of agency.[14]

Gardiner then notes another important aspect of this structure, observing that since actors in the dilemmatic situation do depend on each other's actions, and the aims of the actors are simultaneously at odds with each other, they forfeit not just any goal, but a goal that would be advantageous for both participants in the scenario.[15] Overall, Gardiner explains the unfettered path of greenhouse gas emissions by understanding how individual consumptive behaviour causes a scenario comparable to the Prisoners'

Dilemma scenario. It forfeits a reasonable and desirable goal by simply acting on individual preferences as consumers.

Jamieson also recognizes the close tie between the structure of the atmosphere as a public good and climate change. He also links the scenario to claims of justice in that he sees a "disproportionate appropriation" of the atmosphere by rich people, which produces harm to poor people, as particularly unfair.[16] This picture is made even worse since it is the rich who could produce fewer greenhouse gases by making small changes in their lifestyles. More specifically, Jamieson also gives space for a thorough analysis of the collective action problems involved in the debate on the atmosphere. He first states that the schemes of cooperation tied to the atmosphere provide a collective action problem, even – as already noted – "the world's largest and most complex collective action problem."[17] Yet he also warns that this problem should be compared with textbook examples of collective action problems, by way of emphasizing "the differences of scale that are involved in moving from human action to the climate system, and back to damages."[18]

Jamieson discusses a recurrent theme in the debate on climate change in explaining the discontinuity between individual emissions and the event called "climate change." In order to explain this, he recalls the trajectory of single carbon dioxide molecules emitted by individuals to its effects in the climate system: "The influence of my emission must travel upward through various global systems that affect climate, and then downwards, damaging something that we value."[19] Jamieson finds the assumption that individual emissions are closely related to climate effects particularly implausible. And Jamieson is certainly right in explaining that collective action problems rest on an interpretation of action as causing effects that in turn can be tied to these actions and their actors.[20]

Even when granting that collective action problems as related to common-pool resources differ from textbook examples, Jamieson also highlights two important implications in analysing the causes of environmental degradation more generally. First, he states that environmental problems result from a problem of coordination:

> Some of the most serious environmental problems occur when the same resource is used both as a source and as a sink: for example, when the same stretch of river is used both as a water supply and as a sewer; or when the same region of the atmosphere is used as a source of oxygen to breathe and as a sink for various pollutants. Using the environment as a source or a sink typically degrades its ability to function.[21]

By illustrating the problems that result from using a shared resource simultaneously as a sink and as a substance for use, Jamieson highlights an important problem. He also highlights an equally devastating aspect of the shared use of a resource. From the perspective of allocating a resource among various beneficiaries, another problem becomes obvious. Goods that can be used by every single person, without accounting for the costs, will indeed be used that way. Jamieson again: "It is difficult to allocate such goods efficiently because people use them, diminishing their value to others, without paying the full costs of their use."[22] Overall, Jamieson appears to support an interpretation of collective action problems as part of the causal history of climate change, although "causation" and "action" need to be interpreted with caution.

In order to complete this first assessment of how collective action problems relate to an NCPR, another perspective is important. As I mentioned earlier, Gernot Wagner and Martin L. Weitzman compare the physical causes of climate change to an overflowing bathtub.[23] In their view, it is not an over-exploitation of the atmosphere in terms of excessively voluminous and persistent greenhouse gas emissions that is responsible. Instead, the authors assert that the problem comes down to externalities that do not enter into any coordinated activity scheme. This coheres with another interpretation of climate change, focusing on the destructive effects of high emission rates in terms of negative externalities that tend to destroy climate stability as a public good. This interpretation was also put forward in the Stern report:

> In common with many other environmental problems, human-induced climate change is at its most basic level an externality. Those who produce greenhouse-gas emissions are bringing about climate change, thereby imposing costs on the world and on future generations, but they do not face directly, neither via markets nor in other ways, the full consequences of the costs of their actions.[24]

This interpretation also takes as its backdrop a common-goods interpretation, yet not in terms of the atmosphere as a common-pool resource, but in terms of the climate system as a public good. Stern states:

> The climate is a public good: those who fail to pay for it cannot be excluded from enjoying its benefits and one person's enjoyment of the climate does not diminish the capacity of others to enjoy it too. Markets do not automatically provide the right type and quantity of public goods, because in the absence of public policy there are limited or no returns to private investors for doing so ... Thus, climate change is an example of market failure involving externalities and public goods.[25]

Note that in the latter explanation Stern gives an interpretation of a collective action problem that adds a further interpretation, different from the interpretations already acknowledged. Stern contends that climate stability is comparable to goods provided by markets. He then applies the logic of market failures to climate problems. Although many would agree that the atmosphere has the characteristics of a public good, it is not obvious that markets can play that role in remedying public-goods problems.[26]

To sum up, collective action problems form part of the explanation of the causal history of climate change. Climate change is primarily caused by a high degree of greenhouse gases in the atmosphere resulting from many single emissions, which in turn result from uncoordinated activities of many different actors. The socially desirable goal of climate stability is forfeited because single actors do not consider the effects of their activities accordingly. Although the authors discussed here highlight the differences between textbook examples of collective action failures and climate change, they also draw comparisons. The undesirable social effect called climate change is the result of activities that – because of the correlation with emissions and because of a lack of a reasonable, coordinated scheme for the appropriation of the atmosphere – cause climate change. More specifically, this is the effect of uncoordinated activities that do not consider the characteristics of either (a) the atmosphere as providing only limited space for emissions or (b) climate stability as physically dependent on a certain mix of gases in the atmosphere.

3.2 The non-accountability view

In the view of another interpretation, the presented reconstruction of climate change as the outcome of failings in coordinated activities still misses the point. The interpretation of conflicts surrounding common-pool resources might not only be the effect of uncoordinated behaviour; it might rather resemble an unfortunate event that has not been "caused" in the literal sense. In addition, it is not fair to hold anyone responsible for this situation. This coheres with an interpretation of climate change as being, to a high degree, man-made, yet not willed as such. This is the "nobody's fault–nobody's accountability" view.

Coming back to the notion of "tragedy," an analysis of the meaning of this term as a first step in dismantling the "nobody's fault" perspective is helpful. A tragedy is an event that has three characteristic features: it is caused by actions and not by natural forces, its bad outcome is not intended, and its outcome is particularly disastrous.[27] Tragedies have been reported, staged, and debated since ancient times. People are interested in tragedies because there are some thrilling questions that are almost impossible to answer, but nevertheless intriguing to ask. The first question is, of course, whether that tragedy could have been prevented – and, if so, by what means? Tragedies often result from complex conjunctures that move actions in unintended directions. Moreover, people are interested in whether the actor really was blameworthy. In ancient times, tragedies told the stories and fortunes of people who bore guilt for disastrous events without being responsible for them.

The outcome that has been labelled "climate change" is particularly threatening, even disastrous. It is caused by anthropogenic forces. To address issues of blameworthiness and thereby also of agency in the context of climate change, two preliminary reservations need to be made. First, the tragedies of the natural commons are particularly difficult to judge because often it is not people or a group of actors who cause damage to the resource. Instead, industrial activities as well as the outcomes of collective behaviour need to be included in that picture.[28] Moreover, it is the accumulation of long-term effects that in the end might cause disasters. Second, I shall exempt from the discussion of tragedy Derek Parfit's much-debated "identity problem." This analysis of environmental problems as a specific set of problems supports the claim that it is "nobody's fault."[29] Parfit does not mean, however, that climate change is *not* the outcome of agency or collective agency. Instead, he argues that in the case of climate change it is false to say that activities that contributed to climate change really cause *harm* to people who are living now, and therefore it is also false to blame people who formerly produced high levels of emissions for this result.

After this preliminary clarification, I intend to discuss the question of whether the interpretation of climate change should include the statement that climate change is nobody's fault. To answer this question and to propose an alternative interpretation of "responsibility," I first review some important points that authors in the field of climate ethics have already contributed.

To begin with, many agree with Peter Singer's assessment of responsibility when he claims that the main polluters should also be the first to provide remedies for coping with the effects of climate change and to support mitigation, which is a costly enterprise. He forcefully argues that the ones who spoiled climate stability should now also fix it.[30] Following the logic of the polluter-pays principle, the main emitters in the past should now contribute to climate politics and pay for it. Singer makes an important point in claiming the responsibility of the main emitters. As for my discussion of the drama of NCPRs, it needs to be said that all the talk of "tragedy" should not overlook that some of the causes of climate change are clear-cut. Over-consumption of the atmosphere is not only bad luck or a result of the physical structure of climate and atmosphere alone. It is, in fact, an effect of overconsumption. Yet I do not think that the conclusion that the main past emitters now have to fix it is straightforward.

This point has been argued by authors who think that responsibility also includes the knowledge of doing something wrong. It has been a matter of vigorous debate whether it is right to presuppose knowledge about the effects of emissions.[31] Yet it is not necessary to presuppose that an actor has knowingly willed an outcome in order to hold him liable.[32] The latter distinction blurs another difficulty in addressing polluters as being responsible for fixing the damage that results from pollution. Since climate change is a process that has evolved over decades and centuries, claiming responsibility today means addressing not the contributors of climate change, but the successive generations who are already burdened with the effects of climate change that they have not caused. It has been proposed that instead of focusing on a polluter-pays principle, it is more promising to look towards a beneficiary-pays principle. This principle accounts for the gains derived from emissions.[33] This also provides a bracket between past emitters and the generations currently living. Affluent societies still benefit today from previously high levels of emissions, since economic development is coupled with high levels of greenhouse gas emissions in carbon-based industries. It is therefore right to claim major contributions from affluent societies both for mitigation and to support the adaptation of poorer nations.

This reasoning is appealing, yet it shifts the debate from the question of whether anyone can be held responsible to a reasonable politics of mitigation and adaptation. Such an approach, one that first seeks to name the perpetrators of climate change and then to ask them for action, has been criticized for several reasons. First, and congruent with my arguments, it is not fair to regard climate change as the outcome of particularly bad will. This does not mean that the outcome is bad per se and that acts of continuously endangering climate stability are morally negligible. Moreover, it would, of course, be possible to take the outcome of carbon-based industry in former generations into account. So, although it is someone's fault and although big climate "sinners" remain unchecked today, the language of "moral responsibility" in the sense of "blameworthiness" does not hold either. Instead, it might even contribute to a simple "fatalist" view – when it is already too late, it is also too late to claim responsibility in an altogether different sense.

I argue that instead of looking back, each person and each actor has responsibility for future developments regarding the climate as a natural commons. As beneficiaries of shared natural resources and as parties that are mutually responsible for not spoiling life-sustaining goods, I defend an approach to responsibility that focuses on future-looking accountability and on duties to cooperate to support resilience of environmental goods. This type of accountability is not allocated against the background of egalitarian principles, but differentiated among those actors who are more capable and those who are less so; moreover, the benefit from carbon-intensive industries counts for allocating accountability too.

This section prepares this discussion in one respect. It attempts to underscore the claim that although single actors might be exempt from a model of agency that takes single actors as fully accountable regarding long-term effects, the exploration of that structure does not necessarily exempt actors from accountability and contribute to fatalism. Instead, following the interpretations in Section 3.1, collective agency failures in this particular scenario are tied to a lack of coordination and also a lack of commitment to a type of cooperation that includes constraints regarding vulnerable natural goods. It might be wrong to ask the question of "responsibility" in a backward-looking way alone. Instead, I argue that it is part of our duties today to install patterns of coordination that meet ethical constraints and also to cooperate when these schemes are in place. Yet, for the logic of this chapter, it is important to set aside the problem of whether it is anyone's fault. This might be the lesson that readers take from Section 3.1. Yet this would be a false conclusion about the debate on collective action problems.

3.3 Social dilemmas and their parasitic nature

We are not in a position to develop proposals for effective coordinated activities unless another set of problems has been explored. This section portrays three kinds of dilemmas that might still occur. Note that by ana-lysing these problems, we are already jumping into a scenario in which cooperation in favour of a long-term, prudent, benefit-seeking scheme for exploiting a common-pool resource provides a backdrop assumption. The sce-nario for discussing social dilemmas is the following. Even if a group of people tries to adjust individual consumptive behaviour so that coordinated activities that also take into consideration whether environmental exigencies are in place, their activities will still be hampered by social dilemmas that automatically occur. Their collective endeavour will be undermined by free-riding, by the virtuous-as-the-fool problem, and by scoop-the-poop problems. All three social dilemmas will now be discussed.

3.3.1 *Free-riding*

Assume that some actors are successful in arranging their activities in such a way that they succeed together in a cooperative scheme, this scheme helps to support a fair distribution of services contained in a shared resource. And it is designed to keep the resource intact. Yet free-rider benefits will still destroy that success. Free-riders will benefit from the collective effort and the coordinated scheme of exploitation, without sharing the costs of upholding that coordinative scheme and without actively supporting the scheme. This is precisely what might occur in policies that support climate protection. Wagner and Weitzman think that the climate change problem is, in a nutshell, a problem of free-riding:

> You alone can do little beyond scream to get the right policies in place, which could then guide the rest of us in the right direction. Meanwhile, the overwhelming majority of the seven billion of us on this planet are free riders. We enjoy the going while the going is good. We don't pay for the full cost of our actions. Worse, polluting is subsidized worldwide to the tune of some $500 billion dollars per year.[34]

Whereas Wagner and Weitzman think that the object to which free-riding is parasitic is an already existing distribution of costs of negative externalities, others argue that free-riding is simply the most natural behaviour when it comes to the exploitation of a shared natural resource. Its detrimental effect does not result from over-exploitation alone. Instead, it is also tied to what has been termed the "green paradox." In Hans-Werner Sinn's interpretation, environmental resources that are needed in economically developed countries as resources for industrial activities, and for energy supply in particular, will be extracted as highly desirable goods. Whenever one actor tries to spare a good from exploitation to protect the environment, this just makes room for another likely appropriator. When the latter provides technologies that produce higher emissions, the overall outcome is worse than it might have been.[35]

3.3.2 The virtuous-as-the-fool problem

Assume that single actors are willing to invest in strategies that support climate protection, and that they do so by changing individual behaviour that usually incurs some costs. This endeavour of individuals is – in a way – dependent on an encompassing cooperative scheme. Even assuming the goodwill of individuals, it does not make sense to invest unilaterally in climate-friendly behaviour, for several reasons. This behaviour is both ineffective and, in a way, also "stupid." The former has long been observed. Climate change will not be stopped by individuals not driving their cars or investing in climate-friendly products. It is simply very ineffective. Even the argument of tipping points does not change the interpretation. It works in a scenario in which the single impact really will make the difference; yet in the case of climate change, this is not very likely to happen.

Yet there is another example of the "virtuous-as-the-fool problem." Jamieson argues that riding a bicycle in New York is a behaviour that has this quality, unless supported by a collective effort.[36] The bicycle rider will do some good, but his endeavour is undermined by the SUV drivers who literally spoil her engagement. Unless embedded in a scheme of cooperation which, as a collective scheme, not only promises, but realizes success, cooperative strategies that address climate change are self-defeating. This is a more general problem that might be overlooked. The effect of this structure is not only that people are dissuaded from investing in isolated, highly optimistic actions, but that, in addition, a morality that supervenes on those actions might be undermined. Moral societies need to have systems that pay rewards for the "virtuous" who support social developments. But they need to be cautious so as not to undermine such initiatives.

In discussing rule utilitarianism, Jamieson explains that rules might contribute to collective actions that lead to acting wrongly from a utilitarian point of view:

> This is because when facing collective action problems there are instances when I would do best and indeed contribute the most to the world by defecting from collective purposes. While it may be better for no one to drive or fly than for everyone to engage in these activities, what may be best of all is for me to drive or fly while everyone else refrains. If despite this reasoning I refrain from driving or flying then I risk my

utilitarian credentials, for I knowingly do what is less than the best. Even worse is the case in which I conform to some ideal set of rules to which there is not widespread conformity. In these cases, I might find myself riding my bicycle through the snow while everyone else blows by me in their SUVs. In both cases the only difference my behaviour makes is to reduce the general happiness by reducing my own happiness.[37]

Although an approach aiming to think through a collective endeavour would not necessarily side with rule utilitarianism, Jamieson outlines two more general lessons. First, he addresses free-riding as the optimal choice when many are contributing to achieving a shared climate goal. Second, he discusses a situation in which conformation with a reasonable rule, for instance the rule to reduce greenhouse gases, is ineffective. As for joint action, we could discuss the group of environmentally engaged bicycle riders who think that their behaviour has an effect on climate change, when actually it is helping the most egregious emitters carry on business as usual. As for environmental goals, a more general lesson is not that rule-following provides the solutions, but that the solutions are to be found in a mix of intentional action, the reshaping of the big institutional settings that cause the problems, and of course creative technological fixes. Imagine that the one person serving the environmental goal is not the one on the bicycle being stressed out by SUV drivers, but the creative mind in a laboratory who invents the ultimate "carbon catcher." Today, many environmentally inspired people are working on good and forward-looking solutions. They are part of a movement that, in some important respects, a joint action approach theorizes. More specifically, it adds to a clear view of the conditions under which this movement may gain momentum.

3.3.3 Buck-passing[38] and scoop-the-poop problems

Some authors argue that successful collective action assumes that each actor investing in collective action to bring about a socially desirable goal should also benefit from that shared goal. Yet climate action today means either investing in costly mitigation, whose benefits cannot be allocated to current generations, or it means investing in adaptation, which in turn is particularly costly and provides no advantage except being equipped when severe weather events occur. In addition, no one really knows whether there is anything to achieve through these strategies: mitigation might be too late; adaptation might not work out. Even when mitigation and adaptation are social goods, from the viewpoint of each individual actor, these investments come close to a lose-lose scenario.

The scoop-the-poop problem focuses on one facet of the problems I have outlined. It addresses the practice of transferring environmental costs to someone else by means either of simply not addressing the costs responsibly or by directly making an environmental hazard into someone else's responsibility. In Krantz's and Weber's analysis, the fact of possibly not benefitting from a collective effort plays into the temptation not to cooperate in the first instance. As Krantz has put it,

> [I]f many cooperate, the gain for each person is large, but the portion of that gain that stems directly from any one person's cooperation is too small to repay his or her effort or cost. Therefore each person has an incentive not to cooperate, regardless of whether many or only few others cooperate … Discharge of chemical wastes into waterways, or discharge of polluting or greenhouse gases or aerosols into the atmosphere provide a host of examples.[39]

The underlying rationale is that even when cooperation was in place, the gains will not go to individuals in any amount that could be considered compensation for the efforts.

This still leaves open the possibility of inventing climate institutions that by artificial means contribute to a win-win scenario. Although Posner and Weisbach try to develop a global contract that is good for everyone,[40] the proposals fall short of preventing lose-lose scenarios. As a consequence, not investing in climate change cures, and instead investing in other agendas, is the better strategy for efficient spending.

As different as these social dilemmas are, they all have one important trait in common. They also explain defection in relation to a scheme of cooperation that would provide a desirable scenario, yet one that is destroyed by defection. The diagnosis of social dilemmas does not necessarily assume that there is no solution, but rather that failings in a scenario happen that could have been successful. In this respect, the outlined dilemmas are comparable to "free-riding" as the most common trait of collective action problems in the case of climate change. In the most basic interpretation of it, free-riders benefit from a common scheme for the appropriation of a shared natural resource without paying for the services they derive from it. The presupposed scheme is not a physical scheme alone, but a social arrangement that guides the appropriation of services and assets of environmental goods and the atmosphere. It is a scheme of cooperation that coordinates single activities that relate to each other in a fruitful way. Free-riders benefit from that situation without contributing to its existence. And free-riders potentially also destroy the common effort of those who engage with the scheme of cooperation.

Although it is apparent that collective action problems assume that there could have been successful cooperation, it also has to be mentioned that another group of social dilemmas surfaces as soon as a cooperative scheme is in place, even when agreed-upon non-compliance with a cooperative scheme and behaviour that tries to circumvent the fair share of burdens in cooperative contexts are likely to happen. These problems will not be discussed until Chapter 10. They can only be addressed in a concise way after having explained the theory of joint action in the case of climate change and additionally the duties that various types of actors hold when cooperating with respect to climate goals.

Overall, at this point in the discussion the lesson from an analysis of a variety of social dilemmas is already straightforward: even when cooperation on a shared natural resource is a realistic and helpful alternative to destructive patterns of exploitation, social dilemmas prevent cooperation. Once in place, a cooperative scheme invites free-riders, whose behaviour destroys the possible success of cooperation. The virtuous-as-the fool problem will contribute to minimizing the social engagement of the "virtuous few." Instead of paying costs for environmentally correct behaviour, they will be frustrated by people who benefit from not complying with environmental exigencies. And the scoop-the-poop problem explains that incentives for compliance and cooperative behaviour will, in the case of climate change, be easily destroyed by disincentives, resulting from the fact that the gains made through cooperative behaviour do not go to the individuals who engage in cooperative schemes that also cohere with sustainable practices of the appropriation of a shared natural resource.

Various lessons can be drawn. On the one hand, it is easy to support a fatalistic view that says that an analysis of collective action problems only adds to an already dire view of climate change. It is too late to address it; the only thing that might still be possible to achieve is a prevention of worst-case scenarios. Yet social dilemmas appear to thwart even this. They undermine the collective efforts that would be needed to prevent the worst-case scenarios. But this is only one possible conclusion. In my view, an analysis of collective action problems and of social dilemmas also points in a different direction: in order to benefit from natural resources in accordance with socially desirable goals, it is particularly important to develop schemes of cooperation that do not suffer from the failings described. Obviously, uncoordinated and mutually impeding schemes of appropriation cause over-exploitation and destruction of shared natural resources. More specifically, social dilemmas obstruct cooperative schemes that supervene on the willingness and the motivation of many single actors to contribute to environmentally sustainable schemes of appropriation of natural resources – including in terms of the appropriation of them as a shared sink.

It is the goal of the remainder of the study to provide a theory of a cooperative scheme of appropriation that takes cognizance of the lessons of the reported failings. The remainder of this chapter only provides an initial step in that direction. It distinguishes two types of appropriation of natural commons that will not succeed and explains a third, more promising, alternative.

3.4 Natural commons and cooperative schemes

With regard to natural goods, one argument claims that as soon as people benefit from a good, they are already cooperating in one way or another. People are not free to benefit from natural goods. Instead, every single person and every single performance needs eco-services to be successful.[41] Most basically, people need air and water to survive; they need the services of both the air and of a stable climate. Even when most goods we daily consume are processed and transported goods, they only provide people with resources as converted materials.[42] In the end, our livestock depends on a range of common-pool resources, including, most basically, water reservoirs and the atmosphere.

In this interpretation, the hidden structure of NCPRs is brought to the foreground. Instead of natural goods merely being items that have systemic structures and that possibly offer benefits, they are at the heart of patterns of exchange. Instead of physical resources merely being shared resources that single actors appropriate to benefit from, common-pool resources are surrounded by patterns of exchange and cooperation that become part of a system of exchange. In this interpretation, there is no NCPR, such as a lake or the atmosphere, surrounded by potential beneficiaries; there is instead a system of exchange that supervenes on the physical resource and the social arrangements that regulate access to the lake and the atmosphere and determine patterns of exchange.

This perspective is important since it portrays a much more realistic initial scenario. Yet this argument goes in two different directions. On the one hand, it supports a *dependence approach to cooperation*, which says that people are more or less forced to cooperate, at least when they do not want to put at risk the sustainability conditions of that good. It also highlights that societies have already chosen their patterns of exchange by means of institutions that regulate access to natural resources, most importantly property systems, yet also all sorts of regulations, including pricing systems. Although this might be the case, this assessment includes a very loose sense of "cooperation" at best. The distribution of eco-services is often a matter of sheer political power; it is also regulated by institutions, including the institution of a free market, which are not in a position to provide a fair distribution of assets.[43]

On the other hand, it is also important to note that in many instances NCPRs only provide the tail end of all sorts of exchange mechanisms and processes of transformation, at least in economically highly developed societies. Water, land, and air serve as the backdrop and as resource for all sorts of agricultural and civilian processes. Much more realistic is an acknowledgment of what might be termed *chaotic exploitation schemes*. Currently, both the existing frameworks in environmental law and the institutions that regulate the appropriation of natural resources support this view. Instead of supporting the sustainable use of shared resources, they provide all sorts of opportunities for chaotic exploitation schemes.[44]

Taking the literature on social dilemmas into account not only helps us understand the sources of those problems, it also helps us see that defection and dilemmas are conceptualized in such a way that a "good scenario" is constantly presupposed. Moreover, the traits of the "good scenario" have been a matter of vigorous debate in theories that discuss a fair scheme of appropriation of a shared resource, be it the atmosphere, a water resource, or a public good such as climate stability. More specifically, social dilemmas account for failures to achieve socially desirable – and therefore good – scenarios. Achievements would include (a) a sustainable use of that resource and (b) a fair distribution of benefits and costs to uphold a fair scheme of appropriation or to restore it where misappropriation has already taken place and has already produced environmental harm. Social dilemmas are conceptualized in such a way that they destroy good scenarios that could be achieved – yet only by joint effort and by a scheme of appropriation the constituents of which need to be developed first.

In addition, even when the existence of truly cooperative schemes cannot be assumed, authors who analyse failures of collective action appear to have an idea of what that would consist of. Environmental problems not only occur because some actors over-exploit resources and thereby also jeopardize existing schemes of exploitation that are already well-established; instead, with regard to current concerns regarding climate stability (but also water and soil), the cooperative schemes are being forfeited for obvious reasons. One reason is that natural goods are particularly vulnerable in two respects. First, they usually lack access barriers that cohere with the performance of services that do not contribute to destructive scenarios.[45] And second, the terms of exchange with natural goods – the deposits in nature that civilization produces and the services that beneficiaries take from natural goods – are not integrated into schemes of exchange that produce and sustain fair outcomes. The underlying normative claim is that most basic goods need to be handled with care, and that schemes of appropriation need to be modeled against the background of joint efforts and joint commitments to protect the shared resource.[46]

Yet there appears to be something more in the analysis of failures of social cooperation. It appears as if the authors also have a much more concrete idea of what successful cooperation consists of. This does not amount to a theory of environmental cooperation, but it might lead the way in that direction. More specifically, this type of cooperation is in a certain and unusual way goal driven. The goal is not a static one, nor one that ultimately can be achieved. Rather, it is a pattern of cooperation that fulfils two normative claims. It serves a fair distribution of services and burdens resulting from the shared scheme of appropriation, and it supports the long-term usability of that resource. Basically, this also includes respect for exigencies resulting from what we know about the robustness of natural goods.[47] In the case of climate change, this vision is framed in a pessimistic way. Climate stability has been forfeited, because patterns of exploitation have not been oriented in cognizance of this exigency. Yet despite this special case, there is a perception of a match between affordances of natural goods and the desires and needs of people that guides ideas about cooperation when applied to natural goods.[48] Basically, cooperation is critical not only as a means of providing a fair share of assets and services of a common-pool resource, but also to achieve long-term benefits and to protect or restore a scheme of appropriation that serves a variety of needs and desires in the best possible way.

The approach I develop in the reminder of this book takes these lessons and the lessons of the previous chapter into account. It builds on three insights in particular. First, the drama of the commons is a very severe problem – especially when it comes to the most basic and life-sustaining natural goods. In order to address these multi-layered problems, it is necessary to develop a scheme for coordinating the exploitation and distribution of the natural resource that prevents defection. As a matter of fact, environmental problems have been resolved not by single actors, but by a synergetic effect that includes technological innovation, visible leadership, collective action, and effective politics. Since defection cannot be totally excluded, it is particularly important to develop models of cooperation that strengthen cooperators and help to contain the negative effects of defection. One way of doing this is by developing a scheme of cooperation that brings together all sorts of willing cooperators, independently of their role and their function in society and in nation-states. As a first step, the joint actor is a group of single actors, consisting of all the willing actors who share the goal of benefitting from a shared resource in a fair and sustainable way.[49] This model then needs to be adjusted to non-ideal scenarios.

Second, a theory of joint environmental action also appears to provide a forceful tool for addressing collective action problems. Authors in the field of climate ethics agree with Daniel Jamieson's observation that climate change is the "biggest collective action problem" of all time.[50] According to his assessment of collective action problems as portrayed in this chapter, the failures have many facets. In climate politics, there are loopholes for free-riding that can only be stopped by efficacious institutional means. Many also benefit from existing schemes of cooperation – for instance, from the activities of the virtuous and modest climate protectors – without contributing to climate goals themselves. The goal of this study is far more modest than proposing ideal institutional settings that avoid social dilemmas altogether. A theory of joint action, however, argues that the starting point for alternative practices and for new ideas about successful institutions is provided by an exploration of a new type of cooperation. More specifically, cooperation, when tied to NCPRs, needs to be reasoned as a scheme of appropriation of a shared resource that takes its particular shape into account. When benefitting together from a shared resource, more is at stake than fairness in the distribution of the available eco-services. Foremost, a theory of cooperation also needs to address the special exigencies resulting from that situation: the exploitation of a resource needs to cohere with the exigencies resulting from the vulnerabilities of that resource. In addition, as I argued in Chapter 2, appropriation schemes need to be coordinated schemes.

More specifically, cooperation needs to be reasoned in a way in which the goals cannot easily be destroyed by defectors. As a first step, it is helpful to think of goals as being negotiated and developed as in-group goals. Instead of enforcing all-encompassing schemes of cooperation no one is willing to comply with, the joint action model recognizes that people act in favour of goals to which they knowingly and willingly subscribe. Starting with a group of willing cooperators, the subsequent steps end with the tools to defy defectors efficiently. More specifically, the goals must accord with the existing goals of groups and with desirable goals. Obviously, changing the path of climate developments is not a reasonable goal in this context, yet it is still possible to form goals that are attractive and will have an impact on climate change and on necessary steps towards adaptation.

3.5 The theoretical way ahead

The debates on the failings of climate politics and on the degradation of climate stability have been accompanied by a debate on problems of collective action. This chapter has taken a closer look at this narrative. Three outcomes are particularly important. The first is that, although authors in the climate justice debate emphasize that collective action can have a dramatic impact on climate change, the climate problem is in several respects an exception to the general assessment of collective action problems. The extraordinary structure results not only from the sheer scale of the problems, which extend into the future and are global in magnitude. Collective action problems also provide a structure that differs from the textbook examples. People do not defect out of self-interest, nor is the most desirable social outcome forfeited simply because of mistrust among self-interested actors. Instead, collective action problems play into a structure that in itself is unfortunate. Climate change is the result of many isolated decisions that, on their own, have only a marginal effect, but together have produced the dramatic situation we now face. At its core are problems of a self-defeating structure of exploitation.

A second, and most remarkable, outcome is that, far from analysing just another "tragedy," thinkers consider that lack of motivation for contributing to and upholding schemes of cooperation is also a driver of this unfortunate situation. Some authors support the idea that ethical commitments would make a difference. Although neither people nor institutions are literally responsible for climate change, Jamieson argues that the solution to the current situation can only be worked out by also developing a new approach to ethics. And Gardiner even claims that the most dreadful aspect of climate change is the "tyranny of the contemporary,"[51] which is the unwillingness to stop buck-passing and instead to pay for the negative externalities being produced. Interestingly, the ethics that these authors propose is not exclusively an ethics of justice that claims a fair share in emission budgets for every individual. Instead, they also explore underlying values and virtues, in short, an "ethos" as a set of dispositions and beliefs that play out as a behaviour that differs from short-term consumptive behaviour.

A third outcome is that, in analysing social dilemmas, it appears as if there is an idea of a successful scheme of cooperation lurking in the background. Free-riding, leeching, and scooping-the-poop are types of behaviour that damage an otherwise successful scheme of cooperation. Yet an assessment of collective action problems also contributes to the insight that successful cooperation in appropriating shared natural resources will differ significantly from cooperation in the usual sense of the word. In the remainder of this book, I try to develop this insight further and propose a theory of joint climate action.

Notes

1 Peter Singer, "One Atmosphere," in *One World. The Ethics of Globalization* (New Haven, CT: Yale University Press, 2002), 35.
2 Steven Vanderheiden, *Atmospheric Justice: A Political Theory of Climate Change* (Oxford: Oxford University Press, 2008), 104.
3 Dale Jamieson, *Reason in a Dark Time. Why the Struggle Against Climate Change Failed – and What it Means for Our Future* (Oxford/New York: Oxford University Press, 2014), 61.
4 Stephen M. Gardiner, "The Real Tragedy of the Commons," *Philosophy and Public Affairs* 30, no. 4 (2001): 387–416.
5 Stephen M. Gardiner, *A Perfect Moral Storm. The Ethical Tragedy of Climate Change* (Oxford/New York: Oxford University Press, 2011), 36–38, 160–169.
6 Gardiner, *A Perfect Moral Storm.*
7 Jamieson, *Reason in a Dark Time.*
8 Gernot Wagner and Martin L. Weitzman, *Climate Shock. The Economic Consequences of a Hotter Planet* (Princeton, NJ/Oxford: Princeton University Press, 2015).
9 Gardiner, *A Perfect Moral Storm*, 24–44.
10 Ibid., 143ff.
11 Ibid., 47.
12 Ibid., 54–57.
13 Ibid., 56.
14 Ibid., 58.
15 Ibid., 56.
16 Jamieson, *Reason in a Dark Time*, 148.
17 Ibid., 162.
18 Ibid.
19 Ibid., 164.
20 On the other hand, Jamieson also appears to think that moral behaviour is necessary if collective endeavours are to be successfully addressed. In that respect, he ties the analysis of collective action problems to the concept of social dilemmas that need to be overcome for the same purpose. After having thoroughly analysed the novel aspects of morality needed in times of climate change, Jamieson concludes that a moral theory radically different both from former reasoning and from common-sense morality is needed. At the end of his book, Jamieson explains why virtue ethics is an important part of any ethics of the Anthropocene. In his view, a reasonable approach needs to fulfil two conditions. Ethics needs to be "proximate," which means that it is about "what presents to our senses and causally interacts with us in identifiable ways" (Jamieson, *Reason in a Dark Time*, 185) and it is agent-centred in that it gives individuals not only reasons, but the personal wherewithal to act accordingly. In particular: "Ethics is a collective construction, like morality, but it

seems to allow more individual variation" (Jamieson, *Reason in a Dark Time*, 186). At the cost of oversimplification, the proposal of a virtue ethics can in my view be considered just another interpretation of collective action problems. To address scarce resources in a way that guarantees fair conditions for the appropriation of services, more is needed than a pattern of exchange that helps to do justice to all parties. Rather, coordination problems can only be overcome by schemes of cooperation that differ from market-style behaviour. Jamieson envisions actors who are mindful, cooperative, and who realize simplicity. This provides the backdrop against which fair cooperation can also be outlined – even when addressing a complex resource.

21 Dale Jamieson, *Ethics and the Environment. An Introduction* (Cambridge: Cambridge University Press, 2008), 14.

22 Ibid., 15.

23 Wagner and Weitzman, 15.

24 Nicholas Stern, *The Economics of Climate Change. The Stern Review* (Cambridge: Cambridge University Press, 2007), 27.

25 I do not discuss the remedies for collective action problems that have been proposed as a consequence of this approach. Yet, according to the diagnosis, they need to include public policy interventions on a global scale in order to cure the allocation problems and the problem of free-riding. See Stern, 40–45. And they also need to include cures for collective action problems. Whereas Stern thinks that game theory (including ethical tools such as an ethics of responsibility) can cure the ills of collective action, this book presents an approach that reverses that programme, arguing that ethics is fundamental for overcoming collective action problems and that incentives are necessary to realize an ethical programme. For the remedies Stern proposes, see Stern, 510–529.

26 Here again, one concept needs to be made more explicit. Even when something is regarded as a failure of collective action, alternatives might still be available. This has been the underlying idea of the Kyoto protocol. For a critical discussion of the effects and possible effects of the Kyoto protocol if ratified by the US, see Jamieson, *Reason in a Dark Time*, 43–49. For an all-inclusive scheme of cooperation in terms of a global carbon tax, see John Broome, *Climate Matters. Ethics in a Warming World* (New York/London: W.W. Norton, 2012), 40–43.

27 For an interpretation emphasizing that climate change needs to be attributed to the effects of externalities, see Stern, 27. This includes the presupposition that the outcome is not intended. In this respect, it also argues that climate change is a "tragedy."

28 I come back to this normative problem, which is a problem of allocation of blame as well as of duties, in Chapters 7–9.

29 The "non-identity problem" goes back to a scenario of "harm," which Parfit introduced in *Reasons and Persons* (Oxford: Clarendon Press, 1984), 70–71. The identity problem states that it is logically impossible to think of acts that caused emissions as harming currently living individuals. John Broome interprets it this way: "The identity of a person depends on her origin, which is to say the sperm and egg she originates from. No one could have come from a different egg or a different sperm from the one she actually does come from. Consequently, even the slightest variation in the timing of conception makes a different person. A very slight change in people's lives means that they conceive different people. Had we significantly reduced our emissions of greenhouse gas, it would have changed the lives of nearly everyone in the world in ways which are more than slight. Within a couple of generations, the entire population of the world would have consisted of different people. Call this the 'non-identity effect'" (Broome, 62). This thought leads also to the following proposal: If a person living 150 years from now complains that by causing high emission rates we have done her an

injustice, this claim is false. The reason is given in the quotation. Had we reduced the emissions, it would not be she who would complain, since lower emissions contribute to a difference in conception. For a more sophisticated analysis, see Broome, 61–64. See also Parfit's own explication in Derek Parfit, "Energy Policy and the Further Future. The Identity Problem," in *Climate Ethics. Essential Readings*, eds. Stephen M. Gardiner et al. (Oxford/New York: Oxford University Press, 2010), 112–121. This problem does not affect my reasoning, since I do not argue that individuals need to be blamed for harming a now-living person.

30 Singer explains the argument of the polluter-pays principle by stating, "If the developed nations had had, during the past century, per capita emissions at the level of the developing nations, we would not today be facing a problem of climate change caused by human activity, and we would have an ample window of opportunity to do something about emissions before they reached a level sufficient to cause a problem. So, to put it in terms a child could understand, as far as the atmosphere is concerned, the developed nations broke it. If we believe that people should contribute to fixing something in proportion to their responsibility for breaking it, then the developed nations owe it to the rest of the world to fix the problem with the atmosphere" (Singer, "One Atmosphere," 33–34).

31 The debate on whether former generations could know the effects of high emissions and how this relates to responsibility is described in Simon Caney, "Cosmopolitan Justice, Responsibility, and Global Climate Change," *Leiden Journal of International Law* 18 (2005): 761–762.

32 Liability is distinguished from responsibility in that liability does not presuppose full knowledge of the effects of an action. In jurisprudence, for example, this distinction accounts for the difference between manslaughter and murder.

33 Caney explores this principle in "Cosmopolitan Justice," 2005.

34 Wagner and Weitzman, 97.

35 For an interpretation of the paradoxical situation that underpins the rationality of free-riding, see Hans-Werner Sinn, *Das grüne Paradoxon: Plädoyer für eine illusionsfreie Klimapolitik* (Berlin: Econ Verlag, 2009). [*The Green Paradox. Plea for an Illusion-Free Climate Policy*]. For a critique of the logic underlying the green paradox, see John Barry, *Rethinking Green Politics: Nature, Virtue and Progress* (London/Thousand Oaks, CA: SAGE, 1999), https://doi.org/10.4135/9781446279311.

36 Jamieson, *Reason in a Dark Time*, 173.

37 Ibid.

38 The term "buck-passing" was coined by Steve Gardiner to illustrate the behaviour of a currently living generation that unfairly favours its own interests over those of others. More specifically, it illustrates the behaviour of a group that is exclusively concerned with events that happen during the timeframe of their existence. "Each will secure benefits for itself by illegitimately imposing costs on its successors, and avoid costs to itself by illegitimately failing to benefit its successors" (Gardiner, *A Perfect Moral Storm*, 153).

39 David H. Krantz, "Individual Values and Social Goals in Environmental Decision Making," in *Decision Modeling and Behavior in Complex and Uncertain Environments*, eds. T. Kugler et al. (Dordrecht: Springer, 2008), 168.

40 Eric A. Posner and David A. Weisbach, *Climate Change Justice* (Princeton, NJ: Princeton University Press, 2010). One flaw in their proposal is the claim that since climate contracts will not support a win-win scenario, they should be enriched by support for poverty reduction. This needs to be criticized as adding one more dire task to an already difficult agenda.

41 For the concept of eco-services, see Robert Costanza et al., "The Value of the World's Ecosystem Services and Natural Capital," *Nature* 387, no. 6630 (1997): 253–260, https://doi.org/10.1038/387253a0.

42 Whether eco-services can be replaced with artificially generated products is a matter of debate; in the climate discussion, the hope of generating materials that produce much better outcomes in terms of lower emissions is debated in terms of efficiency and in terms of "discounting." The latter means that because of future prospects for higher eco-efficiency and higher substitution rates, claims of sustainable use need to imply that future generations will not be as dependent on environmental goods as we are. "Discounting" is the appropriate way to account for this. For a detailed discussion of "discounting," including the rejection of the discounting of the future well-being of people, see Broome, 133–155.

43 For proposals to distribute water through market mechanisms – and their drawbacks – see Kristin M. Anderson and Lisa J. Gaines, "International Water Pricing: An Overview and Historic and Modern Case Studies," in *Managing and Transforming Water Conflicts*, eds. Jerome Delli Priscoli and Aaron T. Wolf (Cambridge: Cambridge University Press, 2009), 249–265; Doreen Burdack, "The Australian Water Trade," in *Water Management Options in a Globalised World. Proceedings of an International Scientific Workshop (20–23 June 2011, Bad Schönbrunn)*, ed. Martin Kowarsch (Institute for Social and Developmental Studies (IGP) at the Munich School of Philosophy, 2011), 172–181, www.researchgate.net/publication/308118199_Water_management_options_in_a_globalised_world.
For a thorough debate on a normative scheme of water distribution and water conservation resting on a rights system, see Tracey Skillington, *Climate Justice and Human Rights* (New York: Palgrave MacMillan, 2017), 207–230.

44 On the patchwork of regulatory environmental law and its historical development, see Radkau.

45 One way to explain the shift into destructive scenarios has been proposed by Johan Rockström et al., "Planetary Boundaries: Exploring the Safe Operating Space for Humanity," *Ecology and Society* 14, no. 2 (2009), whose authors explain that anthropogenic pressure on the earth system has reached a scale at which abrupt environmental change cannot be excluded. In order to define safe operational spaces for human activities, they explore thresholds as intrinsic features of planetary systems. The systems they explore include the climate as being exposed to global change and, among others, freshwater supply, with primarily local effects. Further thresholds apply to ocean acidification, stratospheric ozone, global P and N cycles, atmospheric aerosol loading, land-use change, biodiversity loss, and chemical pollution. The concept of planetary boundaries states that the use of NCPRs can reach a critical level when thresholds as intrinsic features of the goods are being reached. Although much is still unknown, they also think that since these thresholds have not been surpassed for too long, there is still room to manoeuvre in a stability domain. Yet they also state that they think that three boundaries have already been crossed, including the climate threshold. See Rockström et al.

46 This is of course a very controversial claim. I come back to it in the chapter on fairness principles with respect to accumulative goals (Chapter 9) and the chapters on climate duties more generally (Chapters 7–10).

47 Concepts of a good situation of a natural good, including the concept of "robustness," will be explained at length in Section 5.3.

48 See Angela Kallhoff, *Why Democracy Needs Public Goods* (Lanham, MD: Lexington Books, 2011).

49 The concept of "joint agency" will be explained at length in Chapter 4.

50 Jamieson, *Reason in a Dark Time*, 61.

51 Gardiner, *A Perfect Moral Storm*, 143–184.

4 Joint environmental action

Cindy: Wonderful. Now I've got the point. Philosophers think that short-comings in collective action are a real and big problem, because they prevent concerted action. This brings me back to the analysis of shortcomings at the beginning. Now I can understand the link between the tragedy analysis and the proposal to investigate collective action. The link is now really obvious: There is no time for and no sense in debating ways to accuse people or collectives of wrongdoing in the past. Instead, an analysis of shortcomings of the past tells a simple story: people have not succeeded in achieving goals that would have been important to achieve by means of collective action. And the reasons are now obvious: As long as people do not have to cooperate in favour of climate goals, they choose alternatives. These alternatives are cheaper and easier for them.

Bert: Yes, this appears to be the message of all these debates. But why do we not force people to work together? Climate change is very dramatic right now. And climate goals are not only obvious, but we have a world-wide scientific committee, the Intergovernmental Panel on Climate Change (IPCC), that reports the dramatic changes and that also gives recommendations.

Cindy: I think the problem is that nobody forces governments to act. But possibly this is not the only problem. I think it would be very difficult to enforce climate change action, because it would end in a complete surveillance of everyday life. And possibly it's not even possible to enforce climate change action. What do you think?

Bert: I think you're right. Moreover, there is no interest in doing that. Scientists say that greenhouse gas emissions have to go to zero. But this means that our lifestyle is not possible any longer – all the things that are fun come with greenhouse gas emissions.

Cindy: I think it's better to discuss voluntary climate change action first.

Bert: Yes, you're right. Why should we always think so negatively about people and their commitment to act?

Cindy: But how do you think collective climate action would work?

Bert: Oh man. This is difficult to imagine. Actually, people pretty much live their lives. And they don't ask what other people are doing.

Cindy: Wait a minute. I've heard about two things that make things look different. First of all, even though not "cooperation" in a literal sense, it is nevertheless true that people are not as isolated from each other as one might think. When it comes to natural resources such as water and the atmosphere, everybody is pretty much connected with other people, possibly even more so when it comes to the atmosphere. There is interconnectedness that no person can actually avoid. And second, I've heard about a revival of theories in philosophy that discuss cooperation on a new level. They speak about "shared agency" and "joint action." And in a way they claim something very similar to the first point, just on another level. They say that working together, acting together, and even achieving goals together is part of the building blocks of life. It is natural!

Bert: Possibly we're much more connected with each other than we think we are. This is fascinating. But let's look into it step by step. My proposal: we first listen to the insights of authors who speak about social collectivity. I really want to know how "joint action" works. And later revisit the outcomes regarding climate change. Possibly there is as strong a link as we thought just a moment ago.

Natural common-pool resources (NCPRs) are at the centre of a range of conflicts. Not only are these conflicts likely to occur, but NCPRs also function as their catalysts. To some degree, this is the consequence of a specific structure of NCPRs. They lack clear-cut entrance barriers and simultaneously serve a multitude of different functions. As a consequence, different beneficiaries with different expectations generate conflicts of interest. In the context of theories of natural common goods, this has been debated in terms of unavoidably occurring "tragedies of the commons."[1] Since this term was introduced by Garrett Hardin for conflicts surrounding NCPRs, it has dominated the debate on natural commons.[2] Still, this metaphor is not particularly helpful as it blurs the differences between a number of various types of conflicts. A thorough analysis shows that various dimensions of conflict need to be distinguished. In addition, tragedies result not only from the very structure of natural common goods, but also from social dilemmas that occur when coordination is needed to make use of shared natural resources successfully.[3] In general, over-exploitation and destruction result from patterns of appropriation of natural services that have two failings: they do not comply with the conditions of the ecological health of that resource, and they do not support a fair distribution of benefits among all beneficiaries.

As a response to the fact that NCPRs are conflict catalysts, this study introduces a theoretical approach to natural goods that sides with a specific concept of cooperation. The central idea is that a theoretical approach needs to take the following insight into account: when multiple parties are utilizing a shared resource, it is helpful when they coordinate their use. Part of this is a clear idea of joint expectations and shared goals. Although conflicts among various users will occur, a joint action approach highlights the option of first defining goals. More specifically, future scenarios sketched out as "environmental goals" are framed as scenarios in which societies not only continue to use natural resources, but also regard nature as supportive in achieving further shared goals. The theory will explain how concepts of sustainability and of what is meant by a "fair share" are built into joint agency. But first the most basic premises of a theory of joint environmental action need to be unpacked.

In this chapter, I discuss a theory of joint action that can provide the theoretical background for environmental cooperation. At its heart is an approach to collective action that differs significantly from other types of cooperation, such as a market-style system of exchange, or cooperation via contract. Specifically, the proposed model of joint action explains how individuals come to act together without losing their identities as single actors who perform their actions wilfully. Individual actors do not dissolve into a collective actor. Instead, they form groups to achieve shared goals by acting together. Simultaneously, each form of cooperation in the use of an NCPR rests on material exchange processes among beneficiaries who use the NCPR, either as a dump for materials or as a resource. In the remainder of this study, I shall refer to an institutional setting that channels these processes of exchange among various beneficiaries as *a scheme for the joint appropriation of a resource*.

To interpret cooperation in terms of "joint action," two steps are necessary. At the beginning of this chapter, I illustrate some key elements of one version of a theory of "joint action." Since the debate on cooperation in terms of "joint agency" has recently gained momentum, the available literature is vast. It is not my goal here to discuss the myriad philosophical problems these theories present for theoretical debate. Instead, this chapter relies on elaborated theories of cooperation that figure as theories of "joint action." The approaches I build on include the concept of social agency put forward by Raimo Tuomela[4] and the theoretical work of Michael E. Bratman.[5] Both thinkers have provided theories of joint action, and both support a view of cooperation that differs significantly from that of cooperation as an exchange process serving the interests of single individuals as part of the cooperative process. I then introduce the various elements of a theory of joint environmental action.

The explanation of joint environmental action does not end there. I go on to use these theoretical insights to present a theory of joint action that focuses on joint practices of the use of natural common goods, in particular of water and of the atmosphere. Basically, joint environmental action aims to reach socially and environmentally desirable goals by working together. A central claim of this chapter is that successful schemes of cooperation can be achieved and sustained by means of joint action. Defending this claim requires several theoretical steps.

This chapter begins with an explanation of basic insights into explaining "joint action." These insights will be used to clarify "joint environmental action." Although joint environmental action is a special case, it relates to key elements of joint action in a more general sense. It includes various individuals who not only intend to cooperate, but who also subscribe to a goal that they intend to bring about together. It also includes an aspect of cooperation that differs from processes of exchange as achieved in market-style cooperation. More specifically, joint action includes processes of planning.

Two provisos are necessary at the outset. First, this chapter does not offer a comprehensive approach to joint environmental action. Instead, it is restricted to the discussion of certain fundamental ideas. It provides a theoretical backdrop for further discussion of the elements explored in subsequent chapters, which analyse the type of action involved in joint agency, a discussion of the relevant actors, an exploration of the normative resources for carrying out joint environmental action, in particular the role of an ethos in collaborative action in favour of shared environmental goals. They also include a debate on normativity in terms of the obligation to cooperate and a discussion of whether individuals are obliged to join climate action. Finally, necessary steps towards an institutional framework will also be discussed.[6]

A second proviso is that the debate on joint action has gained momentum recently and has reached the stage of sophisticated discussions about details of the approach – for instance, the question of whether single actors are obliged to join joint actors. This discussion will be discussed in Chapters 7–9, while the present chapter primarily draws on basic insights that date back to Margret Gilbert's pioneering work in the 1990s.[7] Parallel to this discussion, philosophers also started to explore anew the question of collective actors. This included a debate on the status of organizations and institutions as actors, and the concept of responsibility as applied to collective actors.[8] Finally, the debate on the ontological explanation of groups as actors and of collective actors also became a matter of vigorous debate.[9] This chapter does not engage in these important and fundamental debates. Instead, it starts with the insight that joint action is not only a fact, but also an important alternative model of cooperation as compared to market-style exchange processes. This claim has been elaborated by Bratman and Tuomela.[10] Although the two thinkers do not give the same interpretation and weight to the various elements of "joint action," it is assumed that, regarding the elements that are particularly important for joint environmental action, they do not provide systematic alternatives, but instead explain varieties of joint action by emphasizing different elements. Whereas Bratman stresses the goal-directedness of joint action, Tuomela is particularly concerned with the ethos and commitment of actors who together form a joint actor. In addition, both models are metaphysically modest approaches.

This chapter is organized into six sections. Section 4.1 starts by explaining the theoretical background of "joint action" that has been developed in social philosophy. Section 4.2 introduces the central building blocks of joint action according to a "planning theory" approach. In Section 4.3, I will be in a position to explain what "acting jointly on environmental goods" means.

Section 4.4 highlights a particular characteristic of joint environmental action and explains the theoretical challenges to which a joint environmental action approach needs to respond. Section 4.5 takes the opposite perspective and investigates the opportunities offered by a theoretical approach to joint action. Section 4.6 concludes.

4.1 Theories of joint action: Key concepts

To introduce the background of this theoretical approach, it is helpful to start with a different though closely related family of theories. In addressing collective action problems related to public goods, public choice theory provides the leading paradigms. However, this approach deviates significantly from "joint action" approaches. I first discuss some of its most distinguishing features; after that, I highlight some core concepts in joint action approaches.

Methodological individualism is a fundamental premise of social choice theory. This concept holds that to explain collective actions, it suffices to explore the actions of single actors. In this theoretical context, decisions about common goods rest on the decisions of individuals. Moreover, individuals are endowed with sets of preferences that provide the ultimate building blocks of decision-making procedures. However, the dependence of collective choices on individual sets of preferences causes theoretical failings. More specifically, in acting strategically, individuals may contribute to destructive scenarios regarding public goods.[11] Although theories of collective choice have been significantly overhauled and multilevel models have been introduced,[12] the leading paradigm of social choice theory still receives approval. The rational behaviour of self-interested individuals still provides the backdrop for the analysis of collective choices.

Recently, researchers in the field of social philosophy have developed a systematic alternative to this approach, calling into question the most basic premises in social choice theory. They disagree with methodological individualism and the individualistic interpretation of the plans and actions of individuals. These theorists are particularly interested in explaining how individuals who hold their own beliefs and have particular desires nevertheless succeed in accomplishing things together.[13] The anchor of concern is "joint action." Diverse as these theories are, theorists in this field of research generally subscribe to one central premise: *acting as an individual is radically different from acting as a member of a collective.*

This insight has also opened up a range of new research questions, including the notion of "shared intentions" and research on a variety of concepts of "mental sharing."[14] In addition to debates on the bedrock of joint action, the role of social institutions in framing collective action has also been examined.[15] This chapter will not take up this discussion, but rather addresses the more general point that, as compared to theories of individual action, this research provides new insights about collective action.

One of the leading research hypotheses of this study is thus that "shared agency" approaches are particularly helpful in revisiting and revising theories of environmental cooperation.

Here are some of the features of the explanation of "acting together" in the context of environmental goods that I propose are favourable: groups are not regarded as pre-existing units. Instead, processes of team formation that include processes of "team reasoning" are addressed as critical elements of acting together.[16] This opens space for introducing self-constituting groups of individuals who share a common endeavour, in particular regarding the use of environmental goods. It also provides space for addressing reasons to act together in favour of a goal. Instead of regarding preferences as the ultimate bedrock for explaining collective choices, a theory of joint action builds on the shared goals of various people. Additionally, single individuals who engage in a joint action are regarded as having reasons for doing so. Moreover, research on joint agency opens space for debating group action in terms of collectives of individuals and of stakeholders who work together to achieve goals that can only be achieved together. Theories of joint action are about activities that single individuals cannot undertake on their own. Examples that have frequently been chosen to illustrate this are "dancing" or "walking together," as well as "visiting New York together." The common background for these examples is the insight that these activities cannot be explained other than as joint actions. In these examples, acting together is a basic fact.

In environmental cooperation, coordinating activities is an important imperative. Following the analysis of NCPRs as conflict catalysts and the analysis of collective action problems, it is necessary to address the pitfalls of defection and of uncoordinated behaviour in the utilization of scarce and shared resources. In my view, "joint action" can serve as a theoretical tool for addressing these issues. In working together, both the problem of uncoordinated usage as well as the set of problems of collective action can be addressed – and perhaps even be overcome to some degree.

There is another feature of theories of joint action that is interesting to discuss in relation to joint environmental action. Since the interpretations of joint action resonate with concepts employed in the explanation of single actions, and since single actions cannot be explained without concepts of intentions, plans, and goals, they play an important role in explaining action.[17] Shared agency cannot be explained without also addressing shared goals individuals want to achieve together. Instead of relying on pre-existing sets of preferences of individuals, which in turn are transformed into collective preferences, the interpretation of shared agency starts with a concept of goal-directed action. As a consequence, the transformative process related to shared goals as well as to processes of group formation plays an important role.[18] This is helpful for an application to environmental goals, too. In order to achieve coordinated practices of sustainable use of a shared resource, it makes sense to first address the goals that people share when addressing that resource. This also contributes towards shaping patterns of utilization and perhaps even towards remodelling them when they are not conducive to those goals.

Another interesting feature of shared agency approaches is that they are not constructed as normative theories in terms of an ethical approach, though many of them are loosely related to normative issues or even explicitly relate to normative claims. One example is group agents, which have not only been explored in terms of thorough description and explanation; the responsibility of group agents and normative constraints of group agency are also of major interest.[19] Moral constraints, as well as the values included in acting together, have become an important part of the debate.[20] Although construed along the lines of an explanation of how group action works, this investigation also offers ideas about how normative concepts such as norms and responsibility relate to acting together. For an analysis of joint environmental action, this is a helpful trait. Instead of starting with norms and constraints, this approach first investigates how people come to work together on goals they have and share. It then addresses normative implications and normative constraints.

Moreover, some theorists claim that the sheer fact of individuals acting together already confronts them with obligations as co-actors. They should not cheat, and they should not behave like leeches.[21] Although this claim is controversial, it introduces the debate on how actors relate to each other when acting together. When truly subscribing to a shared goal, this also might make a difference in terms of their likelihood to cheat. At a minimum, it has been claimed that the obligations of people acting as group members differ from those of people acting as single actors.[22] As for environmental action, the binding forces of group membership need to be explored. More specifically, the question of whether they represent a type of mutual obligation that contributes to really motivating people to act needs to be explored.

Researchers have at times mentioned environmental examples, but usually only to exempt these cases from a standard theory of joint action. In expanding the basic account of joint action, Tuomela explains that doxastic elements can be made part of its basic explanation. In addition to jointly intending to perform an action X, to being in a situation to divide an "achievement-whole" into parts, and to a joint intention to perform X, this also includes shared beliefs about the former parts of joint action.[23] Tuomela then goes on to state that standard tasks, such as cleaning a yard together, are included in this extended version of joint action, but one example is excluded:

> Suppose that some people intentionally refrain from polluting the Gulf of Finland with the hope that this would eventually make the gulf unpolluted. They might intend to contribute to "cleaning up" the gulf. However, even if they believed that many other agents also similarly would refrain from polluting and also believed that this collective activity might result in the gulf getting sufficiently clean, all this need not give acting together.[24]

The reason Tuomela gives for this assessment is that there is no "joint intention," but rather that it suffices that every single actor is acting intentionally: "All that we are guaranteed to have here is a kind of contingent, causally connected aggregate of individual actions contributing to the same goal."[25]

In my view, Tuomela is correct in this assessment, to a degree. Nevertheless, this environmental example differs significantly from the examples that cohere with the proposals in this study. Whereas cleaning up environmental hazards is not a joint action, utilizing a shared resource by means of a cooperative scheme is indeed one. Bratman recognizes this when claiming that certain environmental concerns meet a condition of interdependence in policies. In this case, there is interdependence in the strong sense, but it is still only feasibility-based. Tuomela interprets the example of environmental policies as follows:

> We each recognize that if the other did not have this policy [certain environmental concerns] then, other things equal, it would not be feasible for us to deliberate together in relevant ways; that is why our policies are persistence interdependent. But neither of us sees the other's policy as contributing to the desirability of our giving such weights: each favours our giving these weights solely because of what he sees as the importance of these environmental issues, not because of a social value of interpersonal convergence in policy. Nevertheless, what each of us intends (namely, that *we* reason in this way) involves the contributions of both. And we each recognize that the policy of the other is, other things equal, a condition of the feasibility of such joint reasoning. That is why our policies are interdependent.[26]

This example demonstrates that Bratman does indeed consider environmental examples, but only in a very restricted sense.

Despite the former comments, research on "acting together" indeed offers a new spectrum of theoretical concepts and insights. More specifically, these approaches differ from those of theories of collective choice and collective actors in that they do not simply attempt to apply concepts from the debate on the actions of individuals to actions of groups. Instead, they take social action as a unique phenomenon and as the starting point for an analysis of action.[27] The attractiveness of explaining shared agency as a phenomenon that differs from both of these has stimulated research, but it has not – to the best of my knowledge –so far been systematically used for environmental theories.

In the remainder of this section, I start by applying these ideas to environmental issues. More specifically, I discuss the building blocks that, in my view, are particularly fruitful for addressing shared environmental agency. For the basic model, I first focus on Bratman's approach to a planning theory of joint action.[28] I later widen the spectrum and introduce further elements. More specifically, Sections 4.3–4.5 are dedicated to explaining how this theory can be applied to environmental action, and also to addressing its most challenging aspects.

One proviso needs to be made in advance. It is not the goal of this chapter to add new insights to theories of joint action or to cooperation more generally. Instead, I intend to translate key insights into the scenario I am interested in, which is that of an appropriation by a group of actors of a shared natural resource qualifying as an NCPR. Although there are significant differences among various theories, the joint action approach as proposed by Bratman and the theory of Tuomela share enough common ground that the insights of both are useful in reinterpreting joint environmental action. Obviously, this attempt remains controversial.

4.2 A planning theory of joint action

One of the irritating facets of theories of shared agency is that they immediately begin with joint action. More specifically, the concept of "joint action" fills two spaces. In explaining what individuals are doing when they act jointly, the goal they want to achieve together is already classified as a "joint action." Bratman is well aware that this renders the joint action approach close to circular reasoning.[29] The problem of apparent circularity, however, can easily be overcome under two conditions: first, the goal is different from the actions individuals undertake to achieve it. More specifically, each person has the overall goal of realizing a joint action and therefore develops sub-plans she tries to render constantly coherent both with the overarching goal and with the sub-plans of other individuals. Second, although it does not perfectly depend on this person to achieve the shared goal, it is the object of her will, and something that does not yet exist (intention, not mere desires). When intending it, the goal does not exist; however, individuals involved in joint action not only want to see it achieved, but are also prepared to undertake further steps to realize it together.

I need to make another assumption explicit right at the beginning. It concerns sociality. More specifically, what type of social cohesion needs to be assumed to explain "joint action"? Bratman's approach differs from other theories of joint action in that he regards joint action as merely an upgrading of a theory of action whose building blocks are provided by a theory of individual action. Groups are distinct entities; they figure as collectives with shared intentions or goals. An action consists of an event that is a real change in the world, and it is undertaken by a single person. This does not change when individuals act together to achieve a shared goal. Groups are assemblages of individuals. The glue holding them together is provided primarily by the intention to achieve a goal together.

This is not the place to discuss action theory more thoroughly; indeed, the explanation of how mental states relate to an event classified as an action is one of the most intriguing issues in contemporary philosophy. Here, I just address the question of what a theory of joint action adds to the assumption that an event is related to an intention formerly held by an actor.

Bratman explains that for a "joint action" to be joint, two individuals must both have the intention to bring about a shared action. Moreover, each of the two needs to be aware that the other has that intention. A further building block of this joint action approach is the awareness and intention of realizing the joint action by a means other than individual agency. Instead, in a two-person case, both actors also intend to accomplish the joint action by way of acting together. In addition, they know that sub-plans are necessary to explore the steps for accomplishing their overarching and joint goal. More specifically, two individuals need to be ready to "mesh" their sub-plans. This is the process of mutually readjusting plans.[30]

Overall, these building blocks are modest and do not imply anything like a specific notion of a "we" as a collective actor in the strict sense. Nor does this theory employ an unusually demanding idea of how actors should react to each other to achieve a shared goal of joint action. More specifically, it is not assumed that intentions are shared by means of mental acts or mind-reading. Bratman states that, instead, a frequent exchange about the sub-plans is critical to re-adjusting plans accordingly.

Bratman's approach is – in contrast to other approaches – not a particularly morality-laden framework, either.[31] More specifically, he does not see moral principles in place when people start acting together. Moreover, a completely shallow approach to agency is also precluded from the beginning. Purely strategic behaviour that lies at the heart of problems of collective action is rejected as a necessary premise in explaining scenarios of groups of actors. Actors who share only the intention of achieving a joint action without also having the intention that it be accomplished by means of mutual awareness of the intentions and sub-plans of the other participants are not properly engaged in "shared agency."[32] Precluding purely strategic behaviour does not imply a moral obligation not to cheat on the other person. Instead, it is simply unreasonable to cheat when both parties want to achieve a shared goal. Whether or not this assumption can be applied to environmental scenarios needs to be debated at a later point.[33]

To sum up, shared agency in terms of "joint action" explains how individuals succeed in accomplishing a shared goal. It focuses on types of action like "going together to New York" or "walking together," which are themselves cooperative enterprises.

The challenges a theory of joint environmental action faces are twofold. First, it must be explained how environmental goals figure in this approach to acting jointly. Second, it has to be explained whether this model of small group behaviour can also be applied to groups of joint beneficiaries. Groups dealing with benefits from a shared resource are much more incoherent and likely more controversial than are groups of people who engage in a joint enterprise. The latter includes the presupposition that all participants in relatively small groups have an interest or even a pre-existing intention of bringing about a desirable goal together with other individuals. As for the appropriation of natural resources, there are no shared goals at hand – at least not at first glance. Moreover, in the history of environmental regulation, schemes for appropriating natural goods have predominately been the result of sheer power. Joachim Radkau makes the important point that "nature" and "power" are closely related, not only with regard to gains from nature, but also regarding cooperative schemes that have been part of the regulation of appropriation of natural resources, such as the regulation of waterways.[34] This speaks to top-down approaches rather than for flat ones among individuals acting together.

The first reply to these reservations is that both problems can indeed be connected to debates on theories of joint agency. The first problem resonates with the question of which type of goals figure in a joint action approach. What makes a goal work as a joint goal – aside from being desired by all parties? The second problem is a recurrent theme in theories of joint action. What group size is adequate for joint action and how can groups of people be defined when these people figure as a group actor? Both challenges will be addressed in turn. First, however, I outline the fundamental components of constructing a theory of "joint environmental action."

4.3 The model of joint action in a nutshell

In order to render the application to environmental agency more plausible, I first intend to start with an example of small-scale interactions. The example – which has also received much attention in the literature on environmental governance – is environmental cooperation in relation to a hypothetical river. This river is utilized by many different beneficiaries. It supplies water to agriculture, serves as fishing grounds, and is used as a transport medium for ships. Companies use its water for cooling, and it is used for recreational purposes.[35] Taking this example, the question I now want to address is: how can the activities of the participants be explained in the context of a theory of joint agency?

To start with the beneficiaries, while they all benefit from the river, they do so in various ways. They do not share plans or have the same plans regarding these activities. Instead, they all have their own plans. But they also know that they are not only dependent on the river, they are also dependent on each other to benefit from the river. The benefits one person receives do not necessarily undermine the benefits of other individuals, but that can happen. More specifically, they need to coordinate their benefit-seeking behaviour to keep the resource intact. At a minimum, they share the goal of not destroying this resource, which they all need to realize their individual plans and goals. The theory of joint action as presented in Section 4.2 presupposes not that shared goals are the comprehensive goals of individuals, but that they are intended by means of participatory intentions. My proposal to frame environmental goals is related to this insight. The goal not to ruin the common resource is a shared goal. However, it is not an inviting goal for participants. To render that goal more attractive and to transform it into a real joint action goal, it needs to correspond to a positive and more comprehensive vision of the joint achievements of a group of people.

In order to fully understand common-pool-related goals, it is helpful to correctly understand the interface between the NCPR and the activities of groups of people. The shared interest is a pattern in the appropriation of the resource that fulfils two tasks: (a) it serves the interests of beneficiaries and (b) it keeps the resource intact and thereby enables the benefits to continue. However, joint action does not rely on shared interest alone. Instead, it rests on a shared view of something individuals want to achieve together in the future. As for common-pool-related goals, they describe a desirable future scenario in which the common-pool resource plays an important role. To explain it metaphorically, the key question is: "How do we, as a group of people, want to live together with the river in the future?" The answer could be that "we, as a group, want to build a dam" or that "we, as a group, want to protect the aesthetic value of this river." What already distinguishes this interpretation from other theories of cooperation is the type of "joint action" the group intends to achieve. In order to translate the "shared goal" as a joint activity to the environmental action of a group actor, it needs to be equated with a version of cooperation combined with an idea of a worthy development goal.

This proposal rests on several presumptions. A river, like other common-pool resources, continuously performs eco-services.[36] These services are needed and desired by various beneficiaries. The problem of uncoordinated extraction results from a missing scheme of appropriation. At any rate, a scheme of appropriation cannot be developed without also outlining shared goals about the future of the resource. More specifically, this includes decisions about the range of available schemes of appropriation. The decision to keep the river as a recovery area rules out its utilization as a water resource by an industrial facility; it also rules out its utilization as a waste dump. The building of a dam, on the other hand, prevents the use of the river as a habitat or as a resource for irrigation systems. Still, all beneficiaries want to make use of that river. What differs in terms of "joint agency" is the negotiation of a scheme of appropriation that coheres with an overarching vision of the future design of that resource. This type of coordinated plan-making with respect to a shared resource has also been termed "active co-designing with nature."[37]

The same scenario can also be explained the other way around. The joint future scenario of "benefitting continuously and collectively from the river" is not an event that can be accomplished by single individuals alone. Instead, it rests on ideas for future schemes of appropriation that shape possible processes of exchange, both among the individuals involved and among the beneficiaries and the resource. As for the aspect of a joint activity, usage schemes come close to the examples of "dancing" or "visiting New York." Still, the ability to bring about a scheme of usage in the first place is restricted by conditions that all environmental cooperation needs to meet. Once the resource is overused or destroyed by regional pollution or chaotic patterns of exploitation,[38] the goal of shared long-term utilization can no longer be achieved. In addition, the goals are not freely chosen; rather, they exemplify visions of future ways to shape the natural resource. They are an expression of what groups of people consider a good way to "live with the resource" in the future. Yet, on second thought, even the latter can be compared to insights into the theory of joint action. Whether "we" succeed in "dancing together" or "travelling to New York" is also constrained by conditions that are beyond our power to act upon.

One further important component in shared environmental goals is that each person is aware of her role as co-actor in achieving the intended aim. Actually creating "shared goals" is not the only precondition for working them out together. Joint action on a scheme for appropriating eco-services that a group of people wants to bring about is also tied to explicitly shared goals. Note also that common-pool-related goals differ significantly from environmental goals more generally. Usually, environmental goals such as conservationist goals or goals of re-naturation are not part of the set of desires of actors; instead, environmental goals tend to provide challenges and are often seen as negative, but necessary. Yet this interpretation of environmental goals is far from self-interpreting. Instead, environmental goals can be interpreted as particularly important, but also as especially desirable goals.

The next step in adapting joint agency to environmental agency is to explain the process of developing sub-plans and meshing them and integrating them with each other. Let us assume that individuals develop sub-plans of how each of them can contribute to realizing a shared goal – for instance, a green future for the city of New York. People who subscribe to this goal frequently meet to exchange ideas and plans. They also "mesh" their plans – that is, they attempt to render them mutually supportive. At the very least, they need to revise their plans and sub-plans if they prove to be counter-productive in relation to other plans.[39] If one person intends, say, to stop the release of sewage into a river, but another person proposes that investing in filter techniques would be a better solution, they need to discuss more than how their sub-plans can best be coordinated. Obviously, they also have to set priorities for achieving their shared goal. Most importantly, they need to understand how their individual sub-plans relate to that overarching goal. The framing within which these processes happen is still the shared intentions of each participant to contribute to achieving the shared goal.

A final modification that needs to be made relates to the self-interpretation of actors in joint environmental action. Besides the problem that actors in this field often confront negative agendas – rivers need to be cleaned up, people need to be compensated for suffering from severe environmental hazards, climate protection is primarily addressed in terms of adaptation to new living conditions and to cuts in harmful emissions and their consequences in terms of additional costs for the transformative process – the role of agents also needs to be discussed. One severe reservation regarding the interpretation of environmental action as "joint action" concerns the distribution of responsibility with respect to the natural environment. It is perhaps crude and unfair to push people into the role of guardians of a natural resource. As a first answer to this problem, a reservation with respect to practicability needs to be made. To propose a reconsideration of joint action in terms of joint environmental action does not include the proviso that the fulfilment is up to single actors. Instead, I discuss how both the realization of appropriation schemes that cohere with common-pool-related goals as well as their supervision and enforcement is up to institutions. However, the willingness to cooperate and to contribute to these opportunities nevertheless depends on the existence of shared goals that are either justified and possibly obligatory goals on the one hand, and on the other hand are actually desired goals. This desirability goes beyond the continued availability of eco-services, but instead needs to resonate with values to which the group of actors is willing to subscribe.

At this point I want to emphasize that although every single person can be addressed as a possible co-actor, this does not mean that the responsibility for achieving shared goals needs to be determined within an equal-responsibility approach. On the contrary, I maintain that responsibility needs to be allocated by means of various principles of fairness.[40] Furthermore, this also relates to presuppositions about "fair goals." In order to address these and other questions, it is important to develop a thorough reading and interpretation of what co-agency can mean in this context. Each person can and – as I shall discuss later – should, under certain conditions, contribute to achieving common-pool-related goals. However, she alone cannot, of course, bring them about in the literal sense. Instead, nation-states, international institutions, and instruments of environmental law are needed to push forward agendas that ensure the sustainability of common-pool resources. More specifically, a theory of joint environmental action also leaves room for addressing the shared goal of sustainable appropriation of a shared resource not as a separate agenda, but as integrated into a set of further goals. Whether or not actors should regard environmental goals in this light is another question.

4.4 Theoretical challenges

So far, I have explored the theoretical means of applying a model of joint agency to environmental action. More specifically, I have explained some of the building blocks of a theory of shared agency in the context of environmental goals. It is now time to more thoroughly discuss a major objection to this proposal. This reservation has many facets, though, in sum, it posits that acting with respect to environmental goals differs from other processes in that the former are primarily reactive aims. Environmental action is primarily conceived as the process of cleaning up environmental hazards and restoring the environment to a previous or more pristine condition. Still, so the objection goes, these reactions to environmental hazards have nothing in common with joint action. In addition, environmental agendas also react to failings of collective action in terms of undesired negative externalities.[41]

These objections are important. Instead of taking reactive agendas as the standard case, a theory of joint action starts with forward-looking agendas. Concerning environmental action, it appears to assume that it is not too late to modify patterns of exchange that resonate with the functional soundness of natural goods. More specifically, joint environmental action should work as a blueprint for helping to correct existing practices. Furthermore, although the spectrum of environmental action is today often tied to nega-tive agendas, I intend to defend the claim that environmental action should be framed in terms of future-looking proposals for shared goals. This includes a shift from a reactive scheme to an explicitly proactive one. This revision of environmental action is not only challenging as a matter of fact, but also as regards a theoretical approach to environmental action. In this section, I address three challenging aspects of framing environmental action as joint action:

- Is environmental action really a capacity to *act*?
- Are environmental goals really *joint action goals*?
- Are environmental goals reasonable goals in an era of *considerable environmental damage*?

The discussion in this chapter of each of these questions will not yield a comprehensive answer, but it will serve to highlight the challenges as a way to set the stage for further discussion.

Is environmental action really a capacity to act? Individuals as well as agencies who have to work on environmental hazards and who attempt to repair environmental damage might not regard themselves as "actors" in the strict sense. More specifically, the fate of developments in the environment is to an important degree attributable to invisible and long-term effects of pro-cesses of production and consumption. Even when leaders or individuals step into the role of actors vis-à-vis nature, they experience their capacities as being rather limited.

In discussing the capacity to act, there is of course the difficulty posed by long-term effects on the environment. Theorists who favour the concept of the "Anthropocene" – the notion that humankind has reshaped the natural environment in a significant and irreversible way – argue precisely this. The developments that have been caused by civilizations are irreversible and quite uncontrollable.[42] Climate change appears to be a particularly clear example of uncontrollable and unforeseeable developments. Even if the emissions of carbon dioxide were immediately switched to zero, the effects of greenhouse gases in the atmosphere would continue for thousands of years. Under these circumstances, it is difficult not only to regard action as effective, but to think of climate agendas as encapsulated in the concept of action.

The diagnosis of environmental change challenges theories of joint action. When neither individuals nor groups of individuals or political entities are any longer in a situation to regard themselves as effective actors, it is difficult to apply concepts of shared agency. Besides, in another respect, the experience of an inability to act also results from a lack of coordination and cooperation. Even when an intact climate is no longer an achievable goal, there is still room for manoeuvre and also for addressing joint actions that show which groups of people want to live with a common-pool resource in the future.

The Paris Agreement is an example of that.[43] Assuming all states really deliver on their pledges, this will contribute to the achievement of a common-resource-related aim. This is the goal of stopping the overuse of the atmosphere as a waste dump for greenhouse gases in order not to risk unforeseeable climate events. The problem with environmental agendas is not a general incapacity for action, but rather the ineffectiveness of individual actions. A joint action approach provides the tools for achieving effective agency.

Another difficulty is posed by the interpretation of the *goals* that figure in a theory of joint action. Tuomela is right in claiming that even when many people desire an end to the pollution of the Gulf of Finland, this does not amount to a joint goal.[44] Rather, it is a case of many people wanting the same thing. The desire might be strong enough that when everyone does his or her share, the goal will be realized. True, individuals who engage in environmental agendas usually want to bring about a certain situation or condition. Instead of merely desiring certain policies or a certain outcome, the joint environmental action approach assumes that each of the participants in joint action wills that a goal be achieved by common efforts. And the shared goal comes with a distinct shape of future patterns of appropriation and exchange that will be achievable among beneficiaries of a shared resource. Instead of wishing for a condition to come about, the approach takes the more realistic route of recognizing that NCPRs are constantly shaped and reshaped. It proposes to render these practices explicit, to interpret them as a common affair of a group of people, and it proposes to think through the effects of the design processes with respect to the available patterns of appropriation.

Another obstacle to interpreting environmental goals for a theory of joint action is that collective processes might not be regarded as helpful for accomplishing environmental agendas in another respect. Beyond the fact that outcomes are unknown due to uncertainty and a lack of scientific capacities for projecting further developments,[45] the argument is that due to the complexity of the side effects of exploitative practices, collective environmental action is threatened by a range of unforeseeable effects and developments. The only reasonable goal presented in the literature when this uncertainty is considered is precaution.[46] This again overlooks the possibility of agency, even when the "operational safe space" for action is already limited.[47] Of course, societies continuously shape access conditions to natural resources, and the same is true with respect to the atmosphere. Still, the way this happens is often rather invisible. It would make sense to discuss various options to shape the interface of societies and shared natural resources and then to decide upon them.

One more objection needs to be addressed before outlining the theoretical approach to joint environmental action in more detail. Today, many researchers agree that Planet Earth suffers from the enormous environmental impact of an ever-growing world population – a population that is particularly ingenious at developing technologies that effect environmental change. The impact of humankind on our geophysical systems is grave, irreversible, and all-encompassing. Hence the aforementioned eschatological notion of the "Anthropocene."[48] An initial response is that this diagnosis does not necessarily render agency, and environmental agency specifically, superfluous. Nevertheless, what actually is called into question is a naive interpretation of goals. Theorists who support the Anthropocene thesis explain that the idea that humankind is free to benefit from nature as an unlimited resource is an illusion. Instead, nature as a resource that is free from human impact is viewed as just a residuum that serves as a substrate for food and goods that are the outcome of highly industrialized processes. But does this diagnosis also speak against demanding "environmental goals" as such?

In one respect, this is certainly the case. Since people do not live in natural environments, they are no longer in touch with nature. Most of us lead alienated lifestyles when it comes to our relationship with the natural environment. The goods we require and consume are the final products of often global production chains. Even regarding the most basic natural goods, such as water and food supplies, it is illusory to think that they provide schemes of appropriation near our homes. People who live close to a lake, for instance, will possibly never eat a fish from it, but instead receive their food from a supply system that relies on long-range transportation, perhaps even on a global scale.

Furthermore, the consequences of over-exploitation of nature in the Anthropocene might also call for a necessary readjustment. In accordance with the demand to respect the limits to growth[49] and with the alarming situation of safe operational spaces in the geological systems of Planet Earth,[50] it is apparently necessary to reshape the processes of utilization – in particular the utilization of the atmosphere, water, and soil. The thesis of the Anthropocene does not appear to undermine a joint action approach. What it challenges is the framing of environmental goals in a bio-conservative way, which means that environmental conservation of remaining resources is the appropriate answer in the Anthropocene. But this is not the case in the approach to environmental agency I am proposing. Instead, environmental goals are framed as those resonating with "schemes of cooperation" that can only be achieved by means of joint action. More specifically, environmental goals are framed as future scenarios that offer a match between the needs of people who benefit from goods and the affordances of natural common goods, such as a river or a lake.

4.5 Specific characteristics of joint environmental action

After having discussed the challenges of providing a theory of joint environmental action, I now discuss some special characteristics of joint action as applied to NCPRs. The question is, what are the features of joint action as applied to common-pool resources? Some of the answers also contribute to the insight that joint action is particularly attractive in the application of that specific mode of cooperation to forward-looking agendas regarding common-pool resources. Overall, four characteristic traits can be spelled out at this point.[51] First, joint action as presented in Bratman's theoretical approach to a planning theory[52] provides room for individual plans that contribute to bringing about overarching goals. Second, joint environmental actions are conceptualized as processes, not as singular events. Third, joint action is achieved as physical events, not merely as deliberation or as speech acts, and is still tied to the will of individuals. And fourth, joint action as discussed in this study includes a forward-looking, positive goal. I discuss each of these aspects in turn. Together, they already provide a first impression of the distinguishing features of joint environmental action. More specifically, this section presents aspects that are likely to interpret joint action not only as an *adequate* explanation for environmental agendas, but also as an *attractive* one.

Although individuals who are walking together or traveling to New York together are performing a single joint action, they do not achieve it by means of doing the same thing. More specifically, the path from conceiving the shared goal to accomplishing it is marked by steps that together accomplish this objective. Depending on the theoretical details of addressing joint action, this has received different interpretations. In Bratman's theoretical approach, individuals develop sub-plans in order to achieve a shared goal. Success in achieving that goal does not depend on each person actually doing the same thing. Quite the opposite is true. Success depends on the successful sub-meshing of plans that differ from person to person.[53] The thing holding the whole process together is the shared intention to bring about a common goal. The shared intention accounts for the interpersonal coordination of single actions; it simultaneously structures the thinking of how the goal can be achieved as a collective. More specifically, joint action is not accomplished by means of a single event; instead, it has what might be called a "multiple realization," in that several actions together provide for the successful fulfilment of the goal, which resonates with a shared intention.

This characteristic is helpful for conceptualizing the accomplishment of common-pool-related goals. An overarching goal that many individuals intend to achieve together can be accomplished via not only a diversity of single actions, but also by individual actions that differ from each other, yet which jointly contribute to bringing about a common-pool-related goal.

Another important issue that resonates with environmental agendas as encapsulated in environmental goals is that environmental aims cannot be accomplished all at once and forever. Instead, the structure is one of trial and error, of adjustment and re-adjustment. An environmental goal is achieved through a process.[54] Jointly benefitting from an NCPR is not like walking or dancing together – but it is a kind of coordinated behaviour structured by "rules" that prevent, for instance, bumping into each other or treading on someone's foot while moving through common space. It is a type of physical, as well as coordinated, collective action on a timeline.

The exploration of this model of action needs to include a thorough assessment of how processes of re-adjustment can be explained. The important point is that joint action gives space for cooperation in terms of mutually non-destructive practices of exploitation, based on agreement on a shared goal. Another feature is that in acting as joint actors, individuals do not have to change their identities, nor must they give up their personal life plans. Instead, they must simply contribute to achieving a shared goal that they want to accomplish together. This aspect of the theory has received various interpretations, depending on theoretical decisions regarding the underlying theory of action and on the actors. The important part, at this point of the argument, is that a theory of joint action explains the readiness of single actors to adjust their individual actions in such a way that the shared goal can be achieved, though without necessarily abandoning individual interests or desires. What is needed, though, is the intention to really bring about the action as an event that can only be accomplished as something done together.

As for environmental goals, one of the most persistent prejudices is that they are not part of the preference set of individuals, but instead are usually regarded as being *against* the individuals' interests. Individuals do not intend to benefit from a lake in a coordinated way. Instead, each person utilizes it in a self-interested manner. Be that as it may, a theory of joint action need not accept this assumption. Nor does it have to assume that individuals will automatically choose environmental joint goals. Instead, at this point of the theory, it is assumed that common-pool-related goals can figure as joint goals in a comparable way to "dancing" or "walking together". They are objects of the intentions of individuals in the sense of willing to bring them about as a collective endeavour. This suffices to make them part of the goals that individuals desire to achieve. However, they also differ from those processes in that the conditions for realization reach far beyond the capacities of individuals.

I later explain that in order for people to really have the intention to participate in a shared goal, something more must be assumed. People who act in favour of common-pool-related goals subscribe to an environmental ethos that provides motives and reasons for achieving these goals.[55] Furthermore, besides this important element in joint environmental action, it is important to note that within the context of the joint goals of keeping a river intact or of reaching climate goals by a common effort, personal interests are not restricted to that endeavour. Some participants might even hold a range of conflicting goals. But nowhere has it been said that common-resource-related goals cannot be part of a range of goals people actually like to invest in.

An approach to environmental action that highlights ways of acting together to achieve future goals together sets itself apart from an approach to environmental goals in terms of "negative externalities" of civilization and necessary repair mechanisms. It focuses on what people can achieve together. As for NCPRs, individuals who engage in joint action focus on what they can do together, not on what side effects their activities might entail. At a later point, I also explain that the model of voluntary joint action needs to be strengthened for normative reasons and by means of explaining the duty to cooperate in certain cases. Still, for now it is important to note that the model starts with positive agendas that relate to particularly valuable goods, the NCPRs.

In general, joint environmental action includes the following elements. A group of individuals intentionally sets out to realize a shared goal that entails a distinct pattern of appropriation of a shared resource. At a minimum, the participants believe the scheme of appropriation serves their interests and allows long-term sustainable practices. As in theories of joint agency more generally, this approach assumes a positive tie between the individuals who intend to work together and the end result they envision. At a minimum, the preservation of NCPRs as desirable goals is possible, even when the realization imposes constraints on agency.

4.6 A revisionist approach

Assume that environmental goals that protect the NCPRs can be achieved by means of joint action and that the resulting terms of cooperation cannot be developed in the office, but need to be advanced and implemented by individuals and organizations benefitting from a shared resource and coordinating their activities successfully. In this scenario, the model of joint action proposed by Bratman fills an important theoretical gap in debates on climate change politics and climate change action. It explains how people can and do really succeed in accomplishing shared goals by means of joint action.

Authors in the field of social philosophy also discuss the conditions under which people succeed in coordinating sub-plans. They mutually adjust single steps to achieve their final goal, provided they really have the intention to work together for a shared goal. As a consequence, it is also important to reframe environmental goals. They need to be framed in such a way that cooperation in favour of the achievement of these goals is rendered possible. The next chapters discuss the elements of this proposal in detail. It is necessary to recapitulate the ways commitment to climate goals can be enhanced and goals can be adjusted according to the proposed theory of joint action. It is also particularly important to recapitulate the resources for the motivations for cooperating in the first place. This part of the theory will be developed by taking a closer look at an ethos that environmental cooperation entails.

This theoretical model provides not only an idea of the characteristic traits of the actors involved in joint action, but also of the goal itself. First of all, the goal is itself a joint enterprise. As for environmental objectives, this translates into a cooperative scheme of utilization according to which individuals succeed by approaching a shared resource as a collective actor.

Instead of undermining each other's success through profit-seeking beha-viour, this model also highlights prerequisites and elements in the successful realization of a future shared goal. As applied to environmental goals, a cooperative goal also takes into account the fact that natural resources can be allocated so that beneficiaries share "baskets of benefits."[56] Having a clear view of future goals that support this option is a precondition for this.

The theoretical approach is revisionist in that it thinks through the process of appropriating shared natural resources and their services in a new way. It first investigates the conditions of successful joint action according to theore-tical insights into social philosophy. It then explores the prerequisites for applying that model of joint action to the accomplishment of environmental goals by collectives. Still, in accordance with theories of water management, the comparison is not too far-fetched. In appropriating shared natural resour-ces, cooperation is also obligatory. What a theory of joint action adds are the theoretical prerequisites that explain action in the first place and that explain successful coordinated action. As such, these theories offer an approach to cooperation that stands in stark contrast to cooperation as a voluntary exchange among benefit-seeking individuals.

Once the elements are outlined in detail, it is also possible to explain why environmental action so often goes astray. Joint environmental action cannot be accomplished without also assuming that actors actually share the inten-tion of bringing about a shared objective. Nevertheless, these aims disappear when NCPRs are regarded as utilizable resources. Instead, they need to be re-interpreted as systems that offer baskets of benefits, but only under certain conditions. The remainder of this book is dedicated to explaining further elements of that revisionist approach. This includes a thorough exploration of the function of shared values in cooperation. It also includes a translation of the model of joint action into scenarios in which individuals need to be motivated to contribute to joint action, and in which institutions need to accept responsibilities for environmental goals.

Notes

1 See Chapter 2.
2 Garrett Hardin, "The Tragedy of the Commons," *Science* 162, no. 3859 (1968): 1243–1248, https://doi.org/10.1126/science.162.3859.1243. For an exploration of this notion, see Chapter 2.
3 For this diagnosis, see also Chapter 3.
4 Raimo Tuomela, *Social Ontology: Collective Intentionality and Group Agents* (New York: Oxford University Press, 2013); Raimo Tuomela, *The Philosophy of Sociality: The Shared Point of View* (Oxford/New York: Oxford University Press, 2007); Raimo Tuomela, "The We-Mode and the I-Mode," in *Socializing Metaphysics – The Nature of Social Reality*, ed. F. Schmitt (Lanham, MD: Rowman and Littlefield, 2003), 93–127.
5 Michael E. Bratman, *Shared Agency. A Planning Theory of Acting Together* (New York: Oxford University Press, 2014); Michael E. Bratman, "Shared Intention," *Ethics* 104, no. 1 (1993): 97–113.

6 It is important to note that this chapter is dedicated to outlining an explanatory approach to the natural commons. I intend to explain how achieving sustainable water use, land use, and atmospheric use works and which conditions need to be met in this specific case of joint action. This task needs to be distinguished from another one: it is independent from sustainability as a reasonable goal in addressing the natural commons; it also needs to be distinguished from the normative question of whether individuals should engage in joint climate action and in joint environmental action more broadly for moral reasons. I shall discuss the normative claims in Chapters 7–10.

7 See Margaret Gilbert, *On Social Facts* (Princeton, NJ: Princeton University Press, 1989); Margaret Gilbert, *Living Together. Rationality, Sociality, and Obligation* (New York: Rowman and Littlefield, 1996).

8 See Peter A. French, "The Corporation as a Moral Person," *American Philosophical Quarterly* 16, no. 3 (1979): 207–215; Peter A. French, *Collective and Corporate Responsibility* (New York: Columbia University Press, 1984); Larry May, *Sharing Responsibility* (Chicago: University of Chicago Press, 1996).

9 For prominent positions in this field, see Christian List and Philip Pettit, *Group Agency: The Possibility, Design, and Status of Corporate Agents* (Oxford/New York: Oxford University Press, 2011); Philip Pettit, "Collective Persons and Powers," *Legal Theory* 8 (2002): 443–470; John Searle, *The Construction of Social Reality* (New York: The Free Press, 1995); John Searle, *Mind, Language and Society – Philosophy in the Real World* (London: Phoenix, 2000).

10 Bratman, *Shared Agency*; Bratman, "Shared Intention"; Tuomela, *Social Ontology*; Tuomela, *The Philosophy of Sociality*; Tuomela, "The We-Mode and the I-Mode."

11 Examples to support this thesis in the context of group theory are provided by Itzhak Gilboa, *Rational Choice* (Cambridge, MA: MIT Press, 2010); Shmuel Nitzan, *Collective Preference and Choice* (Cambridge: Cambridge University Press, 2010); Christian List and Philip Pettit, "Aggregating Sets of Judgments: An Impossibility Result," *Economics and Philosophy* 18, no. 1 (2002): 89–110.

12 See, for example, Lina Eriksson, *Rational Choice Theory: Potential and Limits* (New York: Palgrave MacMillan, 2011); Jan De Jonge, *Rethinking Rational Choice Theory* (New York: Palgrave MacMillan, 2012).

13 Although this field of research has been significantly augmented in recent years, a list of key references can be given. The "classical" texts include Annette C. Baier, "Doing Things with Others: The Mental Commons," in *Commonality and Particularity in Ethics*, eds. Lili Alanen, Sara Heinämaa, and Thomas Wallgren (London: MacMillan, 1997), 15–44; Bratman, "Shared Intention"; Gilbert, *Living Together. Rationality, Sociality, and Obligation*; Margaret Gilbert, "The Structure of the Social Atom: Joint Commitment as the Foundation of Human Social Behaviour," in *Socializing Metaphysics – The Nature of Social Reality*, ed. Frederick F. Schmitt (Lanham, MD: Rowman and Littlefield, 2003), 39–64; List and Pettit, *Group Agency*; Tuomela, *The Philosophy of Sociality*.

14 The concept of "sharedness" of intentions and the ontology of intentions is one of the key issues in this debate. For different perspectives on the ontological status of shared intentions, see Bratman, "Shared Intention"; Gilbert, *On Social Facts*, 128; Tuomela, "The We-Mode and the I-Mode."

15 See Seumas Miller, *Social Action* (Cambridge: Cambridge University Press, 2001).

16 See Robert Sudgen, "The Logic of Team Reasoning," *Philosophical Explorations: An International Journal for the Philosophy of Mind and Action* 6 (2003): 165–181; Robert Sudgen, "Thinking as a Team: Toward an Explanation of Nonselfish Behavior," *Social Philosophy and Policy* 10 (1993): 69–89; Tuomela, *The Philosophy of Sociality*.

17 The role of shared intentions and potential circularities are examined in Bratman, *Shared Agency*, 40–47.

18 For the processes of group formation, see Amy R. Poteete and Marco A. Janssen, *Working Together: Collective Action, the Commons, and Multiple Methods in Practice* (Princeton, NJ: Princeton University Press, 2010); Axel Seemann, "Why We Did It: An Anscombian Account of Collective Action," *International Journal of Philosophical Studies* 17, no. 5 (2009): 637–655.

19 Normative constraints are made a matter of debate in List and Pettit, *Group Agency*, 153–169; Seumas Miller, *The Moral Foundations of Social Institutions: A Philosophical Study* (Cambridge/New York: Cambridge University Press, 2010).

20 On values inherent in group action, see Monika Betzler, "Valuing Interpersonal Relationships and Acting Together," in *Concepts of Sharedness. Essays on Collective Intentionality*, eds. Hans Bernhard Schmid, Katinka Schulte-Ostermann, and Nikos Psarros (Frankfurt: Ontos Verlag, 2008), 253–272; Hans Bernhard Schmid, Katinka Schulte-Ostermann, and Nikos Psarros, *Concepts of Sharedness: Essays on Collective Intentionality* (Frankfurt: Ontos Verlag, 2008).

21 Gilbert, *On Social Facts*; Gilbert, "The Structure of the Social Atom."

22 For my position on normative claims, see Chapters 7–9.

23 Tuomela, *The Philosophy of Sociality*, 112.

24 Ibid., 113.

25 Ibid.

26 Bratman, *Shared Agency*, 144.

27 This is not precisely true of all texts in the field of social action. John Searle and Philip Pettit both hold views according to which collective action derives from single action. They subscribe to "social ontology" in the sense of interpreting group action in terms of actions of individuals. See List and Pettit, *Group Agency*; Pettit, "Collective Persons and Powers"; Searle, *Mind, Language and Society – Philosophy in the Real World*; Searle, *The Construction of Social Reality*.

28 Bratman, *Shared Agency*; Bratman, "Shared Intention."

29 In this section, I primarily focus on the theory of joint action proposed by Bratman, which, for two reasons, is particularly good for explaining the building blocks: it is metaphysically modest and also ethically modest, nor does it include thick ethical claims.

30 Bratman, *Shared Agency*, 84. Note also that the elements in Bratman's explanation of joint agency are more modest in ethical terms than proposals that view the setting of joint agency as already involving mutual obligations of group members. I comment on this in the next section.

31 It is a matter of vigorous debate whether shared agency presupposes moral commitments and what they look like. Whereas Gilbert defends even mutual moral obligations in a scenario of joint action in *Living Together. Rationality, Sociality, and Obligation*, Bratman appears to think that modest sociality also includes a modest type of morality, expressed in terms of joint commitment. "After all, joint commitment is supposed to be a distinct and unanalyzable phenomenon, not just a familiar kind of moral obligation" (Bratman, *Shared Agency*, 119).

32 This prompts a lot of objections regarding possibilities of nevertheless acting strategically in this framework; for a discussion of some of these objections, see Bratman, *Shared Agency*, 92–96.

33 I maintain that normative claims need to be integrated at two points: first, individuals who engage in joint environmental action subscribe to a certain type of ethos; second, I argue in favour of "a duty to cooperate" once the agenda of joint environmental action is in place. For the latter, see Chapters 7–9. Yet the duties I discuss are restricted to the climate scenario, and they are diverse in that they also include duties of collectives and duties to join collectives.

34 Radkau explores the role of water governance in various cultures, including ancient Egypt and Mesopotamia, China, the Netherlands, and Venice; see Joachim Radkau, *Natur und Macht. Eine Weltgeschichte der Umwelt* (München: Beck,

2002), 107–159 [*Nature and Power. A World History of the Environment* (Cambridge: Cambridge University Press, 2008)]. One more general lesson of his investigations is that water regulation is part of power structures.

35 This example is – of course – in some respects already idealized. It does not take the companies using the river as waste dump into account.

36 On the concept of eco-services, see Robert Costanza et al., "The Value of the World's Ecosystem Services and Natural Capital," *Nature* 387, no. 6630 (1997): 253–260, https://doi.org/10.1038/387253a0.

37 Jerome Delli Priscoli and Aaron T. Wolf, *Managing and Transforming Water Conflicts* (Cambridge/New York: Cambridge University Press, 2009), 121.

38 On this, see Chapter 2.

39 This proposal is close to that of working on water governance against the backdrop of Dewey's philosophy and a model of re-adjustment through cooperation. For a theoretical approach to cooperation on water resources, see Martin Kowarsch, "Beyond General Principles: Water Ethics in a Deweyan Perspective," in *Global Water Ethics. Towards a Global Ethics Charter*, eds. Rafael Ziegler and David Groenfeldt (London/New York: Routledge, 2017), 57–74. The proposal in this book deviates from that proposal by introducing principles of justice and by arguing for a shared ethos underlying joint climate action.

40 See Chapter 9 on principles of fairness for allocating burdens to actors.

41 This idea was worked out in the context of a theory of political institutions by John Dewey, who wrote that once negative side effects become so dramatic that they turn into particularly long-term and threatening events, political institutions need to be generated to cope with the desire of the public to check these events. John Dewey, *The Public and Its Problems* (Athens: Swallow Press, 1991), 47.

42 The term "Anthropocene" was initially introduced by Crutzen and is now a widely shared term primarily addressing the significant impact of humankind on the geological systems of Planet Earth. See Paul J. Crutzen, "Geology of Mankind," *Nature* 415, no. 3 (January 2002): 23. For the opposite claim that responsibility in the Anthropocene does not vanish, but increases, see Klaus Töpfer, "Nachhaltigkeit im Anthropozän," *Nova Acta Leopoldina* 117, no. 398 (2013): 31–40.

43 United Nations, "The Paris Agreement," accessed October 4, 2018, https://unfccc.int/sites/default/files/english_paris_agreement.pdf.

44 Tuomela, *The Philosophy of Sociality*, 113.

45 This is particularly important regarding geophysical structures, which are already close to thresholds that cause uncertainty about further developments. In discussing safe operational space, Rockström thinks that both climate change and the acidification of the oceans match that scenario. See Johan Rockström et al., "Planetary Boundaries: Exploring the Safe Operating Space for Humanity," *Ecology and Society* 14, no. 2 (2009).

46 The literature on the precautionary principle is vast. Here, I shall just mention the basic disagreement debated in the literature about either a strong precautionary principle, considering not only risks, but also unforeseeable risks, versus a more general approach to precaution that works on uncertainty. For the debate on the precautionary principle and its presumptions, see Stephen M. Gardiner, "A Core Precautionary Principle," *Journal of Political Philosophy* 14, no. 1 (2006): 33–60; Christian Munthe, *The Price of Precaution and the Ethics of Risk* (Dordrecht: Springer, 2011); Mariam Thalos, "Precaution Has Its Reasons," in *Topics in Contemporary Philosophy 9: The Environment*, eds. W. Kabasenche, M. O'Rourke, and M. Slater (Cambridge, MA: MIT Press, 2012), 171–184.

47 For this term in the context of a debate on planetary boundaries, see Rockström et al.

48 Crutzen.

49   Donella H. Meadows and Club of Rome, eds., *The Limits to Growth: A Report for the Club of Rome's Project on the Predicament of Mankind* (New York: Universe Books, 1972).
50   Rockström et al.
51   Further qualities will be introduced step by step in developing the content of the theory.
52   Bratman, *Shared Agency*.
53   Ibid., 53–56.
54   For this structure as part of an approach to natural common-pool resources in a Deweyan framework, see the proposal by Kowarsch to shape policies of water supply accordingly: Martin Kowarsch, "Ethical Targets and Questions of Water Management," in *Water Management Options in a Globalised World. Proceedings of an International Scientific Workshop (June 20–23, 2011, Bad Schönbrunn)*, ed. Martin Kowarsch (Institute for Social and Developmental Studies (IGP) at the Munich School of Philosophy, 2011), 38–49, www.researchgate.net/publication/308118199_Water_management_options_in_a_globalised_world; Kowarsch, "Beyond General Principles."
55   Even if this never becomes a realistic case, it does not obstruct the theoretical gains; instead, it is possible to transform the theoretical approach into an institutional account. This accepts as a further premise that the protection of environmental goods and the fair distribution of services of those goods is part of the "telos" of institutions that are obliged to care for the common good. According to Seumas Miller, this can be taken for granted and spelled out in terms of cooperation and collective action. See Miller, *The Moral Foundations of Social Institutions*.
56   See Delli Priscoli and Wolf, 74–78.

5 The ethos in joint environmental action

Cindy: Sometimes philosophers are much faster than one might think. Actually, the last pages were something like a ride through a broad area of different things. Have you understood what they wish to say by addressing "joint environmental action"?

Bert: Yes, I think I got the idea. Philosophers explore "joint action," which is a special type of coordinated action. Usually we think that single persons act because they wish to accomplish something. But these philosophers think that people can also act together. They simply share a goal and have a common intent, and then each single person acts in a way so that they together achieve a shared goal. It appears to be as simple as that. But the difficulties come again with environmental action. The environment is not a goal; nor do people really like to act in favour of environmental goals. Do you think I got it right?

Cindy: Yes, in a way. But I'm not sure that people do not act together in favour of environmental goals. There are so many examples that demonstrate the will and capacity of people to act together. They clean up rivers and landscapes together; they develop green alternatives in cities; they even start building up infrastructure that helps to reduce greenhouse gases. I think part of the message of the last chapter was that there is some hope that environmental action can also be framed as collective action.

Bert: Right. And I think that it is again like in the other chapters: philosophers do not go straight to solutions. Instead, they question conditions for solutions to work. I think this is the plan for this chapter and the next chapter. The question isn't how does collective action work? Theories of joint action have explanations of how it works. But the question is: what are the conditions for environmental action to work as joint action? Do you think this is right?

Cindy: Yes, I think this is right. And this is possibly meant by "revisionist approach". Let's first look what would be needed in order to get joint environmental action to really work according to the proposed model of joint action. And let's then look at how the conditions can best be fulfilled.

Bert: Yes, this is how it works. What do you think the conditions are?

Cindy: I think the model of joint action doesn't work unless people really wish to bring about the shared goal. But I'm not so sure if this is really necessarily the case in order to get joint environmental action done. What do you think?

Bert: I've already learned that the most important thing is keeping different issues apart from each other. You shouldn't mix up too many issues with each other!

Cindy: How do you mean?

Bert: I think it's important to first hear how voluntary joint climate action – or whatever this complex thing is called – could possibly work. And this needs to be explored against the backdrop of existing models. It doesn't help to invent something like cooperative action. Once the conditions are on the table, it's another issue to discuss how the conditions can best be realized.

Cindy: OK, you're right. So first, the conditions for getting joint environmental action off the hook, and then the real-life conditions.

According to theories of joint action, shared goals need to be desired and willed by the actors involved. It is no different with environmental goals. They will not be achieved unless they are shared by individuals who engage together in joint action. Even assuming the theoretical joint action approach can be argued consistently, however, there is still an explanatory gap. The achievement of environmental goals is costly and often not particularly well suited to cohere with our everyday desires. To some degree this fact depends on how environmental goals are framed. Goals of active co-designing with nature might be in the interest of people who benefit from that process. Moreover, local environmental goals such as "greening the city" or "keeping the lake intact" are not necessarily against the interests of the population. But still, the relationship between shared goals and shared intentions to accomplish them still lacks a systematic explanation.

In this section, I intend to outline a normative resource that helps to remedy the inadequacy that results from the gap between the interests and desires of individuals on the one hand and demanding environmental goals as joint action goals on the other hand. Individuals who act together to reach a shared goal act together in a particular way. They share a goal that they regard as the goal of all members of their group. In addition to this presupposition, writers in the field of social philosophy have established another theoretical element for joint agency: an *ethos*. Raimo Tuomela argues the case:

> Basically, acting as a group member is to intentionally act within the group's realm of concern. Such action can be either a successful action or an unsuccessful action. What is required is that the group member in question will intentionally attempt to act in a way related to what he takes to be the group's realm of concern, such that he does not violate the group's constitutive goals, standards, values, and norms (in one word, its "ethos").[1]

The ethos provides the connection that makes the group act as a joint actor.

Tuomela also suggests that a shared ethos is critical for forming a peer group focused on a goal related to the ethos. In summarizing a central chapter in his book *The Philosophy of Sociality: The Shared Point of View*, Tuomela states:

> It is argued that (autonomous) social groups can be partly defined in terms of their constitutive goals, values, beliefs, norms and standards collectively accepted by the members acting. They are collectively called the *ethos* of the group. A group is assumed to try to satisfy and normally also to maintain its ethos.[2]

The ethos thus provides not only a social glue for group members, it also serves as a shared resource for the reasons guiding the action. It also provides something like a group identity over time.

This chapter starts with Tuomela's insights about the function of an ethos in a theory of joint action. It then applies these insights to joint environmental action. I argue that in cases like the appropriation of a common-pool resource according to standards of fairness and according to a shared goal, an ethos of the group members supports successful appropriation of a resource as a group. Before addressing action regarding a global natural resource (i.e., the atmosphere), it is helpful to first inquire a little closer into the theoretical elements of joint action.

One proviso needs to be made in advance. It is not my claim that joint environmental action coheres in every respect with a joint-agency theoretical approach. More specifically, scenarios that are far more complex than cooperation on goals for a lake differ not only with respect to agency, but also with respect to the groups that are involved in appropriating a resource. To inquire as to the steps that are necessary to apply theories of joint action to environmental action, further chapters will be needed; this chapter only takes the first step. It limits itself to explaining that a joint environmental action approach includes a specific normative resource: an ethos. I explore this resource here as it applies to joint environmental action.

Having explained the function of an environmental ethos in groups acting as joint environmental actors, two more lessons can be drawn. First, as a set of norms and values supporting and guiding groups of people to reshape their activities in accordance with "active co-designing with nature," the ethos includes two facets. It puts normative constraints on practices of appropriation of nature in accordance with the convictions and norms of group members. Second, the ethos includes fairness in this appropriation. The ethos guiding groups in the context of active co-designing with nature as related to natural common-pool resources (NCPRs) rests, in particular, on normative insights not restricted to particular situations. Instead, the environmental ethos can to some degree and with respect to distinct NCPRs be generalized – but always in relation to the specific scenario of appropriation of a scarce natural resource by a group of people. At the heart of the environmental ethos are principles of fairness. They not only include basic normative yardsticks for a fair appropriation of a shared natural resource, they also support the motivation of individuals participating in joint environmental action.

The chapter is divided into eight sections. Section 5.1 outlines the function of an ethos in joint action as explained by Raimo Tuomela.[3] Section 5.2 presents an ideal scenario of the appropriation of a shared NCPR by a group of beneficiaries. More specifically, it highlights the environmental ethos as part of that scenario. To take an example, when a group of people benefitting from a lake as a shared natural resource decide to act together to preserve this resource in good condition, they also regard themselves as "active co-designers with nature."[4] More specifically, they share a view not only regarding the goals they wish to accomplish, but also regarding their own commitments or values that are in favour of that activity. This resource for group action is called an "ethos." As for environmental goals, it might include insights about the priority of conservationist strategies, expressed as "respect for the integrity of that resource." It also might include the claim that both burdens and benefits are shared in such a way that people can agree on the outcome. And it possibly includes what I call "moderate eco-centrism" as well. This is the claim that NCPRs should be preserved not only because of the benefits people receive from them on an ongoing basis, but also because they are habitats for non-human organisms and sites for biodiversity. Each aspect of the ethos is outlined in a separate section (Sections 5.3–5.5). Section 5.6 presents the ethos as moderate eco-centrism. Section 5.7 presents an alternative to moderate eco-centrism. It takes up a recent discussion of environmental ethics and explains the demanding claims about the value of nature that have recently been argued in environmental ethics. In the context of this chapter, this helps us to understand that proposals to frame a shared ethos based on integrity, fairness, and respect for non-human living beings and their needs really are moderate. Section 5.8 summarizes the normative baseline of this proposal.

5.1 Functions of an ethos in joint action

Raimo Tuomela's work is dedicated to an explanation of the specific conditions under which individuals act when they act together. It should be noted from the outset that the goal of this section is not to encapsulate the complex picture Tuomela has developed over decades. Instead, I intend to interpret just one facet of his approach to social ontology in order to highlight a specific trait of group action in the context of environmental action. In highlighting this specific insight of Tuomela's, this chapter takes sides with respect to a very basic issue in social ontology. It is a matter of vigorous debate whether normative commitment is obligatory in the explanation of joint action.[5] Tuomela thinks that individuals who engage in joint action share a set of normative content, which he calls an "ethos."[6] The ethos has several roles to play. Before outlining them in detail, however, some key elements of joint agency have to be introduced.

Tuomela explicitly explains a modest, but common form of sociality. In his view, individuals are able to cooperate with each other, and sociality is an element of human nature:

> In contrast to higher animals, humans not only live in large groups and act as group members, intentionally cooperating (and also competing) with each other, but also often cooperate with members of other large groups. Let us call this kind of cooperation ... "general cooperation." This kind of cooperativeness is unique to humans and is probably due to coevolution.[7]

Beyond general cooperativeness, people also perform joint action together, which is more demanding: "Two or more actors cooperate in the full, We-mode sense if and only if they share a collective (or joint) goal and act together to achieve the goal."[8] As group members, individuals do not dissolve into a group subject. Instead, people act as group members when they stick to what the group has decided to do together.[9] This approach also includes peer groups who share specified interests and act together to realize these interests.[10] Individuals are also conceived as free to join several groups simultaneously; joining or leaving a group is always possible.[11]

One key concept in the explanation of joint agency is termed the "We-mode" of acting. In accordance with other theories of joint agency, joint action is theorized as a goal-driven activity.[12] The constitution of the group, however, depends on the capacity of each member to take a stance that is the "We-mode" of acting. This "We-mode" varies with the groups constituted by individuals. In a structured group, individuals act together, acting in the "We-mode." Tuomela explains:

> A We-mode state (or action, for that matter) is conceptually group-dependent and requires that the person functions qua group member in accordance with the group's ethos. There are three central criteria for the presence of We-mode attitudes (or actions): the members function for a group reason, they are on quasi conceptual grounds "in the same boat" and thus satisfy a special collectivity condition with respect to their We-mode attitudes, and they are collectively committed to these attitudes.[13]

It is not my goal here to explain this type of action in the extended sense. What is, instead, critical to our inquiry of joint environmental action are the meaning and the function of an ethos in this context.

Regarding the *ethos*, it is important to note from the outset that it does not necessarily consist of a set of moral values.[14] Instead, it is primarily defined as normative content that has a function for the participants in joint agency.

Three aspects of shared ethos are particularly important for understanding its role in group agency.

First, groups are not pre-existing entities, but instead consist of single agents. Groups are made up of individuals, but they are also able to hold beliefs and goals together and to act as a group. Groups are assemblages of individuals and not a new ontological unit. Nevertheless, they are in a position to have shared intentions in the sense of participating in one and the same intentional content.[15] One crucial difference between acting as an individual and acting as a group member is the relationship to the other group members:

> A We-mode belief [is] expressible by "We, as a group, believe (accept) that p requires that the group in question is collectively committed to upholding and acting on its mutual belief or at least to keeping the members informed about whether it is or can be upheld."[16]

Acting in the We-mode assumes that each participant is aware of a shared goal, and that every single actor is supportive of the commitments inherent in the realization of that goal. As Tuomela puts it,

> Functioning as a group member (relative to a group g) in the positional case, that is, in a structured group, is equivalent to acting intentionally, with the purpose to satisfy or at least not to contradict the ethos of g, in one of the senses 1.–4., or attempting to act.[17]

Tuomela refers to the context of structured groups, in which he distinguishes four types of actions: (1) positional actions related to a group position or role, (2) actions that other group norms require, (3) actions that do not belong to (1) and (2) and are instead based on situational intention formation or agreement making that has not been codified in the task-right system of the group, and (4) freely chosen actions or activities, but still being in the concern of the group.[18] In other words, the ethos is a normative framework that individuals will not contradict when acting as members in a structured group and therefore also in favour of a collective goal.

The second key aspect of a shared ethos, according to Tuomela's proposal, is that it contains constitutive goals, values, beliefs, norms, and standards collectively accepted by the members acting and, as such, defines the very identity of the group. With respect to an autonomous group, Tuomela explains that the ethos includes its constitutive goals, values, beliefs, standards, norms, and the like:[19]

> The content aspect of group g's identity is "defined" by its collectively accepted ethos, E; this is expressible by "We form a group, g, based on ethos E." That this definition holds is a constitutive principle for the group, and its rational collective acceptance already gives a kind of weak

normative reason for maintaining it, based on the rationality of the persistence of maintaining the group's identity.[20]

The concept of an "ethos" thus serves as an umbrella term for the normative self-understanding of group members as such. As a consequence, the ethos is also central to the continued existence of the group: "A group is assumed to try to satisfy and normally also to maintain its ethos."[21] In other words, the ethos not only creates a shared normative bond for a group of actors, but also includes the values and norms that group members share.

The third aspect of shared ethos that should be understood is that individuals who are group members also have reasons to act not only in accordance with their shared goal, but also in accordance with the ethos. It is one crucial function of the ethos to provide reasons for actions when individuals act as group members and in favour of a shared goal.[22] Although Tuomela does not go so far as to relate group membership to a special type of mutual obligation,[23] he nevertheless regards commitment to a shared ethos as essential for mutual reliability in acting together: "Being committed to each other, the group members can better rely on the others to perform their respective tasks, which, especially in the case of interdependent and joint actions, is central."[24] Again, the ethos does not have to be value-laden in a moral sense, nor does it necessarily fit into a moral approach to action. But it serves as a resource for reasons to participate in bringing about a shared goal.

It also should be noted that Tuomela is not particularly clear regarding the strength of the ethos. At some points it looks as if the ethos gives foundation and identity to the group of individuals who act together; it appears to be a reason-giving authority ascribed to the members of the group.[25] At other points, the assumptions about the functions of an ethos are much more modest. At a minimum, Tuomela accepts the view that "proper group responsibility involves a group-external normative standard that in general relates to things like what the group-external social environment ... normatively (e.g., morally) expects or can be taken normatively to expect."[26] In other words, the ethos does not include an outright moral stance. Instead, the ethos serves as a set of norms that cannot easily be overridden by the members of the group.

This is only a very rough sketch of the central elements of Tuomela's account of ethos in group agency. Each group has a realm of concern, which is what group agency is focused on. Tuomela also calls the realm of concern the "intentional horizon of the group,"[27] by which he means the sum of the contents that are of interest to the group and publicly known by all group members.[28] The concerns of group members are connected to a set of shared values and norms. Indeed, the theory Tuomela presents can be applied to a variety of different groups – as long as a common concern can be identified. With regard to the themes in this study, Tuomela's theory is well-suited to discuss environmental action in a new light. More specifically, it explains group action as the action undertaken by individuals who not only identify with some goal, but who share a set of norms and values. More specifically, these norms and values give reasons for action and therefore also support common goal-driven activities.

In the remainder of this chapter, Tuomela's theoretical model of shared ethos will be used to explain a model of environmental action with some parallels to joint agency. Tuomela's view of cooperation, especially, provides an alternative to either cooperation based on a model of voluntary exchange or to cooperation in the sense of exchange processes among self-interested beneficiaries. It is still a goal-driven interpretation of joint action. Still, unlike a "planning theory of action," it adds the insight that people act together to the extent that they share values and norms not only as individuals, but as participants and supporters of a group's ethos.[29]

5.2 Environmental cooperation on a water resource

Imagine a village on the shore of a lake.[30] The lake serves many different interests, including as a receptacle for waste. For years, the practice of utilizing the lake as a waste dump went unquestioned. The amount of waste dumped in it was marginal in comparison to the vast amount of water in the lake. No one had really noticed the problems caused by polluting the lake. But one day the situation begins to cause trouble. The lake begins to stink, and it poisons a river. As a consequence, the practice of using the lake as a waste dump is finally called into question.[31] At that point, a basic problem of using the lake becomes apparent. Although the lake serves many different interests, it is necessary to coordinate its utilization and not to over-exploit the resource. More specifically, it is apparently necessary to prohibit hazardous practices of water use. To prevent the lake from reaching a tipping point beyond which it would no longer be usable, the villagers come together and discuss the problem and debate possible solutions.

A first step in developing a solution is to interpret the problems. One might argue that the problems the inhabitants of the village face result from over-exploitation of a shared resource. This interpretation, however, is only part of a much more complex situation. A lake is a common-pool resource, and common-pool resources have two specific characteristics. They lack effective entrance barriers, so anyone is free to use them. At the same time, many beneficiaries compete over the resource. A frequently occurring situation, as previously mentioned, is "when the same resource is used both as a source and as a sink," as Jamieson has put it.[32] The problem is not only that scarcity of a shared natural resource provokes conflicts and might lead to the degradation of that resource. Instead, uncoordinated use of a resource has the effect that some types of usage are made impossible by other types of usage.

The villagers are well-advised to develop a model of cooperation that can prevent the present degradation of the resource through uncoordinated use. This includes a focus on certain core values and an attempt to translate them into basic normative guidelines of water usage. More specifically, the users of the water should act together to reach future developments as a practice of "active co-designing with nature."[33] Instead of focusing exclusively on the negative side effects of their current behaviour, they should start regarding themselves as actors who can shape the future practices of water usage according to their best knowledge and their values. That means that they not only interpret their shared activities as driven by a distinct vision of future development, but that their vision of future use is based on values that help them to shape and coordinate activities. Together, they consider good guidelines for access that accord with their values. They also know that to succeed in achieving a good situation, they had better work together. They will rely on valuing the environmental conservation of the lake, on valuing a fair distribution of the burdens and benefits of that lake, and on values attached to certain practices of water usage. In subscribing to an ethos that includes these values, water users form a group. They also develop a shared vision of future goals.

This proposal for framing "water cooperation" as a value-based type of joint action is based on two premises. It maintains that cooperation in terms of "joint action" is indeed the best model for resolving and transcending water conflicts. In addition, it assumes that there is some common ground regarding basic values that help to shape the decisions of group members. In accordance with this proposal, the villagers develop the following strategy.

As a group, the villagers first discuss the critical elements of an ethos that helps to shape their goals regarding the common-pool resource. Basically, they acknowledge the fact that they are all in some way dependent on the lake. Moreover, they also agree that shared goals have priority over private goals. In addition, they also reach consensus about the priority of environmental conservation, about fairness in distributing the benefits and burdens of water usage, and about respect for various cultural values attached to the water resource. They also figure out which concrete values are best to guide their common action. They then agree about which cooperative schemes of water usage to implement.[34]

A minimal agreement on shared values has three elements. First, the conservation of the lake is prioritized. All water users agree that heavy pollution and heavy over-exploitation of the lake is a worst-case-scenario. They agree that it is reasonable to keep basic ecosystemic functions of the lake intact. They also discuss how the resilience of the lake can be supported and how its eco-functions, currently under pressure, can best be restored. Second, the villagers agree that the benefit-seeking practices need to be framed and guided by long-term strategies that accord with claims of fairness. This includes a range of principles of fairness. Water fairness has at least three components: (a) a minimum fair share to all users, (b) distributive fairness in sharing burdens from maintaining fair schemes of appropriation of the lake, and (c) respect and support for diverse types of water usage. A third element of a minimal agreement on shared values is that the villagers agree that they will commit to the agenda and achieve their goals as joint actors. This means that each of them supports fair practices of water usage. In addition, they try to figure out in which way and how the agendas relate to their convictions and to values that go beyond the needs of the water supply. Although access to water might not figure among their personal values, they regard recreational use – opportunities to go fishing or running along the lakeshore, as well as the aesthetic experience of the surroundings – as part of a fulfilled life.

This proposal for resolving water conflicts is clearly ambitious. It rests on the idea that potential beneficiaries from a natural resource also endorse fairness and a set of shared values. Water resources are interpreted as shared resources that deserve environmental protection. Fairness requires a range of restrictions on the use of resources. Moreover, the proposed scenario ignores many additional conflicts and problems in water supply.[35] More specifically, this interpretation might appear overly optimistic, if not unrealistic. But taking this scenario as the first step does not prevent further steps towards less ideal scenarios. Indeed, the choice to portray this ideal scenario rests first on a theoretical decision. It highlights the difference between market-style cooperation, essentially motivated by self-interest, and environmental cooperation as a completely different type of cooperation. At the heart of a theory of market-style cooperation is mutual advantage, even when conceived as social cooperation. John Rawls – whose fundamental presupposition is that societies are based on cooperation – puts it this way:

> The idea of social cooperation requires an idea of each participant's rational advantage, or good. This idea of good specifies what those who are engaged in cooperation, whether individuals, families, or associations, or even the governments of peoples, are trying to achieve, when the scheme is viewed from their own standpoint.[36]

Rawls adds that it is of course possible that the plan to achieve something good might fail. And although mutual advantage is not the only relevant factor when one is rationalizing cooperation, it is its baseline. Obviously, this type of cooperation fails in the context of environmental cooperation. One reason is that it overlooks the necessity of a shared ethos for successful cooperation.

5.3 Integrity of the natural resource

The environmental ethos underlying joint environmental action is based on the insight that nature's plasticity has limits. More specifically, the conditions for natural resources to regenerate vary. Some resources are not renewable, while others are, but only under conditions that correlate with a certain amount and a certain quality of impacts on that resource. In order not to overuse a limited resource, it is wise to readjust schemes of appropriation so that the risk of destruction of that resource is limited. Although this insight is critical for assessing activities that are neither destructive nor risky, the choice of an adequate concept is difficult. Recently, various proposals have been made to classify a good state of a natural resource and the appropriate means of protecting that good accordingly.

As for the goals of joint environmental action, it is also important to discuss the effect of actions not only as negative side effects on a natural resource. It is also important to acknowledge the value of systemic functions regarding the good state of the natural system at stake. In that context, an ecosystem service expresses a value that results from distinct benefits people derive from that good: "Ecosystem functions refer variously to the habitat, biological or system properties or processes of ecosystems. Ecosystem goods (such as food) and services (such as waste assimilation) represent the benefits human populations derive, directly or indirectly, from ecosystem functions."[37] The sum of ecosystem functions that a unit of natural entities provides is the "natural asset."[38] Theoretical approaches that respond to this claim evaluate processes of natural goods as systemic functions; they also address certain situations as being "good" and "intact."[39] This qualification depends on the characteristics of the natural entity being addressed. In water ethics, water has been addressed in terms of intact water cycles, and more precisely in terms of ecological functions.[40] Unlike environmental services more broadly, ecological functions are restricted to processes that contribute to maintaining NCPRs in an intact and healthy condition. A term that also describes the health of a resource is the notion of "resilience." Falkenmark and Folke explain: "Resilience is a measure of the amount of disturbance which can be absorbed before the ecosystem moves from one stable state … to another."[41] A water cycle that is resilient is able to cope with incidents of stress without losing its capacity to provide eco-functions in the long term. Whereas the interpretation of units of natural goods as "ecosystem services" is widespread, a major point of debate is the tight relation of these systems and their value to human welfare. In listing ecosystem functions, writers include "gas regulation," "climate regulation," "disturbance regulation," "water regulation," and "water supply" among functions that directly serve human populations, as, for instance, "food production," "recreation," and "cultural functions."[42]

Another way to address the good condition of a river or a lake is by employing the concept of "health." It has been claimed that "health" can serve as an inspiration for cooperation on a river.[43] The researchers who explore the conditions of cooperation on a "healthy river" begin with the idea that water resources can be described as items performing a range of services. These services are not restricted to services to people or to civilization. Eco-functions include services to ecosystems, services to the climate, and services to animals and plants. Man-made influences contribute to changing the patterns of eco-functions as well as narrowing down the range of eco-functions. This is not in itself bad or regrettable; water systems have always been reshaped by civilizations. Nevertheless, an interest in the "healthy river" is a shared interest of all potential beneficiaries of the river in not damaging its vital eco-services.

Both proposals – to frame a good state of the natural resource – rest on an assessment of ecosystemic functions that are essential for the robustness of that NCPR. More specifically, when focusing on the value of ecosystem services for humankind,[44] it is most apparent that protection of vital functions is not only reasonable, but a prerequisite for continued life on earth. In Daily's words: "Unless humanity is suicidal, it should want to preserve, at the minimum, the natural life-support systems and processes required to sustain its own existence."[45] And these systems are provided by ecosystems that confer benefits, notably from the assets that are NCPRs.[46]

A concept that can bundle the different insights into the "good state" of an NCPR, which includes a normative component, is "integrity." It is particularly helpful (a) for addressing a "healthy" state of the resource with respect to the limits of its plasticity and (b) for combining this insight with normative exigencies.

5.4 Fairness of a scheme of appropriation of services

To sum up, the basic element of an ethos as an integral part of joint environmental action is respect for the integrity of the natural resource that is appropriated by means of that joint action. Another element is distributive fairness regarding the partitioning of benefits and burdens. To emphasize these insights, I wish to introduce another debate on water justice: the discussion of the "right to water."

Water is not only a shared natural resource, it is also a truly critical good. It has life-sustaining qualities. Assuming that each person has a right to all goods needed to meet basic needs, this includes the right to enough water to meet those needs. Recently, this claim has been embedded in a United Nations declaration of every person's right to water. This includes, among others, the right to receive sufficient supply of clean water within a reasonable reach.[47]

Conditions of fair appropriation of a life-sustaining natural resource include, as a minimum, the fulfilment of a claim of fairness when it comes to subsistence rights. This "guaranteed minimum approach" appears to be reasonable when access to a scarce natural resource is needed to fulfil basic subsistence needs.[48] And this approach leaves space for further additional criteria of allocation beyond the minimum. One additional principle is the *priority to the least well-off* principle, in which fairness in allocating a life-sustaining natural good includes a priority for the least well-off. It is based on the view that it is fair to claim that a scarce but life-sustaining natural resource should be allocated in such a way that the worst-off deserve special recognition, in particular when a bad situation is not something they are responsible for.[49]

Further principles can be debated against the background of special relationships with a shared natural resource. As discussed elsewhere, when local natural resources are integrated into the culture and practice of a people, and this group has also performed services of guardianship, the arrangement of the distribution of services should take this into account.[50] Moreover, it is fair to acknowledge sacral values and specific aesthetic experiences.[51] Another special situation is one in which a group of people needs a resource as a means in their economic systems, and simultaneously is not in a situation to protect against depletion or destruction of that resource. Agrarian societies and societies that literally live with the land or with a water resource are in that situation.[52]

Yet, researchers in the field of environmental justice also explain that it does not suffice to discuss fair schemes of appropriation of benefits. Instead, the burdens that either result from environmental harm or the costs that result from protection, sustenance, or recovery of a joint natural resource also need to be shared in a fair way.[53] Fairness as related to a common-pool resource includes the appropriation of a fair share of benefits from that resource along with a fair share of the burdens resulting from either managing, protecting, or remedying that natural resource or directly from environmental harm.[54] As an outcome of this debate, philosophers have distinguished between an ideal scenario that is perfectly justified, but not applicable, and a non-ideal scenario that requires compromises to achieve applicability. This leads to a "hybrid approach," combining several principles of fair appropriation.[55] Note that this debate is a distinctly moral debate. Fairness does not simply mean equity or a distribution according to already fixed access rights. Instead, the debate is focused on the question of which type of responsibility needs to be shouldered by which actor.[56] Most of the debate, however, is not focused on the eco-services a natural resource performs, but on a fair appropriation either of the remaining space that can serve as a waste dump, or a fair appropriation of the costs of coping with environmental harm.[57]

5.5 Negotiated fairness

In discussing fairness as an element of an ethos in joint action, it is important to note that fairness plays two different but equally important roles. First of all, it addresses the justified claim that in acting jointly on environmental goals, appropriation of shared resources needs to be fair. The most basic intuition regarding life-sustaining NCPRs is that every single person deserves the share of them needed to satisfy her most basic needs. Beyond this, additional principles help to link the fair share of individuals and groups of individuals who relate to a natural resource in a variety of ways.

Second, as part of a shared ethos, fairness plays another important role. With regard to cooperation, empirical evidence states that an ethos that includes norms accepted by all members of the group is critical for the motivation to cooperate.[58] Moreover, a values-based identity of groups supports climate action, as empirical studies demonstrate.[59] Individuals are motivated to subscribe to a scheme of cooperation and to continuously engage in cooperation when they see this scheme as fair. Nevertheless, fairness should not be conflated with equal shares or with equity. Instead, fairness also responds to different positions in the group of beneficiaries.[60]

An example can help clarify these claims. It is derived from the writings of Elinor Ostrom. In harmony with our bid to criticize the theory of the "tragedy of the commons," Ostrom argues that Hardin's interpretation of the necessity of a tragedy of the commons[61] falls short of exploiting all possible scenarios. More specifically, the tragedy story underrates the role of normative frameworks that support beneficiaries of a shared resource in finding an optimal setting. Ostrom's example of an irrigation system can help explain this point of comparison.[62] Basically, an irrigation system needs to fulfil two tasks: it needs to resolve an appropriation problem in that it defines the allocation of water for various processes of agricultural production; and it needs to resolve a provision problem in terms of the maintenance of the irrigation system.[63] The initial scenario is one of asymmetry resulting from the fact that an irrigation system derives water from a river. Some people live upstream on a river, and others live downstream. Those who live upstream are obviously in the favourable situation of being able to allocate for themselves the share of the water reservoir they need; those who live downstream must make do with the remainder. Ostrom explores this situation in the context of the example of an irrigation system in Nepal, located close to the Thambesi River. I shall describe this example and explain how it resonates with a joint environmental action approach.

At the Thambesi river, users of the irrigation system meet annually and discuss the rules affecting appropriation and provision activities in the irrigation system. They agree on rules governing the system, including rotation rules that help to balance the different needs of those who live upstream of the river and those who live downstream.[64] Ostrom and Gardner report that farmers are indeed capable of designing and complying with rules as long as it is left to them to bargain the rules.[65] These negotiations can result in solutions that accommodate the interest of individuals from all points of view. Nevertheless, participants in the process of bargaining do not achieve a situation in which efforts and payoffs are distributed equally. Instead, they try to come closer to an equilibrium by revisiting the existing allocation schemes frequently. Even without equitable solutions, they do not choose a situation in which cooperation comes to a standstill. Instead, they try to continuously enhance the efficiency and equity of the institutional setting. During the process, they comply with the negotiated rules of allocation.[66]

Ostrom asserts that compliance with rules keeping the system intact results from an acknowledgment of dependence, in two ways. Beneficiaries are cognizant of their mutual dependence on the appropriation of services as well as their dependence on the natural resource. Ostrom's arguments emphasize that if these conditions hold, cooperation on common-pool resources will be favourable. The very same example can, however, be reframed as an example of joint environmental action in favour of a shared goal. The integrated environmental goal can be spelled out as a situation in which all potential beneficiaries are satisfied, because a scheme of appropriation is in place that satisfies their needs in the long term. At best, this scheme encapsulates the values enshrined in an environmental ethos: a fair appropriation of the resource, respect for integrity of the resource, and accordance with the values of individuals who regard the river not only as a resource, but also as living space for species and for natural evolution. The scheme of appropriation presupposes a balance between respect for the integrity of the river and the fulfilment of needs and desires of its users – in this case, the farmers who need the irrigation system for their crops.

When framed as an example of joint environmental action, the process of negotiation can also be framed as such. Instead of defining the rules from scratch, the process is now focused on the achievement of integrated environmental goals – in this case, successful farming under the circumstances of sharing a natural resource. If shortage of water supply is a problem, the cooperative scheme needs to be readjusted. This does not, however, undermine the need for all three dimensions of the ethos to be met: the distribution is fair, the resource's integrity is protected from harm, and, overall, an appropriative scheme serves the most basic interests of all users.

The example helps to understand that joint environmental action is special in that the means for achieving the integrated environmental goal need to be modelled against the background of concrete affordances of a shared resource and with respect to concrete needs of users. More specifically, a fair share of the resource is dependent on their agreement – there is not an absolute concept of justice operating in the background. This does not rule out a debate on principles of fairness in relation to specific scenarios, including water supply and climate change. This proposal just denies the ability to fix a distributive scheme of shares in a natural resource that is comparable to a scheme of distributive justice when distributing a cake to several people.

In sum, it is important to notice that fairness as an element of the ethos in joint environmental action has two meanings. On the one hand, it supports the long-term cooperative scheme in a fair way; that is, it does not burden some individuals and distribute the benefits to others. On the other hand, it needs to be adjusted to concrete scenarios and needs to be negotiated. Nevertheless, fairness remains a necessary and basic element in a scheme of cooperation as cooperation with respect to shared natural resources. However, the arguments in this chapter amount to the claim that fairness is only one principle among others in addressing normative resources of cooperation. More specifically, the negotiation of a fair share is already based on a framework that includes normative proposals.

5.6 Moderate eco-centrism

A third element of the ethos in environmental cooperation as joint action is much more open to discussion and critique than the other two. Joint action aims to accomplish shared goals. Principally, individuals are free to choose the goals that they want to reach together. But joint environmental action is qualified as a type of joint action whose goals are not chosen arbitrarily. It is instead focused on goals that provide direction and orientation regarding the future shape of activities that have an immediate or an indirect impact on NCPRs. In discussing what I am calling "moderate eco-centrism," NCPRs are not only interpreted as sources of services and as entities that are vulnerable, but also as living space for other species. More specifically, protection of this living space is regarded as a valuable enterprise. In order to explain this claim, the exploration starts with a rather radical position and moves forward to a more moderate eco-centric view.

Far from merely providing justification for environmental practices, environmental ethics is also dedicated to exploring the relationship between humankind and nature.[67] One key question is whether the life of non-human beings is regarded as valuable in and of itself. This question has been framed by Bernhard Williams in the following way: it is one thing to ask whose questions we are posing when discussing environmental issues, but it is altogether different to ask whose interests should play a role in the answers. The interests that environmental theories account for are not necessarily exclusively interests of people, but the interests of non-human living beings as well.[68] Obviously, people need to discuss environmental challenges against the background of the human language, human epistemic instruments, and shared knowledge. But this does not mean that their arguments necessarily focus exclusively on human interest. Rather, challenging moral anthropocentrism as a far too narrow approach to values in human life is perfectly reasonable.[69]

An answer to the question "Whose interests count?" is provided by a biocentric ethics.[70] Biocentrists assert that organisms have an inherent worth as related to their capacities to function as "centres of their lives" – in particular, each living entity has a "good of its own." As Paul Taylor explains it,

> One way of knowing whether something belongs to the class of entities that have a good is to see whether it makes sense to speak of what is good or bad for the thing in question. If we can say, truly or falsely, that something is good for an entity or bad for it, without reference to any other entity, then the entity has a good of its own.[71]

More specifically, Taylor does not think that living beings are self-authenticating sources of value. Instead, he is aware of the fact that to define the value of living beings accordingly, it is important to use an approach significantly different from an anthropocentric worldview. A biocentric worldview emphasizes the conviction that the lives of people should not – prima facie – be granted more moral weight. Instead, it is important to regard life on earth as a dense web of life-forms, in which every single being figures as a centre of its life.[72] To develop normative power, however, this ecological approach needs to be combined with normative premises.

This element of the environmental ethos has two sides. In a more modest view, according to Nussbaum's appraisal of "flourishing" in nature, the capacity of living beings to evolve and to struggle for a good life on its own demands respect.[73] Nussbaum explicitly claims that it is the capacity to flourish that deserves respect and that this even allows for determining entitlements of animals:

> And yet, if we feel wonder looking at a complex organism, that wonder at least suggests the idea that it is good for that being to flourish as the kind of thing it is. This idea is at least closely related to an ethical judgment that it is wrong when the flourishing of a creature is blocked by the harmful agency of another.[74]

On the other hand, respect for nature is also a perfectly reasonable approach to entities whose development and whose qualities humankind cannot control. Environmental ethics researchers remind us that respect and humility are the appropriate attitudes – not only as moral attitudes deserving of appreciation, but also as reasonable attitudes towards nature, which is ultimately not under our control.[75]

In the context of a discussion of NCPRs, natural goods also need to be addressed as goods serving the interests of non-human animals and of other living beings.[76] Natural goods and NCPRs, in particular, serve as habitats and as living space for creatures and ecological organisms that are different from humankind. Reasons to appreciate natural goods transcend arguments focusing on the benefits of ecosystemic functions.

This is not the place to argue that this approach to nature is more reasonable or better than a pragmatic or functionalist approach. When addressing joint environmental action, though, the assessment of the many functions of nature and the value of living beings therein appears not to be too far-fetched. Moderate eco-centrism acknowledges this interpretation of nature, even without necessarily supporting the stronger claim of bio-centrism, that all living beings have an equal moral worth. It is compatible with moderate eco-centrism to claim that humans are unique and different from animals and other life-forms, which are also uniquely valuable.[77] As for shared natural resources, moderate eco-centrism acknowledges the fact that, in order to preserve diversity on Planet Earth, these resources should also be protected as habitats, as living space, and as resources for the evolution of new species and forms of life.[78]

These proposals for defining the ethos included in joint environmental action can now be summarized. As related to the appropriation of scarce natural resources that qualify as NCPRs, the scheme of appropriation needs to reflect respect for the integrity of the resource; it needs to implement patterns of a fair appropriation of the benefits from that resource; and it should cohere with moderate eco-centrism. One remaining question is whether the rationale for this ethos really is altruistic or particularly nature friendly. Couldn't it simply express the self-interest of users of a shared resource who are seeking to benefit in the long term?[79] Obviously, assuring the long-term capacity to appropriate a natural resource is in the self-interest of humankind; accordingly, rules that support this goal are not necessarily particularly "ethical." Both of these points are valid. Then again, the difference between an ethos that mirrors self-interest and an environmental ethos is twofold: first, the latter emphasizes non-human species and their needs; second, it supports fairness even when it is not perfectly aligned with the most efficient appropriation of the shared resource.[80]

5.7 Alternative proposals for an environmental ethos

An ethos for joint environmental action rests on three normative elements. First, the need to be fair in order not to appropriate an overly egoistic portion of natural resources, one that undermines the fair shares of other appropriators. Second, people who share the ethos understand that Planet Earth is not made exclusively for human beings, but is also living space for other living beings that are trying to survive and whose evolution is part of life on earth. Third, and foremost, people are dedicated to respecting the fact that geophysical systems come with a safe operational space, but beyond that, they are vulnerable, and can be exhausted and destroyed. It is reasonable to not test these limits.

This is not the only attempt to think through the underlying normative claims of collective action for environmental aims. Since an environmental ethos can have many different forms, I shall briefly summarize the alternatives and also explain what is "moderate" about "moderate eco-centrism." This section discusses three alternative moral framings. It starts with the view that protection of natural resources is owed to future generations; it explains the concept of "new harm" in the context of a cosmopolitan ethics; finally, it addresses environmental rights as subsistence rights. This list is not exhaustive, nor does it represent the current debate in environmental ethics. Instead, it represents positions thinkers have taken towards fundamental claims in climate ethics.

One frequent issue that comes up when discussing a fair appropriation of natural goods is the question of what we as people living today owe future generations. This is, of course, also motivated by the prominence of the concept of sustainability, which has been positioned as being needed "for the sake of future generations."[81] The question of what current generations owe future generations when it comes to environmental concerns has been a matter of vigorous debate. On the one hand, philosophical puzzles have avoided making a general and consistent claim that it is necessary to honour the needs of future generations. More specifically, population growth generates the most serious issues in environmental terms. Over-exploitation of soil, water, and the atmosphere is worsened by a growing world population. In spite of this, the argument that population numbers ought to diminish is difficult to defend.[82] In addition to this dilemma, the non-identity problem has also generated confusion. This dilemma posits that harm to future individuals cannot be accounted for in a theory for analytic reasons that provide a close tie between the existence of people and the then existing climate conditions.[83] All this puts severe limitations on future-generation claims. In addition, it is not easy to claim that entitlements of people need to be respected when these individuals do not even exist yet.

Overall, the recent development of climate ethics pays limited attention to the problem of future generations. One of the sad truths in the background is that climate change is developing so rapidly that the generations currently living will already suffer the effects, so concern for future generations may be moot. More specifically, current generations should respect "planetary boundaries" in order not to put at risk the ability of future generations to shape their environmental surroundings and to act on their ambitions. One way to establish this claim recalls conditions of sustainability. In that context, the broad notion of sustainability is broken down to specific incidents of "natural capital," which include water resources and the atmosphere.[84]

Another alternative moral framing of an environmental ethos is based on the claim of fairness among strangers in the context of ethical cosmopolitanism.[85] Philosophers have classified causal harm resulting from environmentally hazardous private behaviour as a class of "new harms." These harms are "new" in various respects: the causal nexus between the act causing the harm and the occurrence of harm is not straightforward; the harm results from an accumulation of negative side effects of activities taking place in common-pool resources; it is neither willed nor can it be easily avoided; both the spatial and the chronological distance between the source causing the harm and the effect can be extremely large. Although this class of harm differs from more straightforward types of harm, and although it impacts "strangers" who are spatially distant,[86] it can be stated that individuals who cause this harm are nevertheless obliged to take responsibility for its consequences for other individuals. Recently, Simon Caney has highlighted the claim that the "harm-avoidance justice" approach works on premises that differ from the "burden-sharing justice" approach:[87] "Its focus is primarily on ensuring that the catastrophe is averted (or at least minimised within reason). This perspective is concerned with the potential victims – those whose entitlements are threatened – and it ascribes responsibilities to others to uphold these entitlements."[88] It is important to note that although Caney thinks that both approaches might coincide at some point,[89] he thereby also identifies a new type of responsibility. More specifically, the difference between them is not that responsibility concerned with burden-sharing is a backward-looking responsibility, whereas harm-avoidance responsibility is forward-looking. Instead, he states that both approaches have a different scope, the first working on entitlements, the second on burdens; and responsibilities need to be distinguished according to the object of responsible behaviour. Whereas the avoidance-approach is about first-order responsibilities, the latter is about second-order responsibilities, which are responsibilities for ensuring compliance with the first-order responsibilities.[90]

Philosophers who assert duties to make good on "new harms" maintain that a person is at least to some degree responsible for the suffering of individuals at a particular place when she has contributed actively and willingly to the causes of those harms. It is reasonable to suggest that those who caused the harm also need to repair it.[91] Even when not fully aware of the negative effects, a person who spoils a shared natural resource or who over-exploits it is liable for the harm she has contributed to the destruction.

This claim remains controversial. Among others, Walter Sinnott-Armstrong defends the view that since climate change is no one's fault, no one is liable for making good on the harm other people suffer.[92] Another objection comes from Frances Kamm. She maintains that we need to distinguish between incidents of harm caused by accidents and simple justice cases.[93] Distance, in particular, plays out differently in each case. As for accident cases – which can be seen as climate events where more or less randomly distributed incidents of severe harm are caused by an accumulative effect – a stronger duty to rescue people near us than to rescue those far away can be claimed. It is not distance per se that matters, but the distance or closeness of the threat to the victim. Climate harm, instead, needs to be interpreted in the theoretical framework of standard justice cases. In that scenario, distance does not matter to the claim for assistance. As a consequence, there is no reason to claim duties to protect climate victims or even to rescue them.

If a duty to support climate victims exists, it not only differs from the general duty not to inflict harm on people, in that the causal chains are less straightforward; it also differs in that the victims of my potential harmful acts are "far away."[94] In one respect, this reservation can be repaired. As Violetta Igneski argues, if a duty is justified, it does not matter whether the person who is the recipient of acts committed according to moral duties is in one's own neighbourhood or at a distance.[95] This insight is only valid with regard to perfect duties, however, according to the common Kantian understanding that duties do not infringe on basic entitlements that individuals hold. Whether or not people really need to support the well-being of climate victims in non-rescue scenarios needs to be defined on separate grounds. I shall return to duties in the context of joint climate action in Chapters 7–10.

Before we get there, this section addresses a third way to argue in favour of environmental action based on a moral framework. In water ethics, the proposal to establish water rights has not only received much attention, but the rights to access clean water within a reasonable distance and to water sanitation have been accepted in the context of the addenda to the Charter of Human Rights.[96] A sketch of the debate on water rights helps us to understand the complexity of that moral frame. More specifically, fairness as part of the environmental ethos can also be founded in a theoretical proposal for individual environmental rights. Environmental rights do not have to be established as free-standing rights. Instead, it suffices to regard them as part of the subsistence rights of people.

Each person should receive a minimally sufficient share of ecologically intact land, water, and air to be able to live a basically good life. This claim has been made on grounds of fundamental human rights. Henry Shue maintains that rights are basic when "enjoyment of them is essential to the enjoyment of all other things."[97] Basic rights include, inter alia, subsistence rights, where subsistence means unpolluted air, unpolluted water, adequate food, adequate clothing, adequate shelter, and minimal preventive public health care.[98] According to Shue, a right to unpolluted water is a basic human right. Moreover, it includes "justified demands for social guarantees against standard threats."[99] Since the time Shue presented these arguments, a right to water was enshrined in the canon of human rights.[100] The idea of a human right to water consists of two components. First, it implies that the right to water is a human right in precisely Shue's interpretation of something each person is justified to claim for herself. A human right expresses universal but still personal entitlements. The second component of a right to water correlates with the duties of political institutions; although classified as a moral right, its consequences are political.

Recently, environmental rights have been strengthened in that they are interpreted as free-standing rights. There are three main arguments for interpreting the right to water instead as an autonomous right. The first is that, although it is right to say that some of the most basic human rights, such as a right to life and to food, already include a right to water, this "inclusion" might not reach the moral weight of demanding access to water. Like a right to an adequate environment, a right to water meets the criteria of a genuine human right: it is of paramount moral importance.[101] The second argument is that an autonomous right is justified when its content can be made explicit without reference to further claims expressed in other rights. This formal criterion also appears to be fulfilled. It has been maintained that the content of a right to water consists of three claims: (1) water needs to be *accessible*, that is, water resources need to be within the safe reach of all, need to be affordable to all, and need to be accessible to all in law and in fact; (2) water needs to be available in *adequate quality* (it must be safe); and (3) water must be *accessible in a certain quantity*, in sufficient and continuous supply for personal and domestic use.[102] In short, the content of a right to water can be made explicit in a precise way without reference to further rights. The third main argument for an autonomous right to water is that if a right to water is justified as an autonomous right and not as an appendix to other human rights, its moral consequences can also be fully accounted for. At the very least, in view of the shaping and enforcement of the social conditions that cause avoidable global poverty, a negative duty not to cooperate in the imposition of a coercive institutional order that unnecessarily leaves human rights unfulfilled has been justified.[103] Moreover, within a framework of a global ethics, everyone can legitimately be asked to contribute in some reasonable proportion to duties claimed by the rights of others.

Regardless of whether a right to water has been defended as an autonomous right or as a precondition of established rights, two arguments *against* a human right to water also need to be acknowledged. To begin with, specialists call into question whether a right to water effectively contributes to protecting water resources. Although rights are not justified by citing the effects of corresponding institutions, deleterious effects would be an argument against defending a right to water.[104] Moreover, the declaration of a right to water is not very helpful unless "water" has been provided with further definitions and qualifications. Even outlining the three aspects of accessibility may not suffice to explain the nature of a right to water.

These comments contribute to a critical view of a right to water, and possibly also to environmental rights more generally. One severe limitation is the fact that environmental rights are tied to the desires and needs of individuals. Granting these rights does not support the claim of the need to protect NCPRs from over-exploitation. In addition, the concept of rights plays into approaches that underline either a justified minimum subsistence to every single person or a view according to which "subsistence emissions" – which are tied to the needs of individuals – need to be distinguished from "luxury emissions."[105] As a consequence, rights to natural resources – even when restricted to minimum shares – neither respect variances in the lifestyles of people nor their different capacities in terms of an efficient use of resources. Moreover, when taken as an absolute baseline for distributive schemes, they might undermine the efficiency and feasibility of a distribution scheme.

Overall, this section addresses alternative ways of defining an environmental ethos that have been at the centre of recent debates. Unlike an ethos resting on claims of fairness, moderate eco-centrism, and respect for the integrity of a resource, they make one of three claims: first, normative claims are founded on respect for the needs and desires of future generations; second, a moral duty to make good on "new harms" – even when victims are strangers and are "far away" – was maintained; and third, environmental rights have been presented as a backdrop for environmental agendas. Although each of these approaches is valid in its own way, they have also received criticism. More specifically, the applicability to climate ethics and to a fair appropriation of common-pool resources remains questionable.

5.8 Three objections

This chapter has established that an ethos is critical for joint action. It serves as a resource for reasons to act; it also helps participants in joint action identify with the shared goal. Such an ethos can also be applied to joint environmental action.

What distinguishes the values enshrined in an ethos from the values of individuals is not only their character as normative standards to guide the appropriation of a resource. Instead, it is an interpretation of nature as valuable, the claim that each person has a right to a fair share of natural resources as basic as water, air, and land, and the duty to respect "planetary boundaries" that together characterize the ethos. In addition, a shared ethos also supports the motivation for cooperating in the first instance. It serves as a resource for reasons for reaching a shared goal by means of participatory intentions. A thorough characterization of the environmental goals themselves is still not appropriate. It will take another chapter to explain the content of the goals introduced by joint environmental action. Before we get there, objections that are likely to be raised at this point also need to be addressed. Since this chapter is restricted to one element of the theoretical approach, some of the reservations need to be postponed until further elements are on the table. Nevertheless, two standard reservations will be presented. Each of them helps to shape the specific function of the ethos.

To begin with, one might think that the proposal needs as an additional element a balance between investments in cooperation and the fruits of cooperation. Obviously, fairness goes much farther than just the acquisition of a fair share in a natural resource. One of the most important insights into the drama of the commons is the imbalance between gains from natural resources on the one hand and environmental harm to victims of environmental hazards on the other. The logic of benefits from natural resources appears to undermine the kind of win-win scenario seen in the example of the lake.[106] When applied to the atmosphere, the problem is even worse, since the possibility of "passing the buck" and harvesting the fruit without paying the costs – and instead passing them off to future generations – is tremendously tempting.[107]

This is a serious problem. However, a quick answer shows that buck-passing might also result from not receiving a fair share in a common resource. A lengthier answer needs to take those problems seriously. The question of how fairness in terms of receiving a fair share in a natural resource can be combined with fairness in terms of shouldering the costs and burdens that need to be distributed is a recurrent theme in this book. In my view, both need to be mediated by an understanding of goals, which I will introduce shortly. Schemes of fair appropriation are the baseline regarding distributive fairness with respect to shares in a natural resource. Regarding the burden of attaining goals that rely on these patterns or that even imply a rectification of existing patterns, these burdens need to be distributed according to principles of ability to pay, of liability, and of second-order responsibilities. It is particularly important to detach the issue of a fair share of benefits in terms of eco-services from the issue of a fair share of burdens and costs. In addition, the transition from an ideal scenario in which "enough" is still available needs to be distinguished from a scenario in which a "safe operational space" is no longer available. Still, the very possibility of fair practices in appropriating shared natural resources should not be rejected from the outset.

It should also be noted that this approach stands in stark contrast to an economic approach to addressing the natural commons. Instead of "selling indulgences,"[108] and instead of arguing for economic regulation in favour of environmentally friendly practices,[109] this chapter contends that what counts are benefits from natural goods, not primarily the costs of environmental hazards. The general assumption, however, is a joint commitment that in turn cannot be achieved without sharing some normative resources. They include environmental fairness, moderate eco-centrism, and respecting the integrity of natural resources.

A standard critique is that this kind of environmental ethos simply does not exist. Although it is reasonable to propose educating people towards such an ethos, it cannot be assumed that people, much less governments, share values that support acting reasonably regarding shared environmental resources.[110] In two respects, however, this critique misses the point. Examples from water ethics demonstrate that people are willing to adopt an ethos – particularly when it is in their long-term interest to subscribe to shared values.[111] Yet the theoretical approach to joint action taken in this chapter is not yet ready for application. Instead, the discussion is focused on the prerequisites of joint environmental action according to the theoretical insights into joint action. In this chapter, I have portrayed an ethos that fits joint environmental action. However, the idea that joint action relies on a shared ethos is not new; instead, it is seen in theories of joint action as a critical element. Whether this resource really exists is another issue. More specifically, I later maintain that joint environmental action is not only a model for real group action; it also has two additional functions. It helps to determine the moral foundations of successful environmental cooperation. As such, it can also be translated into norms to guide environmental action on an institutional level. The proposals for an institutional setting at the end of the book provide the grounds for this application.

Another perspective can also support the view that a claim for collective action in favour of environmental goals and the idea of a shared ethos are not too far-fetched. Similar insights are mirrored in agreements about shared environmental goals – for example, the Rio Declaration on Environment and Development of the United Nations. This declaration confirms the claim that "states have common but differentiated responsibility."[112] The normative content of environmental action with respect to a shared commitment has also been re-emphasized in the Paris Agreement.[113] The preamble acknowledges that "climate change is a common concern of humankind." It then goes on to describe a shared ethos, including respect for the integrity of natural resources and the protection of biodiversity. It also describes how some cultures consider integrity to be particularly important and that climate justice is an important notion for certain nations in this context. The existence of international agreements is not proof of the existence of a consensus about an environmental ethos. Rather, these documents highlight the need to transform general moral statements into a more concrete form.

The ethos is also modest – not overly demanding – in another respect. It does not claim that current generations need to prioritize conservationist environmental goals as a way of acknowledging the justified claims of future generations. Instead, it suffices to formulate an ethos that restricts appropriation to (a) principles that enable further usage of the resource and (b) principles that regulate a fair scheme of appropriation and protection of a shared resource. This can be stated as a variation of John Rawls' approach to justice, the core idea of which is to introduce an original position. That scenario assumes that participants who are authors of principles of justice are bereft of knowledge about their future position in a society that relies on the principles being argued.[114] Accordingly, the proposal to construe a moderate eco-centric ethos can take the following route. Let's assume that individuals who are all dependent on a shared natural resource, perhaps a lake, did not know their position as a beneficiary of that resource in the future. Which normative guidelines would they choose to establish not only a distribution of eco-services to every single person, but also an institutional framework for the appropriation of that resource? At a minimum, they would take the lessons from the "tragedy of the commons" story and vote against open-access policies. They would also learn the lessons of the failure of privatization as well as from the failure of pure economic institutions. Instead, they would choose a practice that gives credit to basic claims of fairness, including an equal share of basic access to that resource; they would also choose a method that would not destroy the shared resource; and they would possibly also be interested in nature as a space that is not designed – a space not of man-made development – and a place for the existence and evolution of natural organisms. The reasons why they choose this is a recognition of the diversity and multiple functions of a particularly valuable resource.

Finally, the third standard objection is that this approach is simply tautological. People who share an ethos are willing to cooperate in favour of a goal that this ethos supports. An example is an environmental peer group that is moved by an ethos to invest in the protection of a lake or a river. But people who do not share that ethos will simply not join the group. This is true, but it misses the point. It was my goal to establish an ethos that represents an appropriate set of values for addressing and benefitting from a shared natural resource. This attitude is not reducible to the self-interest of users in the long run. Instead, it is the right attitude when collectively harvesting fruit from a resource that offers a rich diversity of eco-services. The claim is that no one has a prior right to that resource; instead, each type of usage is justified as long as it is compatible with the types of usage that other users prefer. The limits are set not by quantities – remember that there is enough water for everyone! Instead, the constraints resonate with the conditions of the integrity of the resource, with the right of each person to satisfy his or her desires as long as they are compatible with the desires of every other person, and with the justified claim to respect non-human life and the irreplaceability of natural resources for their survival and for the natural evolution of species. Goals that individuals and groups of people reach by appropriating natural resources are justified as long as they do not undermine these schemes of appropriation.

Notes

1   Raimo Tuomela, *The Philosophy of Social Practices: A Collective Acceptance View* (Cambridge: Cambridge University Press, 2010), 39.

2   Raimo Tuomela, *The Philosophy of Sociality: The Shared Point of View* (Oxford/ New York: Oxford University Press, 2007), 44.

3   Tuomela, *The Philosophy of Sociality*; Tuomela, *The Philosophy of Social Practices*.

4   This term is derived from integrative water management. See Jerome Delli Priscoli and Aaron T. Wolf, *Managing and Transforming Water Conflicts* (Cambridge/New York: Cambridge University Press, 2009), 121.

5   The main opponents are, on the one hand, Margaret Gilbert, who holds the view that group action is based not only on shared values, but also on the commitment not to cheat, but rather to honour obligations of group membership. This is also applicable to action in the political sphere. See Margaret Gilbert, *Living Together. Rationality, Sociality, and Obligation* (New York: Rowman and Littlefield, 1996); Margaret Gilbert, "Group Membership and Political Obligation," *The Monist* 76, no. 1 (1993): 119–131. On the other hand, some researchers hold the much more modest view that shared agency results primarily from processes of planning and coordination. See Michael E. Bratman, *Shared Agency. A Planning Theory of Acting Together* (New York: Oxford University Press, 2014).

6   Ethos is mentioned in various contexts; for the variety, see Tuomela, *The Philosophy of Social Practices*, 39, 178–180; Tuomela, *The Philosophy of Sociality*, 16, 32–36, 44, 129–133.

7   Tuomela, *The Philosophy of Sociality*, 217.

8   Ibid., 180.

9   Ibid., 128.

10   Modest sociality is a trait that theories of joint action support more generally. It means that joint action is not necessarily tied to groups as organized collectives.

11   Tuomela, *The Philosophy of Sociality*, 15.

12   "If a group accepts doing something, then its members will participate in doing it (or at least try to). Thus, if the group members jointly intend to do X, and will do it, then necessarily (because of this) all members will relevantly participate, that is, to be 'involved' or 'play a role,' possibly in a passive sense, with respect to X" (Tuomela, *The Philosophy of Sociality*, 49).

13   Raimo Tuomela, "Group Reasons," *Philosophical Issues* 22, no. 1 (2012): 405; Raimo Tuomela, "The We-Mode and the I-Mode," in *Socializing Metaphysics – The Nature of Social Reality*, ed. F. Schmitt (Lanham, MD: Rowman and Littlefield, 2003), 97.

14   Tuomela, *The Philosophy of Sociality*, 59.

15   Note that groups have been framed in very different terms in social philosophy. The starkest contrast is provided by Margret Gilbert's concept of groups as "plural subjects" in "Group Membership and Political Obligation." According to that theory, the ethos is not only constitutive of actions, but of the group itself. For a helpful explanation of the meaning for political contexts, see Francesca Raimondi, "Joint Commitment and the Practice of Democracy," in *Group Process, Group Decision, Group Action*, eds. Robert S. Baron, Norbert L. Kerr, and Norman Miller (Pacific Grove, CA: Brooks/Cole, 1992), 285–299.

16   Raimondi, 82.

17   Tuomela, "The We-Mode and the I-Mode," 100.

18   See ibid., 99.

19   See Tuomela, *The Philosophy of Sociality*, 32.

20   Ibid.

21   Ibid., 44.

22   Ibid., 23.

23 This is a prominent and controversial thesis advanced by Margaret Gilbert. See Margaret Gilbert, *A Theory of Political Obligation* (Oxford: Oxford University Press, 2006).

24 Tuomela, *The Philosophy of Sociality*, 37.

25 Tuomela distinguishes thoroughly between "ought-to-be norms" of the members and "ought-to-do norms"; as for the ethos, it gives resources for both, but the most important element is the first, whereas the latter appears to be mediated by the joint intentions that are supported by "ought-to-be norms." For a detailed explanation, see Tuomela, *The Philosophy of Social Practices*, 175–180.

26 Tuomela, *The Philosophy of Sociality*, 238.

27 Ibid., 23.

28 Ibid., 15.

29 It is not my intention to compare Bratman's and Tuomela's approaches to joint action here. However, I do want to underscore Tuomela's interpretation of Bratman's approach as a form of very modest We-mode-actions, which he terms "I-mode we-intention" (see Tuomela, *The Philosophy of Sociality*, 70–73). This interpretation also underscores the systematic possibility of exploring elements of both theories and of bringing them together at chosen points.

30 This chapter draws on a model of cooperation that has been discussed in the context of water ethics. See Angela Kallhoff, "Addressing the Commons: Normative Approaches to Common Pool Resources," in *Climate Change and Sustainable Development: Ethical Perspectives on Land Use and Food Production*, eds. Thomas Potthast and Simon Meisch (Wageningen: Wageningen Academic Publishers, 2012), 63–68; Angela Kallhoff, "Water Justice: A Multilayer Term and Its Role in Cooperation," *Analyse & Kritik* 36, no. 2 (2014): 367–382.

31 I have chosen an example that is very close to the example used by Peter Singer in "One Atmosphere," 27. Whereas Singer speaks of a "giant sink" and compares this to the atmosphere as a waste dump for greenhouse gases, I take the lake as a literal example for a joint natural resource.

32 Dale Jamieson, *Ethics and the Environment. An Introduction* (Cambridge: Cambridge University Press, 2008), 14.

33 Delli Priscoli and Wolf, 121.

34 This claim is, of course, rather challenging. But regarding water resources, there has always been some backing both in the theory of water laws and in the practices of sharing water resources. For examples, see Thomas Naff and Joseph Dellapenna, "Can There Be Confluence? A Comparative Consideration of Western and Islamic Fresh Water Law," *Water Policy* 4 (2002): 465–489; Wilhelm Ripl, "Water: The Bloodstream of the Biosphere," *Philosophical Transactions of The Royal Society of London, Biological Sciences* 358 (2003): 1921–1934; Carol M. Rose, "Energy and Efficiency in the Realignment of Common-Law Water Rights," *Journal of Legal Studies* 19, no. 2 (1990): 261–296; Claudia W. Sadoff and David Grey, "Beyond the River: The Benefits of Cooperation on International Rivers," *Water Policy* 4 (2002): 389–403. Falkenmark and Folke go so far as to claim that their investigation of water conflicts "clearly showed that at the international level, water appears to pose a reason for transboundary cooperation rather than for war, often preventing escalation instead of causing it." Malin Falkenmark and Carl Folke, "The Ethics of Socio-Ecohydrological Catchment Management: Towards Hydrosolidarity," *Hydrology and Earth System Sciences* 6, no. 1 (2002): 1–9.

35 For a more comprehensive approach to an ethics of environmental cooperation in the context of water ethics, see Kallhoff, "Water Justice"; Angela Kallhoff, "Klimakooperation: Kollektives Handeln für ein Öffentliches Gut," in *Klimagerechtigkeit und Klimaethik*, ed. Angela Kallhoff (Berlin/Boston: De Gruyter, 2015), 143–167; Angela Kallhoff, "Transcending Water Conflicts: An Ethics of

Water Cooperation," in *Global Water Ethics. Towards a Global Ethics Charter*, eds. Rafael Ziegler and David Groenfeldt (London/New York: Routledge, 2017), 91–106.

36 John Rawls, *Political Liberalism* (New York: Columbia University Press, 1996), 16.

37 Robert Costanza et al., "The Value of the World's Ecosystem Services and Natural Capital," *Nature* 387, no. 6630 (1997): 253, https://doi.org/10.1038/387253a0.

38 Ibid., 254.

39 Falkenmark and Folke, 6.

40 Martin Kowarsch, "Ethical Targets and Questions of Water Management," in *Water Management Options in a Globalised World. Proceedings of an International Scientific Workshop (June 20–23, 2011, Bad Schönbrunn)*, ed. Martin Kowarsch (Institute for Social and Developmental Studies (IGP) at the Munich School of Philosophy, 2011), 45, www.researchgate.net/publication/308118199_Water_ma nagement_options_in_a_globalised_world.

41 Falkenmark and Folke, 6.

42 Costanza et al., 254.

43 Sadoff and Grey.

44 For an interdisciplinary debate and illuminating papers on the manifold meanings of the concept of an ecosystemic function, see Gretchen C. Daily, ed., *Nature's Services: Societal Dependence on Natural Ecosystems* (Washington, DC: Island Press, 1997). This volume also includes a critical discussion of the reduction of ecosystemic functions to humankind and of the possibility of supplementing the value of markets by monetary means.

45 Daily, 365.

46 Ibid.

47 On July 28, 2010, following an intense negotiation, 122 countries formally acknowledged the "right to water" in General Assembly (GA) resolution (A/64/292, based on draft resolution A/64/L.63/Rev.1). In September 2010 the UN Human Rights Council adopted a resolution recognizing that the human rights to water and sanitation are a part of the right to an adequate standard of living, www.un.org/Waterforlifedecade/Human_right_to_water.shtml, accessed March 14, 2019. For the philosophical controversy over the right to water, see Wolfgang Bretschneider, "The Right to Water from an Economic Point of View," in *Water Management Options in a Globalised World. Proceedings of an International Scientific Workshop (June 20–23, 2011, Bad Schönbrunn)*, ed. Martin Kowarsch (Institute for Social and Developmental Studies (IGP) at the Munich School of Philosophy, 2011), 87–94, www.researchgate.net/publication/308118199_Wa ter_management_options_in_a_globalised_world; Bryan Randolph Bruns and Ruth S. Meinzen-Dick, "Negotiating Water Rights: Implications for Research and Action," in *Negotiating Water Rights*, eds. Bryan Randolph Bruns and Ruth S. Meinzen-Dick (London: ITDG Publishing, 2000), 353–380; Peter H. Gleick, "The Human Right to Water," *Water Policy* 1 (1998): 487–503.

48 One theoretical worry is that, regarding life-sustaining goods, it is difficult to define a bare minimum. Moreover, it might be difficult to determine a fair share of a good that is already over-exhausted, such as the atmosphere as a waste receptacle. Instead of guaranteeing equal access, it is better to prioritize the worst off, a proposal that is defined as an answer to the shortcomings of the guaranteed minimum approach in Stephen M. Gardiner, "Ethics and Global Climate Change," *Ethics* 114 (2004): 586–588.

49 Problems with this view do not result primarily from the difficulty of justifying principles of prioritization accordingly, but also from practicability. See Gardiner, "Ethics and Global Climate Change," 586–588.

50 See Kallhoff, "Normative Approaches to Common Pool Resources."

51 See Kallhoff, "Transcending Water Conflicts."

52 For an example of a community that is perfectly adapted to life with a water resource, see Edward L. Ochsenschlager, *Iraq's Marsh Arabs in the Garden of Eden* (Philadelphia: University of Pennsylvania Museum of Archaeology and Anthropology, 2004).

53 For a concept of environmental justice that includes concepts of distributive justice regarding benefits, but also pays particular attention to the burdens of environmental harm, see Gordon P. Walker, *Environmental Justice. Concepts, Evidence, and Politics* (London/New York: Routledge, 2012).

54 A further complication results from the fact that high emissions of harmful gases are usually related to high living standards. The *beneficiary-pays principle* applies to prior hazardous behaviour, but focuses on benefits already received from a resource. The beneficiaries should also pay for the damage that results from reaping the benefits. Simon Caney addresses the "beneficiary-pays principle," stating: "Put more formally, this claims that where A has been made better off by a policy pursued by others, and the pursuit by others of that policy has contributed to the imposition of adverse effects on third parties, then A has an obligation not to pursue that policy itself (mitigation) and/or an obligation to address the harmful effects suffered by the third parties (adaptation)." Simon Caney, "Cosmopolitan Justice, Responsibility, and Global Climate Change," in *Climate Ethics. Essential Readings*, eds. Stephen M. Gardiner et al. (Oxford/New York: Oxford University Press, 2010), 128.

55 In addition, it also has to be said that a hybrid approach not only results from compromises meant to enhance applicability, but also addresses several principles, each of which is justified in and of itself. For a distinctively hybrid account, see Simon Caney, "Cosmopolitan Justice, Responsibility, and Global Climate Change," *Leiden Journal of International Law* 18 (2005): 747–775.

56 For an outline of two types of responsibilities involved, see Simon Caney, "Two Kinds of Climate Justice: Avoiding Harm and Sharing Burdens," *Journal of Political Philosophy* 22, no. 2 (2014): 125–149.

57 The most well-known principle for a fair allocation of burdens for hazardous practices is the *polluter-pays principle*, which states that actors who are responsible for environmental harm are also liable for remedying that damage. Singer has chosen the expression "you broke it, you fix it" to paraphrase the principle. Peter Singer, "One Atmosphere," in *One World. The Ethics of Globalization* (New Haven, CT: Yale University Press, 2002), 41. Still, this principle has two major shortcomings. First, Singer interprets this principle not as a principle of fairness, but rather as a principle that creates incentives to be careful about causing pollution (Singer, "One Atmosphere," 41). It is a pragmatic principle, not a principle of justice. Caney states that the principle is incomplete, "for it requires a background theory of justice and, in particular, an account of individuals' entitlements" (Caney, "Cosmopolitan Justice," 2010, 134). To see this, is suffices to discuss cases in which people have exceeded their entitlements. In that case, the polluter-pays principle is useless unless we already know in advance what rights, if any, people enjoy to emit greenhouse gases (ibid.). Second, regarding natural resources that cannot easily be repaired and regarding damage that cannot easily be fixed, even when the perpetrators may be qualified, it is still difficult to define remedies. Some types of environmental damage are irreversible – the case of climate change appears to be of this sort. At a minimum, the principle supports cautious practices in using the remaining goods and also supports responsible use. In addition to these theoretical problems, the discussion of the polluter-pays principle automatically leads to the complex debate on "historical responsibility." For a useful summary and illuminating comments, see Simon Caney, "Justice and the Distribution of

Greenhouse Gas Emissions," *Journal of Global Ethics* 5, no. 2 (2009): 133–135. For the connection of the polluter-pays principle and historical responsibility, see also Caney, "Cosmopolitan Justice," 2005, 752–763.

58 This fact is sometimes framed and explored as the readiness of group members to punish defection in terms of free-riding. See Ernst Fehr and Simon Gächter, "Cooperation and Punishment in Public Goods Experiments," *American Economic Review* 90, no. 4 (September 2000): 980–994.

59 In the context of climate research, psychological research highlights the importance of group identity for collective efforts towards climate goals. "Indeed, social psychological research shows the powerful, positive effect of increasing perceived similarity, shared identity and superordinate goals on helping behaviour … Though the analogy is not perfect, the psychological research does suggest that framing the victims of climate change in ways that underscore shared goals and identities should similarly increase their moral standing, and with it, motivation to help them. Further research should confirm that such psychological tendencies can scale to these global levels." Ezra M. Markowitz and Azim F. Shariff, "Climate Change and Moral Judgement," *Nature Climate Change* 2 (2012): 246, https://doi.org/10.1038/NCCLIMATE1378.

60 A similar point is made by Weisbach when criticizing the idea that distributive justice with respect to natural common-pool resources (NCPRs) could ever mean "equity." Instead, access rights to NCPRs are also shaped by territorial theory, which justifies a priority to people living in a certain area, and by other principles of legacy. See Stephen M. Gardiner and David A. Weisbach, *Debating Climate Ethics* (New York: Oxford University Press, 2016), 218–221. The important point for the current exploration is that unequal distribution is not in itself unfair – something that writers who defend a per-capita approach usually take for granted.

61 Garrett Hardin, "The Tragedy of the Commons," *Science* 162, no. 3859 (1968): 1243–1248, https://doi.org/10.1126/science.162.3859.1243.

62 Elinor Ostrom and Roy Gardner, "Coping with Asymmetries in the Commons: Self-Governing Irrigation Systems Can Work," *Journal of Economic Perspectives* 7, no. 4 (1993): 93–112.

63 Ibid., 97.

64 Ibid., 99–100.

65 Ibid., 109.

66 Ibid.

67 This claim has been specifically dealt with by writers who explore a biocentric approach. For the discussion of a "biocentric outlook" or a "biocentric worldview," see Paul Taylor, *Respect for Nature: A Theory of Environmental Ethics* (Princeton, NJ: Princeton University Press, 1986), 99–168.

68 Bernard Williams, ed., "Must a Concern for the Environment be Centred on Human Beings?," in *Making Sense of Humanity: And Other Philosophical Papers 1982–1993* (Cambridge: Cambridge University Press, 1995), 233–240.

69 The debate on "anthropocentrism" is vast. In our context, I take "anthropocentrism" as an approach to moral principles that rests on the assumption that all that counts in a moral assessment of action are the interests of people.

70 For a theory that argues for a biocentric view, see Taylor. For an assessment and a critique of biocentrism, see Angela Kallhoff and Michael Bruckner, "Biozentrismus," in *Handbuch Tierethik. Grundlagen-Kritik-Perspektiven*, eds. Johann S. Ach and Dagmar Borchers (Stuttgart: Metzler, 2018), 161–166.

71 Taylor, 61.

72 "The assertion that an entity has inherent worth is here to be understood as entailing two moral judgments: (1) that the entity is deserving of moral concern and consideration, or, in other words, that it is to be regarded as a moral

subject, and (2) that all moral agents have a prima facie duty to promote or preserve the entity's good as an end in itself and for the sake of the entity whose good it is" (Taylor, 75).

73 Martha C. Nussbaum, *Frontiers of Justice. Disability, Nationality, Species Membership* (Cambridge, MA: Harvard University Press, 2006), 306.

74 Ibid. For an exploration of "flourishing" as a concept of the good life of plants, see Angela Kallhoff, "Plants in Ethics: Why Flourishing Deserves Moral Respect," *Environmental Values* 23, no. 6 (2014): 685–700.

75 For a call to human respect in terms of humility regarding nature, see also Rosalind Hursthouse, "Environmental Virtue Ethics," in *Environmental Ethics*, eds. Rebecca L. Walker and P. J. Ivanhoe (Oxford/New York: Oxford University Press, 2007), 155–172. More arguments for and against environmental virtues are provided in Philip Cafaro and Ronald Sandler, eds., *Virtue Ethics and the Environment* (Dordrecht: Springer, 2010); Jason Kawall, "Reverence for Life as a Viable Environmental Virtue," *Environmental Ethics* 25, no. 4 (2003): 339–358; Louke van Wensveen, *Dirty Virtues. The Emergence of Ecological Virtue Ethics* (Amherst, NY: Humanity Books, 2000).

76 Interestingly, Steve Gardiner has recently enriched his assessment of the "perfect moral storm" with respect to climate change with an additional element, called the "ecological storm." This element emphasizes the fact that moderate eco-centrism is not only a reasonable attitude, but also widely shared. See Gardiner and Weisbach, 35–37.

77 According to empirical studies that test anthropocentrism versus eco-centrism with respect to the motivations of people to act, there is evidence that (a) eco-centrism is part of the motivational system of individuals and (b) that eco-centrism is dependent on situational factors – not so much on the overall motivational character of individuals. On both, see Katherine V. Kortenkamp and Colleen F. Moore, "Ecocentrism and Anthropocentrism: Moral Reasoning about Ecological Commons Dilemmas," *Journal of Environmental Psychology* 21, no. 3 (September 2001): 261–272. In addition, an inquiry into the ethical attitudes of people with respect to climate change comes to a comparable conclusion. See Markowitz and Shariff. Note that both articles defend an empirical claim, whereas my claims in this chapter are conceptual.

78 For arguments that support this claim in the context of plant ethics, see Angela Kallhoff, Marcello Di Paola, and Maria Schörgenhumer, eds., *Plant Ethics: Concepts and Applications* (Abingdon/New York: Routledge, 2018).

79 This question is also of more fundamental interest when addressed in the context of ethical versus economical claims in adjusting patterns of fair distribution in climate studies. Weisbach comments rightly that insights into the short-sightedness of self-interest do not have to rely on ethical reflection. See Gardiner and Weisbach, 152–153.

80 Weisbach is also right in claiming that an ethical outlook should not undermine the feasibility of a scheme of appropriation, as, in his view, some theories of distributive justice include. See Weibach's claim in Gardiner and Weisbach, 141.

81 For a discussion of the orthodox notion of sustainability and its recent extension, see Bryan G. Norton, "Sustainability as the Multigenerational Public Interest," in *The Oxford Handbook of Environmental Ethics*, eds. Stephen M. Gardiner and Allan Thompson (New York: Oxford University Press, 2017), 355–366.

82 For a comprehensive exploration of the future-generations argument, see Edward Page, *Climate Change, Justice and Future Generations* (Cheltenham, UK/Northampton, MA: Edward Elgar, 2007).

83 A critique of the identity problem is presented in John Nolt, "Future Generations in Environmental Ethics," in *The Oxford Handbook of Environmental Ethics*,

eds. Stephen M. Gardiner and Allen Thompson (New York: Oxford University Press, 2017), 344–354. This publication also explains why further reservations against equal respect for people living in the future are not well founded.

84   For an overview of recent frameworks of sustainability, highlighting its inter-generational meaning, see Norton, "Sustainability as the Multigenerational Public Interest."

85   For a moral theory of cosmopolitanism, see Caney, "Cosmopolitan Justice," 2010.

86   In the context of an exploration of duties among strangers, Lichtenberg argues that the concept of harm presupposes a counterfactual test that is not available when causal chains are long and when many different events play a role in determining them. "Our paradigm of harm is one person committing aggression against another ... But when harm occurs over an extended period and involves long causal chains and many intervening people and occurrences, it will generally be impossible to establish the baseline or counterfactual: to know *how things would have been* if the allegedly harmful event had not occurred." Judith Lichtenberg, *Distant Strangers: Ethics, Psychology, and Global Poverty* (Cambridge: Cambridge University Press, 2013), 36.

87   Caney, "Two Kinds of Climate Justice."

88   Ibid., 126.

89   Ibid.

90   Ibid., 134ff. I take up this distinction at greater length in Chapter 10.

91   Singer, "One Atmosphere," 33–34.

92   Walter Sinnott-Armstrong, "It's Not My Fault: Global Warming and Individual Moral Obligations," *Advances in the Economics of Environmental Research* 5 (2005): 298. Although I think harm is not an absolute term and needs to be placed in a concrete context, the argument of relationalism that Sinnott-Armstrong makes appears to be flawed.

93   See Frances M. Kamm, "Famine Ethics: The Problem of Distance in Morality and Singer's Ethical Theory," in *Singer and His Critics*, ed. Dale Jamieson (Oxford: Blackwell, 1999), 162–208.

94   It is important to note that distance in this context is not the same as mere spatial distance. As McGinn explains, duties to children do not disappear when a child is living in a distant place. Colin McGinn, "Our Duties to Animals and the Poor," in *Singer and His Critics*, ed. Dale Jamieson (Oxford: Blackwell, 1999), 150–161.

95   See Violetta Igneski, "Distance, Determinacy and the Duty to Aid: A Reply to Kamm," *Law and Philosophy* 20, no. 6 (2001): 605–616.

96   For a recent discussion of a charter of water rights and the debate on water ethics, see the contributions in *Global Water Ethics: Towards a Global Ethics Charter*, eds. Rafael Ziegler and David Groenfeldt (Abingdon/New York: Routledge, 2017).

97   Henry Shue, *Basic Rights. Subsistence, Affluence, and U.S. Foreign Policy* (Princeton, NJ: Princeton University Press, 1996), 19.

98   Ibid., 23.

99   Ibid., 34.

100   www.un.org/Waterforlifedecade/Human_right_to_water.shtml.

101   See Tim Hayward, *Constitutional Environmental Rights* (Oxford: Open University Press, 2005), 47.

102   For the discussion on water as a human right, see also John Scanlon, Angela Cassar, and Noémi Nemes, *Water as Human Right?*, IUCN Environmental Policy and Law Paper 51 (IUCN – The World Conservation Union, 2004), 28.

103   A similar argument is made by Thomas Pogge, *World Poverty and Human Rights. Cosmopolitan Responsibilities and Reforms* (Cambridge/Malden, MA: Polity Press, 2008).

104 For this critique, see Dinah Shelton, "Human Rights, Environmental Rights, and the Right to Environment," *Stanford Journal of International Law* 28, no. 103 (1991): 109 and 117.

105 For this proposal, see Henry Shue, "Subsistence Emissions and Luxury Emissions," *Law and Policy* 15, no. 1 (1993): 39–59. Note that Shue now rejects one interpretation of this proposal, according to which people have a right to emit greenhouse gases up to the levels of their subsistence. Instead, Shue now strongly advocates the shift towards green technologies, including support for this shift for poorer nations. See Henry Shue, "Climate Hope: Implementing the Exit Strategy," *Chicago Journal of International Law* 13, no. 2 (2013): 381–402; Henry Shue, "Deadly Delays, Saving Opportunities. Creating a More Dangerous World?," in *Climate Ethics. Essential Readings*, eds. Stephen M. Gardiner et al. (Oxford/New York: Oxford University Press, 2010), 146–162.

106 For a discussion of environmental victimhood and the divide between beneficiaries of environmental goods and the poor who suffer from environmental harm, see Gordon Walker's discussion of "hazard export" and the "race to the bottom," in Walker, 95–100.

107 See Stephen M. Gardiner, "The Threat of Intergenerational Extortion: On the Temptation to Become the Climate Mafia, Masquerading as an Intergenerational Robin Hood," *Canadian Journal of Philosophy* 47, no. 2–3 (2017): 368–394, https://doi.org/10.1080/00455091.2017.1302249.

108 This is the title of a well-received article by Goodin in which he mourns the flawed environmental politics not only in terms of monetarization, but in terms of incentive-structures that in the end support polluters in buying out. See Robert E. Goodin, "Selling Environmental Indulgences," *Kyklos* 47 (1994): 573–596.

109 Note that this does not mean that economic proposals are not helpful. Instead, economic solutions are critical, but as instruments to reach goals, not as instruments that set goals. For the importance, as well as the restricted relevance, of economic instruments in climate politics, see Ottmar Edenhofer, Christian Flachsland, and Steffen Brunner, "Wer besitzt die Atmosphäre?: Zur Politischen Ökonomie des Klimawandels," *Leviathan* 39, no. 2 (June 2011): 201–221, https://doi.org/10.1007/s11578-011-0115-0.

110 One way to express this is given by Gardiner, who appears to regard an ethos as an important element in ethics – one that has been neglected in contemporary ethics. In Jamieson's view "unless a duty of respect for nature is widely recognized and acknowledged, there will be little hope of successfully addressing the problem." Dale Jamieson, *Reason in a Dark Time. Why the Struggle Against Climate Change Failed – and What It Means for Our Future* (Oxford/New York: Oxford University Press, 2014), 443.

111 Currently, much effort is being given to preparing and undergirding existing principles of a water ethics. See Ziegler and Groenfeldt.

112 Principle Seven of the Rio Declaration on Environment and Development: Application and Implementation, Report of the Secretary General, April 17–25, 1997, reads: "States shall cooperate in a spirit of global partnership to conserve, protect and restore the health and integrity of the Earth's ecosystem. In view of the different contributions to global environmental degradation, States have common but differentiated responsibilities," accessed April 10, 2019, www.un.org/esa/documents/ecosoc/cn17/1997/ecn171997-8.htm.

113 https://unfccc.int/process-and-meetings/the-paris-agreement/the-paris-agreement, accessed April 10, 2019.

114 See Rawls, *Political Liberalism*, 304–310.

6   Climate goals

Cindy: And how are you doing now? Are you now convinced of joint environmental action?

Bert: I think there are some points in it. I think the water examples are good. And if people succeed in cooperating with respect to rivers, lakes, and even the seas – why should this not also work regarding the climate?

Cindy: Yes. And not only the analogies, but also the social insights are convincing. I think it's true that people do not only act for reasons of self-interest. This is so wrong. But they don't work in favour of other people, either. I think fairness really is something like a meta-value. When people think that they get a fair share of something, they are also motivated to contribute something. Fairness is a real motivator.

Bert: I think you're right. But do you also think that people care about the integrity of resources? I think they usually don't.

Cindy: Possibly. But I think this is not the point here. I think the message is rather that the integrity of a resource should be in the interest of people. It should be part of the "ethos" that theories question as a motivational resource for collective goals.

Bert: But what does this "should" mean? So far I thought that the model of joint action that has been presented is a model of voluntary action.

Cindy: Precisely. I think the goal of the last chapter was much more modest than already presenting something like a perfect solution. Remember the context: The whole book is about the question of how people come to work together in order to achieve climate goals. Or – put negatively – it is about the failures to achieve climate goals so far. And the diagnosis is that the reason for this failure lies in the incapacity to act together. We are now in the part of the book where the aspects of joint environmental action are parcelled together. And I have understood that an ethos is important because goals can only be achieved when people are motivated to act. Right?

Bert: Yes. But doesn't the ethos also depend on the goals and vice versa? And I have learned in a class on philosophy of action that goals as well as actions are not simply out there in the world. Instead, they are represented by words. Actions are always "under a description." Goals also need to be framed and clarified by words. Would it not be reasonable to start with goals and then explore the motivation to realize them and not the other way round?

Cindy: Oh man, this is perfectly true.

Bert: And think about another complexity. This book is about collective action, so wouldn't it be necessary to also distinguish various types of collectives?

Cindy: Yes and no. Theories of joint action actually state that their truth does not depend on the constellations of specific groups. This is one of the interesting features of the theories. But on the other hand, you're right. It is important to learn more about the goals. Obviously, the goals and the motivation to act for goals relate to each other. But I think the last chapter just wanted to explain that there is a way to frame environmental goals according to the conditions of the theories of shared agency. What do you think?

Bert: Possibly, there's even more in it. Possibly, it's important to know that the ethos that underlies joint environmental action includes some issues that are very important. I think something else is already working in the background. When goals are not simply goals, but when the integrity of a resource is at stake, one might also argue that the achievement of that goal is not purely voluntary. Possibly, that goal should be obligatory. And the same with fairness: possibly, people should act equitably with respect to resources that each single person needs in order to lead a good life, possibly even to survive. An intact environment appears to be something like that.

Cindy: OK. This is tricky, but you might be right. Let me recall it in order to see if I got it right. Joint action is goal-driven action of people who work together in order to achieve that goal. But it doesn't work unconditionally. Instead, some elements are needed in order to get joint action to work: people who are willing to act for a specific shared goal, an ethos that supports their goal-driven activities, and a group of people that is focused on that goal. We have learned that there are ways to translate this model into environmental joint action. But we still have to see how this works regarding climate goals. Right?

Bert: Yes, you got it. So the author hopefully explains now how joint climate action works.

Cindy: Yes, hopefully.

One of the most important steps in normative theories that address climate change is the movement from an ideal theory to a non-ideal theory.[1] This step cannot be equated with merely applying a theoretical proposal to real situations. Nor should it be conflated with the fact that politics does not obey moral principles. In that interpretation, the best we can get in ethics when applied to reality is a kind of "messy morality."[2] Instead, in this study, the trajectory from ideal to non-ideal theory is interpreted as an adjustment of theoretical models in climate ethics and in theories of joint action to more specific scenarios. The aim of offering a non-ideal theory will be achieved in two different and consecutive steps. The first step is an adjustment of the theory of joint environmental action and the joint agency approach to the current situation regarding climate change. Although it is not the intent of this approach to develop a comprehensive concept for achieving climate goals, it aims to provide theoretical insights that in the end help to overcome failings in collective action when addressing climate goals. The second step will be presented in Chapter 10. It describes the implications of this approach in terms of institutions that enable cooperation and in terms of institutions that also react to defection and non-compliance once climate change systems have been established.

The discussion of joint environmental action and of nature so far relies on a model that includes two assumptions: first, common-pool resources perform a variety of services, including ecosystemic functions; a rich variety of services and some key eco-services are only performed when the natural common-pool resource (NCPR) is intact. Second, an ideal approach can best be described as joint environmental action of a group of individuals who realize together a pattern of appropriation of a resource that coheres both with normative claims and with the continued robustness of the resource. Yet both also assume a clear vision of the goals that group actors wish to achieve. Joint action is fundamentally goal-driven action.

In order to refine and remodel the theoretical proposal of joint agency with respect to the climate debate, several steps are necessary. As outlined at the beginning of this study in Chapter 1, the joint environmental action approach relies on the premise that climate change is not an isolated phenomenon, nor can the impact on climate change be comprehensively quantified. Instead, climate change is one incident of a misappropriation of a resource that has distinct characteristics as an NCPR. Climate change, as well as over-exploitation of shared resources more generally, results from the misappropriation of shared natural resources by users. The fact of misappropriation can also be explained against the background of the claims in the preceding chapters. Misappropriation occurs when respect for the integrity of that resource is missing, when fairness among all users is not an issue, and when nature is regarded as an exploitable resource, not as a space for nature to unfold. In sum, climate change is the symptom not only of a massive over-exploitation of the atmosphere as a sink for greenhouse gases. Instead, it results from misguided practices of utilization.

As for the current situation, this outcome is particularly bad, because as it stands misappropriation regarding the atmosphere cannot be cured with respect to a heavy overload of greenhouse gases that are in the atmosphere but cannot be taken out of the atmosphere.[3] According to the prognoses of the scientific community, the best the world population can currently achieve is a prevention of worst-case scenarios by an immediate cut of emission levels. Some authors argue that an immediate turn to zero emissions is necessary.[4] In addition, humans need to adapt current living conditions to the dramatic changes in the natural environment that are expected to happen and that are already taking place as an effect of climate change. These include rising sea levels, more intense and extreme weather events, including mega-storms and heavy rain, desertification of land, and many risks resulting from new and unpredictable changes in climate cycles and severe weather events.

This chapter proceeds in six sections. The first provides an exemplary case of joint environmental action. It illustrates the theoretical steps needed to transcend self-interested appropriation of an NCPR such as a water resource by developing a cooperative scenario. The lesson from cooperation on a shared water resource is that users start to work together; they interpret themselves not only as beneficiaries, but as co-designers with nature; and they share a common goal that is not only a realistic and shared vision of how to appropriate the resource in the future, but also a vision of shared values and concepts of justice. This first section serves as an illustration of joint environmental action. It provides the background for exploring the differences that result from the dramatic situation regarding climate change.

Sections 6.2 and 6.3 discuss climate goals. Section 6.2 explores the climate goals that have been at the centre of concern in theories of climate change justice: mitigation, adaptation, and geoengineering. It recalls the interpretation and prepares the discussion in Section 6.3. Instead of reiterating claims of fairness, this section explains how climate goals relate to a theoretical approach to joint environmental action. It argues that the goals can be translated into joint action goals, but that they differ in nature. Section 6.4 explains that three different types of climate group action have to be distinguished. It makes a difference whether a group acts as a peer environmental group or as co-actor. Fortunately, the model of joint action gives space for both. Yet one more systematic problem still has to be discussed. This is the dissociation of actors and beneficiaries. Actually, regarding climate goals that cannot be achieved even by single nation-states, this is a real problem. One proposal to cope with it can be offered at this point. Further and more far-reaching proposals can only be argued after also having explored the normative implications of joint environmental action and climate duties in particular. Section 6.5 summarizes the insight that climate goals can – to some degree – be broken down to joint action goals. It also denies the conclusion that climate ethics needs to be framed as an emergency ethics, although the achievement of climate goals is today one of the most pressing issues for keeping our planet intact and for securing humankind's ability to lead a good life.

6.1 Cooperation as active co-designing with nature

In order to explain the way goals are related to future practice and to joint action in particular, it is helpful to start with a model of goal-envisioning that has been discussed in water ethics. Water resources serve multiple functions for human society as well as for the natural environment. At least three functions of freshwater resources are distinguished: first, water resources are used for *human consumption* in terms of sanitation, washing, bathing, and cultural or religious rituals; second, they are used for *economic purposes*, as for instance agriculture, livestock, industry, tourism, and transportation; and third, they perform *ecosystemic functions*, those that are, for instance, particularly visible in wetlands, coastal areas, mangroves, and in humid, arid, and semi-arid areas.[5] More specifically, the services that freshwater resources perform do not limit themselves exclusively to human society. Water is also particularly important for balancing ecosystemic functions of larger units.

Malin Falkenmark and Carl Folke highlight the fundamental balancing functions of water in the natural landscape by physical services in terms of evaporation and condensation, by chemical functions such as crystallization and solution, and by biological functions, including water molecule splitting and re-assemblage through respiration.[6] "In addition," they write, "there is a whole group of largely water-dependent, yet hidden, ecosystem services ... of decisive importance in the functioning of the life-support system ... physical, chemical, as well as biological."[7] The researchers emphasize that the functions of visible water in terms of direct use for households, irrigation, hydropower, or navigation, are only a small part of the whole set of functions. The description of water functions needs to be expanded by acknowledging the hidden ecosystemic services of water – for instance, in water cycles and as transport mediums for nutrients.[8] An assessment of the manifold ecosystemic functions of water is complicated by a range of factors, including growing pressure on water resources from climate change.[9]

The first step in discussing shared goals as related to water resources is an interpretation of the natural good in terms of ecosystems. The second step is a focus on the integrity of that resource and an exploration of the effects of actions that either support integrity or jeopardize it. As for water resources, Wilhelm Ripl's analysis of water as the "bloodstream" of the planet highlights this interpretation.[10] After a thorough analysis of the many functions of water, some adverse impacts of society on water cycles, and the possibility of achieving sustainability, Ripl recommends:

> The difficulty with solving our sustainability problems seems to be the transition from that of a net-production society into one whose strategy is characterized by the maintenance of the steady state, where water and matter cycles are to be closed and the function of ever-continuing "growth" reduced to that of improving relations within society, life quality and sustainability. Planning which does not take into account temporal phases and system-immanent cycles has to be replaced by coupled spatio-temporal-related planning, which means re-implementing dynamic structures which have minimized irreversible losses. These goals can be only achieved by respecting life cycles and adaptive resource management on managed land areas: locally providing the water cycles, the energy, the food and necessary service functions of nature.[11]

This claim can be summarized thus: "Man has to be reintegrated in ecosystems as the most suitable intelligent controller of water cycles and matter cycles."[12] Still, the idea of the re-integration to local affordances of natural goods is only one interpretation of how common-resource-related goals can be interpreted. There is a range of further joint goals, including goals of restoring a resource that is already suffering from environmental degradation; another goal could be to generate an ecologically intact city; or it could be to enhance air and water quality in a distinct area. What is important, though, is the assessment of various options to design the interface of society and environmental resources, and then a decision for future planning that also considers the effects on the availability of a range of eco-services.

The third step in developing goals that resonate both with the affordances of nature and with the shared vision of a group of people is the exploration of the normative elements of an ethos that the users share in appropriating water resources. Specialists in water management explain that it does not suffice to resolve water conflicts that always exist when parties with various interests share a natural resource. Instead, it is critical to embed the process of conflict-solution into a framework that (a) enhances cooperation and (b) relies on values that the beneficiaries subscribe to and share.[13] Delli Priscoli and Wolf explicitly claim a new ethics in addressing water conflicts: "[T]he new ethic we require is not simply one of preservation. It is one that should be built teleologically, on a sense of purpose and on an active co-designing with nature."[14] More specifically, the path towards the shared goal needs to be guided by values. The researchers claim:

> Today, our technology tells us that there is enough water – if we cooperate. One of the most important elements for cooperation is something negotiations experts call "subordinate values." These are values beyond immediate utilitarian values to which competing parties can identify.[15]

Even though recognizing values already represents an important step forward, this alone does not suffice. Instead, the concept of values should not be reduced to the overall utility of a resource. Values that are important are constitutive of an environmental ethos, yet they might also include cultural values and further non-instrumental values.

Moreover, once a goal has been formulated – as in, for example, "We will clean up the river; and we will build that dam in order to prevent frequent flooding" – sub-goals and ways to achieve this scenario can be worked out. More specifically, *transformation goals* stand in stark contrast to reactive environmental goals. True, many environmental goals are dictated by the need to react to dangers and growing threats; they are reactive rather than proactive. However, one and the same situation can be reinterpreted by means of transformation goals whenever there is some safe operational space left.

It is not my aim here to portray the complexity of defining transformational goals calibrated to water-systemic functions, which would also demand an assessment of the functions of groundwater and water cycles in addition to freshwater. Still, it is important to discuss the underlying complexities. A river is an ecosystem that is embedded in water cycles, in landscape qualities, and so on. Furthermore, entities such as rivers and lakes can be defined as common-pool resources. And groups of people can interpret joint actions that shape access to common-pool resources as processes of active co-designing with nature.[16] It is co-designing because it refers to and responds to the eco-functions performed by NCPRs. Processes of active co-designing presuppose that goals and strategies for achieving goals are rendered visible. The minimal requirement for groups to be in a position to actively co-design with nature is that they are made up of individuals.

Accomplishing a joint goal includes cooperation as a process. This process includes a re-adjustment of groups of people to natural living conditions. It also means that the match between nature and society is re-evaluated not only in terms of the integrity of the natural good, but also in terms of conditions of fair access to natural resources. The proposal to act collectively in favour of "active co-designing with nature" requires actors to interpret their roles not as masters of natural processes, but as co-agents. However, there is another interesting point of similarity between an "active co-designing with nature" approach in proposals to integrative water management and a joint environmental action approach. Researchers in integrative water management claim that it does not suffice to evaluate the costs and benefits of a proposed action, but also to adjust future goals to the values that individuals relate to the common-pool resource.

Active co-designing with nature is also part of the vision of theorists who defend the view that climate change action is unavoidable, irreversible, and has a deep impact on other life-sustaining natural resources. Nevertheless, instead of focusing on the negative externalities of human activities, it is necessary to start by revising the way humankind interacts with Planet Earth. Active co-designing with nature as discussed in integrative water management can serve as a model for this new way of practicing the appropriation of scarce natural resources and to think about fruitful ways to do so in the future. More specifically, this vision is helpful for reinterpreting the goals of adaptation and of mitigation that have been put forward in the climate debate.

6.2 Climate justice goals reconsidered

Theories of climate justice address goals that have been established in the community of scientists as "climate change goals." In that context, the goals are framed in the light of normative claims.[17] The discussion in this section focuses on three goals climate ethicists regard as particularly important; realizing them prevents additional and severe harm to "climate victims" resulting from climate change. They also regard them as closely related to claims of fairness. These are *mitigation*, *adaptation*, and *geoengineering* as technological solutions to climate change, as well as precautionary goals. I will discuss them in turn. Note that the interpretation of these aims in this section coheres with the presumption that they are (a) particularly important goals with respect to scientific prognosis, but also (b) normative goals whose realization serves claims of climate justice.

At first glance, there is one predominant aim for climate change: *mitigation*. Mitigation means the reduction of greenhouse gas emissions. It is not controversial that mitigation is imperative if we are to reduce the incidence of climate catastrophes resulting from an excess of greenhouse gases in the atmosphere.[18] What is controversial is, first, the overall approach that is best suited to achieve effective mitigation, and, second, a fair scheme for the distribution of mitigation burdens. Whereas the question of who should pay what is at the heart of the debate on climate justice, Dale Jamieson takes a standard position when explaining that both efficiency and equality are central for a fair scheme for the appropriation of mitigation costs.[19] Another recurrent theme is that some have to carry the biggest burdens, either because of their former benefits or the burdens they generated in former times.[20] The exemption for poor people has also been discussed.[21]

Adaptation is a concept for the necessary adjustments to be made to ecological-social-economic systems in response to actual or expected climate stimuli, their effects, or impacts.[22] Adaptation also causes costs. "These are the costs to persons of adopting measures that enable them and/or others to cope with the ill-effects of climate change."[23] What is controversial, though, is the content and the moral implications of adaptation. Moreover, the relationship between adaptation and mitigation is also under discussion. On the one hand, a prioritization of adaptation helps those who suffer most from climate change – the poor and the vulnerable. Adaptation has not been conceived as a technical term in the debate on climate justice, but is related to the claim of the need to provide funding for adaptation processes on a global scale.[24] On the other hand, if adaptation has a high priority, mitigation is no longer the centre of concern. Jamieson articulates this concern when stating "What is in question is not whether a strategy of adaptation should and will be followed, but whether in addition there will be any serious attempt to mitigate climate change."[25] The solution to this concern is not to regard both goals as competitive. Instead, both goals need to be equally important.

With respect to both goals, the most important problem philosophers in the climate justice debate address is an appropriate policy scheme reflecting claims of justice. Theories of climate justice are not only interested in mitigation as an exigency to prevent climate catastrophe or to also mitigate the effects of climate change. Instead, the focus is on "mitigation burdens" as well as "adaptation burdens":

> Mitigation burdens ... are the costs to actors of not engaging in activities that contribute to global climate change ... To make this concrete, mitigation will involve cutting back on activities like the burning of fossil fuels and, as such, it requires either that people cut back on their use of

cars, electricity, and air flight, or that they invest in other kinds of energy resources. Either way, mitigation is, of course, a cost for some.[26]

Adaptation is also debated as a type of costs. Consequently, the focus on theories of justice is precisely what Caney states: "My focus in this article is on the question 'who should bear the costs caused by climate change?'"[27] As for adaptation, it is also important to note that adaptation is sometimes possible, but not always. Mitigation instead is always possible, but not yet implemented to a sufficient degree. What both goals have in common when spelled out in the way presented so far is that they are *reactive goals*. By supporting adaptation and mitigation, actors try to cope with a situation that may have dreadful implications for human societies.

In addition to adaptation and mitigation, climate ethicists also discuss *technological goals*, primarily *geoengineering* as an instrument to bring about impactful and rapid changes. Geoengineering is "the intentional manipulation of the environment through grand technological interventions at a global scale, such as stratospheric sulfate injection (SSI), a variety of solar radiation management (SRM)."[28] Whereas technologies of geoengineering are not new, the debate on geoengineering as an instrument for tackling climate change is only a recent development.[29] These technologies are controversial, not only as particularly risky technologies. Most recently, they have also included carbon-removal technologies designed to catch carbon dioxide from the air and support the effects offered by natural sinks, primarily vegetative areas and forests.[30]

Ethicists also worry that political bodies might be tempted to use these technologies as a substitute for a serious and costly approach to mitigation.[31] In addition, framed as the "lesser evil" or "second best" solutions,[32] a range of further moral arguments needs to be included. Gardiner provides a long list of those arguments, including the severe problem that many forms of SSI involve, inflicting extreme harm on innocents, that they are not legal, and that they result from moral vices such as hubris or utopianism.[33] In that frame, climate justice addresses responsibility for *climate victims* and their rights to compensation and to special support.

Recently, emphasis has also been laid on a broader range of *precautionary goals*. Gardiner explains, however, that this needs further clarification,[34] because precaution is a notoriously minimal goal and not easy to understand when framed as in Article 3.3 of the United Nations Framework Convention on Climate Change:

> [W]here there are threats of serious or irreversible damage, lack of full scientific certainty should not be used as a reason for postponing [precautionary] measures [to anticipate, prevent or minimize the causes of climate change and mitigate its adverse effects].[35]

Since Gardiner offers proposals to render the precautionary principles more precise and applicable to the climate,[36] the debate on reshaping the principle to apply it to the climate scenario has not come to an end. As for the scope of this section, it suffices to state that the list of climate goals also includes more unspecific goals that stem from the assessment of uncertainty regarding the subsequent effects. This also includes theoretical models that calculate the costs of unabated risks as opposed to the costs of current investments in climate protection.[37]

6.3 Climate goals as joint action goals

The first step in the process of reframing climate goals has already been covered. They have been characterized, not as monolithic end states, but as part of transitional strategies and collective endeavours. The next step is a reconsideration of those goals as being even more closely tied to joint agency.

It is not my aim in this section to go over the whole range of the goals I have explained and that serve as a backdrop for theorizing climate justice. I shall simply note that these are indeed demanding and far-reaching goals. Above all, they are *reactive goals* in that they present proposals for coping with a situation in which there is very limited operational space.[38] Nowhere has it been said, however, that this is an exclusive range of climate-change goals. Instead, I propose to reframe climate goals to some degree so that they fit into the frame of joint environmental action as proposed in the preceding chapters.

Mitigation, as framed in the current debate, is not only a reactive goal, it is also an *accumulative goal*. Effective mitigation can best be achieved when many different ways to reduce greenhouse gases are being combined. Above all, an accumulative goal is a quantifiable achievement that results from the addition of many single achievements. In the context of the environmental debate, accumulative goals usually relate to a quantitative analysis of an environmental hazard. There is too much carbon dioxide in the atmosphere, therefore the emissions need to be reduced. There is also too much pollution in the cities; to protect the health of the citizens, it, too, needs to be reduced. Notably, accumulative goals are goals to reduce existing levels of emissions. The negative externalities that count in accumulative goals are usually side effects of activities that have a negative impact on natural resources.

Accumulative goals are critical in the climate discussion. For developing future-oriented environmental agendas, including more far-reaching visions of active co-designing with nature, however, they are not sufficient. Above all, their negative outlook is not helpful for translating them into a concept of joint action. When formulated as forward-looking and truly desirable goals – as in, for instance, the goal to have clean air in a city or plastic-free and healthy seas – accumulative goals can nevertheless be attractive. Much more has to be said about fairness conditions and justified constraints in terms of fairness regarding mitigation, which is the goal deserving highest priority.

Some of the accumulative goals are also *threshold goals*. Unlike unstructured accumulative goals, the reaching of a threshold goal depends on the total achievement of a predefined sum of reductions. In environmental scenarios, an important threshold is the two-degree aim formulated by the IPCC. The result of a complex calculation that draws on the probabilities given by climate models, the threshold of "not more than two degrees" has been broken down to an overall emissions budget. Thresholds are not only important for defining safe operational spaces (and tolerable ones),[39] they are also important in two other respects. When preset as a threshold that is – by any measure – never realistically achievable, threshold goals have a frustrating effect. On the other hand, thresholds that can *realistically* be achieved have an impact on the motivation of potential contributors.

The goal of Chapter 9 is to present not only a theoretical approach to accumulative goals that coheres with claims of fairness, but also to discuss the question of responsible actors in mitigation goals. I shall therefore postpone the detailed discussion of mitigation and instead proceed with further climate goals.

As for *adaptation*, its relationship to joint action is already obvious. Adaptation includes many actors who also engage in a mutual adjustment of their practices. Yet again, the negative and reactive outlook is a problem. In addition, discussion of adaptation is usually restricted to the discussion of the financial means of realizing it. As integrated into practices of active co-designing with nature, adaptation can count as an important joint goal. In any case, it does not appear to be the most important goal in joint environmental action. Instead, in developing goals in terms of active co-designing with nature, the overall goal is broader. Foremost, it addresses ways of coping with local environmental factors and the future prospects of their development; the community then addresses ways to realize community goals by integrating them into a model of joint environmental practice.

Regarding *technological solutions* to climate change, this is not the place to discuss geoengineering and other proposals at length. After all, it should be noted that once the goal of active co-designing with nature is accepted, technological solutions providing more efficient practices to benefit from local resources and paving the way for a carbon-free energy supply are important. As for the achievement of environmental goals, a common lesson is that success is neither one-dimensional nor linear. Instead, creative solutions that count on the ingenuity of many different actors have proven the best way to address problems as complex as the depletion of the ozone layer or the need to supplement engines that have a negative impact.

Still, the most important re-interpretation of climate goals starts with the attempt to render the underlying practices explicit. Neither adaptation nor mitigation can be accomplished exclusively by single actors. Similarly, actions that contribute to reaching those goals are multifaceted. In outlining the practices needed to bring about mitigation according to the responsibilities of various actors, Simon Caney lists "lowering greenhouse gas emissions," "creating and maintaining greenhouse gas sinks," and "developing and transferring safe and clean energy."[40] As for adaptation, activities include "designing natural and social arrangements so that people are able to cope with climate-related threats and exercise their legitimate entitlements without loss."[41] This diversity is noteworthy, because this also opens space for introducing concepts of joint action. Primarily, those underlying practices that are conducive to mitigation and adaptation are at least in part desirable and worthy as goals of groups of people. The creation of greenhouse gas sinks can also be interpreted as the conservation of valuable forests or as "greening the city"; a safe and green energy supply is worthwhile because it is a way to also prevent air pollution and to foster the energy autonomy of households and cities. To sum up, when translated into sub-goals, climate goals are truly joint environmental goals.

So far, I have argued that the goals discussed in theories of climate action can be interpreted as joint action goals, even when not all of them are literally joint action goals, but rather accumulative goals. Yet climate action is also still dependent on more comprehensive goals that collectives are willing to support. The theoretical approach to joint environmental action was framed as the goal-directed cooperation of various actors to achieve a shared goal. At a minimum, the broad agenda of climate action needs to be reframed in a clear vision of sub-goals that groups of contributors wish to realize.

As a further step, it is also important to get a clear grasp of transformation goals. The bottom line of those goals is that they are *goals of active co-designing with nature with respect to a concrete set of common-pool resources.* Whereas mitigation remains an overarching and pressing goal, a community of citizens needs to focus on goals that acknowledge mitigation and adaptation, but also to respond to local ideas about transformation. They can formulate a transformation goal, for example, of achieving a better water-supply infrastructure within three years. Transformation goals are sufficiently concrete to be achieved together. More specifically, they are related to the concrete setting of a conglomerate of common-pool resources that together build up the environmental conditions for types of action.

Taking the interdependence of life-supporting systems into account, practices of active co-designing with nature are also important as activities on the micro-level that have an impact on the macro-level. An approach to energy supply that includes decarbonization, conservation of forests as sinks, protection of the seas from additional environmental harm, and so on, are all ways to realize overarching environmental goals. In addition to resources from theories of joint action and empirical evidence of the interconnectedness of macro- and micro-agendas, the emphasis on shared goals and on a clear vision of overarching goals also resonates with the psychological insight that the scale of action needs to be modelled against the background of capacity to act. The willingness to contribute to goals depends on more than close ties with others who are also doing their share. What is critical, in addition, is the conviction that the individual contribution really matters and makes a difference to a specific goal.

In order to develop aims that are neither all-embracing climate goals nor tied to local agendas of active co-designing with nature in a local setting, transformation goals need to be articulated and expressed neatly. Accumulative goals also need to and can be embedded in more comprehensive goals – to enable joint action and also to motivate individuals and group actors to invest in their goals. But two remaining problems with respect to the huge goals of mitigation and of climate goals more generally still have to be addressed: the distinction between agency and co-agency and the possible dissociation between resource and beneficiary.

6.4 Agency and co-agency

In addition to a re-interpretation of environmental goals in the way I have proposed, types of agency also need to be distinguished according to their level of involvement in joint action. Obviously, the situation of villagers who benefit together from a lake and develop future-oriented agendas by negotiating future resource-related goals is just a model for explaining the elements and theoretical presuppositions in joint environmental action. In particular, it explains that processes of goal setting in terms of joint action goals are important for getting people involved in those types of agency.

To begin with, *environmental peer agency* is accomplished by collectives that focus on distinct environmental goals and that go forward with really acting in favour of neglected environmental goals. Participants in those groups might share the goal of working in favour of a climate-friendly transport system, the goal of "greening the city," or the goal of protecting a forest from deforestation. They explicitly focus on environmental agendas and spend their time fighting in favour of them. Environmental peer agency is particularly important not only for providing role models, but also as a leadership model.[42]

Second, and differently, joint environmental action can be actuated through various types of engagement in cooperative actions. More specifically, individuals can support joint environmental action by means of participation in institutional settings that offer ways to co-act in favour of joint environmental goals. This is *environmental co-agency*. Examples include participation in activities that support an environment-friendly transport system or simply to change the energy supply system from a carbon-based to a green energy supply. In that way, and by means of participation in opportunities to achieve environmental goals together, individuals as well as collective actors can support climate goals in many more respects. They can try to reduce their ecological footprint; they can create pressure to shift the opinions and practices of officials and other stakeholders. Overall, I shall argue in Chapters 7–9 that – under specified conditions – environmental co-agency is actually obligatory, whereas the responsibility to arrange opportunities to participate in practices that support environmental goals need to be debated on separate grounds.[43]

Environmental co-agency is also particularly well suited for accumulative goals. Here again, the debate about duties to accumulative co-agency needs to be distinguished from second-order responsibilities.[44] Whereas single actors may possibly be obliged to contribute to accumulative goals, institutions and organizations may be in another situation. They also need to support accumulative goals and to empower individuals to fulfil their own accumulative goals.[45]

Third, environmental agency is realized by means of building up a joint environmental actor; this is *environmental transformation agency*. This type of co-agency can be realized in different ways. One way to build up a joint actor is involvement in processes of active co-designing with nature on a local level. When citizens decide to green their neighbourhood or when farmers decide to build up new strategies to protect their land from overuse, they also build up environmental actors. Thinkers have also argued in terms of "collectivization" for acting efficiently together as another way to build up a group as a joint environmental actor.[46] Examples also include groups of farmers who together build an association to care for sustainable agriculture practices.

Overall, three different ways to get involved in supporting goals as shared goals of joint actors can be distinguished. Whereas it is the aim of the subsequent chapters to argue for types of co-agency in more detail and also to address the normative claims, it is important to note at this point that environmental action is not restricted to environmental peer groups, or to city planners who develop future scenarios including proposals for water supply and infrastructure. Instead, the model of joint environmental action can be realized in different ways and by means of different types of agency. It also allows us to integrate concepts of representation and to interpret the resulting groups not only as responsible actors, but also as themselves building up groups and developing shared views of climate goals. In short, joint environmental action is a theoretical model that can be realized in many different ways.

6.5 The dissociation of resource and beneficiary

When it comes to translating this theoretical approach to joint climate action to a more practical level, the acknowledgement of the goals that have been argued so far is an important step. Yet this is only part of the theoretical application of theories of joint agency to environmental agendas. One basic reservation remains. Can joint environmental action as described in the example of active co-designing with nature really be applied to the far more complex situation of climate change? And does the model fit the current situation at all? Obviously, most people today do not live in rural communities that are close to a lake serving as life-sustaining natural resource. Although local water resources should not be underestimated as a necessary *in situ* resource – water being far too heavy to really be transportable as a commodity – modern life is in many ways alienated from life with nature. What does joint environmental action mean when people are nourished by a food market, when water supply and energy supply is organized by institutions as complex as water supply systems in modern megacities, and when food production is carried out by an agrarian industry that is only to some degree completed by local food production? In addition, the diagnosis regarding the current relationship between nature and societies comes close to catastrophic scenarios.

As a way to begin addressing these issues, three arguments help to support the view that the theoretical model of joint environmental action can be applied to NCPRs, including the atmosphere. The first is that local environmental action and climate action are tied together as a matter of fact. When describing natural resources not as physical resources, but as living systems, the interdependence of the systems is evident. Thus, a first helpful step to explore joint environmental action starts with the insight that it focuses on interconnected life-supporting natural systems. More specifically, the life cycles cannot neatly be separated from each other. Instead, they work together.[47]

A second argument on applying a theory of joint environmental action to NCPRs is that the misappropriation of shared resources is not only a practical problem, but also a normative problem.[48] Regarding fairness in distributing the natural life-sustaining resource, a model of joint appropriation that recalls goals people share is an important element. Moreover, if the arguments in Chapters 2 and 3 are stated correctly, it is also important to interpret the effects of unfair appropriation accordingly. A fair distribution of ecosystems might even be necessary to prevent future water wars and any kind of resource war.[49] More specifically, the theoretical model of joint environmental action sets itself apart from other ways of appropriation in that it focuses on shared goals and on shared interests. If conceived in the right way, the goals are in everyone's interest. The price for this starting point is the restriction to people who really are willing to participate in groups with distinct goals.[50]

Thus, although the model of joint environmental action needs to be translated into various contexts and needs to be modelled against the background of institutions that transform eco-services into supply systems and goods beneficiaries enjoy, it is not far-fetched to interpret forward-looking environmental action against the background of theories of joint agency. This also requires rendering the eco-services visible, discussing the fair distribution of these services anew, and supporting systems that protect shared resources efficiently. In order to serve as goals of joint environmental action, the goals not only have to align with the affordances of natural resources, they also have to align with the values enshrined in the ethos of people acting together.

Still, the problem of scale is significant.[51] It is a recurrent theme in environmental action that value-based practices are best applied in cases of small-scale, if not face-to-nature encounters. Two preliminary replies are possible. On the one hand, joint environmental action can be realized and supported not only by individuals, but by groups with different characteristics. In addition, groups can also be made up of representatives of other collectives. The Paris Agreement is an example.[52] At least regarding the global scale, a model of group action helps to reach that level. On the other hand, the most important elements in a model of joint action are (a) the will of all participants to achieve a goal and (b) the goal itself. For climate action, this proposal needs to include a revision of climate goals. They need to be remodelled so that they cohere with the exigencies that the international scientific community claims, and so that they really fit into joint action. But in addition, the question needs to be raised of whether engagement for joint environmental action and for joint climate action in particular is morally obligatory. I shall discuss this issue in the remaining chapters of the book.

On the spectrum of common-resource-related goals, climate goals are in various respects particularly urgent. Although scientific knowledge has been enormously strengthened and has been extended in the last decade, the effects of an accumulation of greenhouse gases in the atmosphere are still to some degree unforeseeable. In order to frame climate goals as mitigation goals, the IPCC has proposed breaking the goals down to an average rise of temperature that would still be tolerable in terms of not risking catastrophic scenarios.[53] Catastrophic scenarios include a dramatic rise in sea levels, a steep increase in severe weather events, and a possible derailment of the big geophysical systems, including the ocean currents essential for our current climate. Yet the correlation between greenhouse gases and the rise in temperature, and even more so the development of rising temperatures, are uncertain as possible tipping points that cannot be foreseen.

Assuming the prognoses are right and we are close to a climate catastrophe that can only be possibly prevented by rapid, effective, and collective action, the question of whether we are in the situation of an emergency ethics also needs to be posed. According to Michael Walzer's interpretation, emergency ethics should be restricted to incidents of "supreme emergency," where

> "Supreme emergency" describes those rare moments when the negative value we assign – that we can't help assigning – to the disaster that looms before us devalues morality itself and leaves us free to do whatever is militarily necessary to avoid the disaster, so long as what we do doesn't produce an even worse disaster.[54]

In the context of a war ethics, Walzer refers to situations "when our deepest values and our collective survival are in imminent danger,"[55] although certainly not a "permissive doctrine."[56] It proposes that in that exceptional situation, moral rules can and possibly even should be overridden.[57] In this discussion, Walzer grants highest value to the community itself, not the individual.[58] Since climate change threatens community life in various ways, in particular by undermining the capacity to act as a community, the thought that climate ethics could be framed as an emergency ethics is not too far-fetched.

Yet two insights speak against this proposal. The first is that the impression that the world community is already incapable of acting results from an interpretation of climate change that does not cohere with the proposals of this study. When framed as achievable by joint action, climate goals are still within the reach and attempts are obligatory. The constraints that result from an ever-shrinking window of opportunity are real. Yet this does not necessarily lead to paralysis, but should animate us to rethink the relationship between social life and natural resource on a more basic level. One proposal has been offered in this study. The second is that an emergency ethics opens the doors for all sorts of behaviour as legitimate behaviour. From the outset, it is not clear which rules are allowed to be overridden and which ones should still be in place. More specifically, it might be the case that claims of fairness and of moral integrity can also be overridden. This has the implication that climate goals that are related to claims of fairness and to moderate eco-centrism can also be overridden. For both reasons, an emergency ethics is neither adequate nor particularly helpful.

Therefore, I propose to resist the temptation to frame a climate ethics as an emergency ethics. Yet instead of relying primarily on principles of fairness, it is right to claim in a situation of moral exigency that more urgent moral claims than distributive fairness need to be met. Duties to cooperate are much more important, particularly when climate goals can only be achieved by joint effort.

This chapter has been dedicated to explaining one step in the process of producing a theory of joint climate action; the focus has been on climate goals. At the centre of environmental cooperation are shared visions of a goal individuals try to bring about together. Yet most obviously, jointly acting with respect to benefits from a common-pool resource differs significantly from "dancing together," "walking together," or "travelling together to New York City." Although I have already elaborated many of the steps needed to translate those ideas into joint environmental action, it is also important to discuss the concept and the function of goals anew. As for environmental goals, as joint goals they enshrine not only visions of fair and sustainable practices for exploiting natural goods, but also desirable end-states in terms of fair and sustainable schemes of benefit-seeking behaviour. Instead of interpreting climate goals as just another sort of constraint to enhance sustainability, I have argued for a translation and transformation of the big climate goals into more appropriate joint action goals. This includes an examination of a subtle layer of co-agency that is already part of the discussion; it also includes a revision of goals as umbrella terms and an adjustment towards a more detailed picture of transformation goals. This chapter was also dedicated to explaining one element in the transition from an ideal to a non-ideal scenario, including a re-description and reassessment of climate goals in the context of joint environmental action.

Notes

1 The distinction between "ideal" and "non-ideal" reasoning in political philosophy goes back to John Rawls, *A Theory of Justice* (Cambridge, MA: Belknap Press, 2005). In line with a rather broad interpretation of that transition, it has also been interpreted as a transition from a theoretical to a more strategic-political level of argumentation. In the debate on climate ethics, a third interpretation has surfaced. This is an attempt to focus on the translation of moral themes at the core of climate ethics to the question of practicability of the application of solutions. For this, see Derek Bell, "How Should We Think About Climate Justice?," *Environmental Ethics* 35, no. 2 (2013): 191.

2 C. A. J. Coady, *Messy Morality: The Challenge of Politics* (Oxford: Clarendon Press, 2008).

3 Despite research on technologies that recapture the greenhouse gases from the atmosphere, progress is still not groundbreaking. For the current situation of these technologies, see Fred Krupp, Nathaniel Keohane, and Eric Pooley, "Less Than Zero. Can Carbon-Removal Technologies Curb Climate Change?," *Foreign Affairs* 98, no. 2 (April 2019): 142–152.

4 For a particularly clear statement in this regard, see Weisbach in *Debating Climate Ethics*, eds. Stephen M. Gardiner and David A. Weisbach (New York: Oxford University Press, 2016), 170–200.

5 Irma van der Molen and Antoinette Hildering, "Water: Cause for Conflict or Co-Operation?," *Journal on Science and World Affairs* 1, no. 2 (2005): 134.

6 Malin Falkenmark and Carl Folke, "The Ethics of Socio-Ecohydrological Catchment Management: Towards Hydrosolidarity," *Hydrology and Earth System Sciences* 6, no. 1 (2002): 1.

7 Ibid.

8 Ibid., 1–2.

9 Nigel W. Arnell, "Climate Change and Global Water Resources," *Global Environmental Change* 9 (1999): 31–49; David Lewis Feldman, *Water Policy for Sustainable Development* (Baltimore, MD: Johns Hopkins University Press, 2007).

10 Wilhelm Ripl, "Water: The Bloodstream of the Biosphere," *Philosophical Transactions of The Royal Society of London, Biological Sciences* 358 (2003): 1921–1934.

11 Ripl, 1929.

12 Ibid.

13 For an exploration of this method of "managing and transforming water conflicts," see Jerome Delli Priscoli and Aaron T. Wolf, *Managing and Transforming Water Conflicts* (Cambridge/New York: Cambridge University Press, 2009). The researchers identify some trends that push towards cooperation (3–4), inquire into the main drivers for regional instability (9–28), propose a transformation from "allocating water to sharing baskets of benefits" (74–78), draw conclusions for the international community (78–81), and claim a "new ethics for water management" (121–122). In this chapter, emphasis will be placed on the last point.

14 Delli Priscoli and Wolf, 121.

15 Ibid., 122.

16 Ibid., 121.

17 This interpretation of goals coheres with Gardiner's claim that climate goals include normative observations. See Gardiner and Weisbach, *Debating Climate Ethics*, 51–53.

18 The Potsdam Institute of Climate Impact Research also highlights the gaps between the mitigation that is needed and mitigation as it stands now. See The Potsdam Institute of Climate Impact Research, accessed April 3, 2019, www.pik-potsdam.de/en.

19 Dale Jamieson, "Adaptation, Mitigation, and Justice," in *Perspectives on Climate Change*, ed. Richard Howarth (Amsterdam: Elsevier, 2005), 221–253.

20 The additional costs that need to be shouldered because of responsibility for previous environmental harm are framed as the "polluter-pays principle." For a discussion of this principle, which also includes a closer look at the debate on historical emissions, see Simon Caney, "Cosmopolitan Justice, Responsibility, and Global Climate Change," *Leiden Journal of International Law* 18 (2005): 752–756. For the positive correlation of current responsibility because of having benefitted from former emissions, see the "beneficiary-pays principle" discussed in Caney, "Cosmopolitan Justice" (2005), 756–761.

21 Henry Shue, "Subsistence Emissions and Luxury Emissions," *Law and Policy* 15, no. 1 (1993): 39–59. It should be noted that Shue has withdrawn from this judgement in Henry Shue, "Climate Hope: Implementing the Exit Strategy," *Chicago Journal of International Law* 13, no. 2 (2013): 381–402. One reason is that the exemption generates opportunities for poor countries to over-exploit the atmosphere as a waste dump, which contributes to an already dramatic situation.

22 This definition is used in Jamieson, "Adaptation, Mitigation, and Justice," 265.

23 Caney, "Cosmopolitan Justice" (2005), 752.

24 For proposals to develop an appropriate funding scheme, see Paul Baer, "Adaptation to Climate Change. Who Pays Whom?," in *Climate Ethics. Essential Readings*, eds. Stephen M. Gardiner et al. (Oxford/New York: Oxford University Press, 2010), 247–262. See also Allen Thompson and Jeremy Bendik-Keymer, *Ethical Adaptation to Climate Change: Human Virtues of the Future* (Cambridge, MA: MIT Press, 2012).

25 Jamieson, "Adaptation, Mitigation, and Justice," 266.

26 Caney, "Cosmopolitan Justice" (2005), 751.

27 Ibid., 752.

28 Stephen M. Gardiner, "Climate Ethics in a Dark and Dangerous Time," *Ethics* 127, no. 2 (January 2017): 455, https://doi.org/10.1086/688746.

29 Gardiner mentions that the term was introduced by Jamieson in 1996 and that it was regarded then as a "lesser evil" (Gardiner, "Climate Ethics in a Dark and Dangerous Time," 455).

30 An overview of the current state of affairs is provided in Krupp, Keohane, and Pooley, "Less Than Zero. Can Carbon-Removal Technologies Curb Climate Change?"

31 For more detailed moral arguments with respect to this proposal, see Stephen M. Gardiner, *A Perfect Moral Storm. The Ethical Tragedy of Climate Change* (Oxford/New York: Oxford University Press, 2011).

32 This theoretical frame is chosen by Darrel Moellendorf in *The Moral Challenge of Dangerous Climate Change: Values, Poverty, and Policy* (New York: Cambridge University Press, 2014), 202.

33 For a full list, see Gardiner, "Climate Ethics in a Dark and Dangerous Time," 460–461.

34 Ibid., 452ff.

35 United Nations, "United Nations Framework Convention on Climate Change," 1992, 4, https://unfccc.int/resource/docs/convkp/conveng.pdf.

36 Stephen M. Gardiner, "A Core Precautionary Principle," *Journal of Political Philosophy* 14, no. 1 (2006): 33–60.

37 See, most prominently, The Stern Report. Nicholas Stern, *The Economics of Climate Change. The Stern Review* (Cambridge: Cambridge University Press, 2007).

38 For this concept, see Johan Rockström et al., "Planetary Boundaries: Exploring the Safe Operating Space for Humanity," *Ecology and Society* 14, no. 2 (2009).

39 See Rockström et al.

40 Simon Caney, "Justice and the Distribution of Greenhouse Gas Emissions," *Journal of Global Ethics* 5, no. 2 (2009): 127.

41 Ibid.

42 For a decisive claim to leadership in this way, see Henry Shue, "Face Reality? After You! – A Call for Leadership on Climate Change," *Ethics & International Affairs* 25, no. 1 (March 2011): 17–26, https://doi.org/10.1017/S0892679410000055.

43 On this, see Chapter 10.

44 Simon Caney, "Two Kinds of Climate Justice: Avoiding Harm and Sharing Burdens," *Journal of Political Philosophy* 22, no. 2 (2014): 125–149.

45 For a discussion of these claims, see Chapter 10.

46 This claim is outlined in Chapters 8–10.

47 An approach to nature that explains the interdependence of living systems and simultaneously takes nature as an ever-evolving, changing lineage of regeneration and remodelling is provided by scientists who support development system theory. See P. E. Griffiths and R. D. Gray, "Developmental Systems and Evolutionary Explanation," *Journal of Philosophy* 91, no. 6 (June 1994): 277–304.

48 For a discussion of the list of failings, including normative failings, see Chapters 9 and 10.

49 For an informative assessment of "water wars" as possibly severe water conflicts and for strategies to avoid escalation, see Delli Priscoli and Wolf, 9–32.

50 This limitation is important, yet it is restricted to cases in which people act voluntarily together in favour of chosen goals. Later in this study, the model of joint action will be translated into a normative model. In that scenario, cooperation in favour of environmental goals is not only the right thing to do, but under restricted circumstances is also obligatory. For a climate-duties approach, see Chapters 7–9.

51 The problem of climate change as a problem of the misappropriation of an NCPR is particularly difficult to resolve because it has an intergenerational and a global dimension. For an exploration of the resulting particularly dramatic failures of collective action, see Stephen M. Gardiner, *A Perfect Moral Storm*; Stephen M. Gardiner, "The Real Tragedy of the Commons," *Philosophy and Public Affairs* 30, no. 4 (2001): 387–416.

52 https://unfccc.int/process-and-meetings/the-paris-agreement/the-paris-agreement, accessed April 29, 2019.

53 IPCC, "AR6 Synthesis Report," accessed April 3, 2019, www.ipcc.ch/report/sixth-assessment-report-cycle/.

54 Michael Walzer, *Arguing about War* (New Haven, CT/London: Yale University Press, 2004).

55 Ibid., 33.

56 Ibid., 50.

57 Ibid., 34. Here Walzer also discusses the bombing of German cities at the end of WWII commanded by British leaders.

58 Ibid., 42–43.

7   Climate duties

Cindy: Incredibly, that's it! Now we have all the pieces of the puzzle. This has been a long way. But that's philosophy.

Bert: Yes. But insights that are difficult to get are also a special joy.

Cindy: Fine, let's have lunch together. This was hard work.

Bert: I'd like to. But now I also want to learn the complete lesson. You see, there is one tiny problem left – and actually this problem is a huge problem on a second look. Do you remember the section on "the dissociation of resource and beneficiary"?

Cindy: Yes, the last one.

Bert: Yes, but actually also the first one. This tiny problem is the biggest problem I can imagine when it comes to collective action. The whole model of collective action appears to depend on the simple assumption that the benefits from collective action are also reaped by the actors. And this is absolutely not the case regarding a truly global resource. Imagine you have put all your efforts into a collective action that in the end is completely useless or even senseless. This is really like going back to Chapter 3; and if I got it right, and since Chapter 3 depends on Chapter 2, we actually have to go back to the start: just a new level of fallacies and tragedies!

Cindy: But wait a minute. This is not really the case! The problems at the beginning were so vast because there wasn't even an idea of how to come to work together. We now have a sound model of joint climate action. I think we have to invent another way to get to grips with this tiny, possibly big, problem of a misallocation of benefits from acting together.

Bert: But how would that work? Even when most people of the world would work together, one single nation-state could spoil the outcomes – not only be free-riding on the efforts of the all the other people, but also by simply emitting so much greenhouse gas that all the other efforts are eradicated.

Cindy: Stop, there is an error in your thought! Remember the framing of climate change action. Chapter 1 argued against isolationism and quantification; and the whole model of joint action was focused on group action. And think about action and goals "under a description." One easy way to circumvent the problem of a misallocation is a new way to describe the goals. And I think something like that has already been argued. Climate goals can have different descriptions. And it is important to frame them in accordance with the ethos of groups – and also of groups that have many different goals.

Bert: Oh yes, you're right. Thank goodness that the whole effort has not been in vain.

Cindy: And there is another possibility to resolve the problem. Even if philosophers cannot literally force people to do something, they still have a very forceful instrument: they can talk about "duties." Duties are artificial things that rest on the presumption that people are not completely free to act, but that they should do certain things for moral reasons. Parents have duties regarding their children, teachers have duties regarding pupils, etc. Why should citizens of the world not have duties as inhabitants of Planet Earth?

Bert: This sounds plausible. But do you really think that philosophers have an idea regarding this next twist?

Cindy: Actually, they have. I've heard about discussions that really address duties to cooperate – even in this special case.

Bert: Oh man. I'm tired. There have already been so many complications, complexities, concepts, etc.

Cindy: You're right. But remember: climate change action is actually a really important thing to do right now. And I would like to hear more about duties to act. Come on, two more chapters and we have the whole story.

Bert: OK, I'm with you and with these enervating philosophers. But after that we should really have some fun together. And possibly I will skip the next two chapters. As for me, I have already read enough about climate change action. And indeed, I have the strong intuition that everybody should contribute to joint climate action. What do you think?

Cindy: I think as you do. But I'm sure that philosophers will have some arguments for climate duties. It's helpful to hear them when people challenge the view that everybody should do his or her part. And I am also sure that philosophers detect new fine-grained complexities. Actually, I want to hear more about climate duties. Let's see.

So far, I have interpreted climate action in the context of a theoretical approach to "joint action." The elements of a theory of joint environmental action include an explication of the elements of joint action, an exploration of the role of a shared ethos in environmental action, a revision of environmental goals, and a discussion of the concept of active co-designing with nature. This interpretation of joint environmental action is a theoretical model of *voluntary* action. This chapter starts by investigating the option of transforming a voluntary model of cooperation into an *obligatory* model of acting together. In short, this chapter and the following chapters are about *climate duties*. The discussion on climate duties has recently gained momentum in a variety of contexts, some of which will be presented shortly.[1]

The interpretation of climate duties that will be offered in this chapter and in the following chapters is based on the concept of joint environmental action I have already developed at length. As applied to climate change, it is important to note right from the outset that the question is not whether climate obligations exist at all. Instead, the discussion in this chapter focuses on the more specific issue of whether individuals have obligations, in the context of joint agency, to contribute to climate goals. If climate duties[2] exist, it is not only necessary to specify their character and to discuss the issue of duty-holders, but also to keep in mind the theoretical prerequisites.

This chapter begins by investigating climate duties, though only a single aspect of them. It explores whether the claim that individuals have a duty to support climate goals is a justified claim. This is only one element of the shift towards a normative model of climate action. Further elements, including the question of whether individuals should team up as effective joint acting teams or whether they should join pre-existing joint actors will be examined in the remaining chapters. More specifically, this chapter builds on the foundation of the previously explained model of joint environmental action. It simply investigates that model of cooperation under a new premise. While the theoretical model of joint action explains why people join groups or create groups voluntarily to participate in the realization of shared goals, it still leaves room to inquire whether people are also morally obliged to either join existing group actors or to create new groups. What can be assumed as a general backdrop is the fact that the outcomes of joint action on the processes of climate change also relate to goals that are in the immediate interest of humankind.

Several systematic problems preclude a straightforward answer. One problem is that small groups are unlikely to be able to bring about change. Individuals may not even be in a position to establish groups that can achieve effective change. This leads to a first question: would it not be more conclusive to discuss duties of collectives that are capable of achieving climate goals right from the start? Although this question might be answered in a positive way, it is far from clear whether collectives are the only responsible actors in climate scenarios. Instead, it might also be arguable that individuals have a duty to support climate goals; yet this claim might be a conditional one. It may rest on the assumption that duties belong to everyone – and, as part of a huge collective, individuals are also able to discharge their duties to accomplish environmental goals, and climate goals in particular.

This chapter is organized in seven sections. It starts with a brief sketch of two theoretical proposals to establish the climate duties of individuals. Section 7.1 explores the concept of climate duties as established in a rescue scenario by Peter Singer.[3] Section 7.2 summarizes an argument developed by Simon Caney, who justifies climate duties in the context of a human rights approach.[4] Section 7.3 starts with an alternative line of arguments. It addresses climate duties as duties of individuals, but also emphasizes the special characteristics of climate duties. This section also stresses the fact that in the interpretation of joint environmental action, the type of cooperative relationship among the cooperators is also unique. Section 7.4 gives the main argument of this chapter. It explains why individuals do have a duty to cooperate on climate goals – under specified conditions. These conditions include the existence of a fair scheme of cooperation and a situation of mutual dependence in participating in that scheme. Section 7.5 provides further support for a general duty to cooperate once a scheme of cooperation is in place. Section 7.6 explores further conditions in the debate on climate duties: demandingness and effectiveness. Both are regarded as critical in the arguments in favour of climate duties. Section 7.7 summarizes the argument.

The main argument in this chapter draws on theories of joint action and gives this interpretation a theoretically different twist. I contend that whenever a fair pattern of cooperation that supports climate goals is available, individuals as well as groups and other collective actors are obliged to cooperate. Yet the duty to cooperate is restricted to cases of mutual dependence on shared natural goods and is, in particular, restricted to cases in which fair schemes of appropriation are available and are sound means for reaching integrated environmental goals as well as environmental exigency goals. Note that the duty to cooperate is *not* established as an intergenerational duty, nor is it established as an obligation not to forfeit the prospects of strangers to live a decent life. After having presented arguments in favour of a duty to cooperate, I shall also address some potential reservations about this proposal. They include the discussion of over-demandingness, of the reasonable claim of effectiveness, and the problem that duties rest on false assumptions regarding joint action. Before that, however, I will depict two arguments that have been made to give a foundation of climate duties as rescue duties.

7.1 Individual rescue duties for climate victims

A theory of moral duties rests on assumptions that need to be made explicit from the outset. As for climate duties, the discussion also needs to integrate the peculiarities of the scenario in which duties are defined. Therefore, philosophers start with the statement that climate change threatens the lives of many people.[5] Climate change causes considerable harm.[6] When a person's life is under immediate threat, and there are people witnessing this situation (possibly even partially responsible for the threat), people who are able to rescue that person are under a moral obligation to help.[7] This argument has been applied in a variety of different ways to the climate scenario.[8]

In the context of a utilitarian ethics framework, Peter Singer has defined climate duties as rescue duties.[9] His normative claims rely on an interpretation of climate change that has two important implications. First, Singer interprets the atmosphere as a common natural good, and access to the atmosphere should be regulated using fair guidelines. More specifically, the atmosphere serves as a dumping ground for greenhouse gases. As long as no other claim is established, every single person deserves the same piece of that atmospheric resource as everyone else.[10] In Singer's view, there is no prima facie reason why a single person deserves a larger part than every other person of a global and natural good. Second, people are morally obliged to help individuals in desperate situations, in particular when the rescue effort provides only marginal costs to them, yet has an enormous positive impact on the person who receives it. "In practice, utilitarians can often support the principle of distributing resources to those who are worst off, because when you already have a lot, giving you more does not increase your utility as much as when you have only little."[11] Yet additional contributions need to be arranged according to further principles of fairness, including principles that take historical emissions into account.[12]

The principle of rescue duties is thus at the heart of climate duties. Singer writes that he frequently prepares his students for climate debates by discussing an example. Imagine you come upon a lake in which a child is drowning: would it not be fair to ask each person passing by the lake to rescue that child if it were possible? He then asks his students:

> [W]ould it make any difference if the child were far away, in another country perhaps, but similarly in danger of death, and equally within your means to save, at no great cost – and absolutely no danger – to yourself? Virtually all agree that distance and nationality make no moral difference to the situation. I then point out that we are all in that situation of the person passing the shallow pond: we can all save lives of people, both children and adults, who would otherwise die, and we can do so at a very small cost to us: the cost of a new CD, a shirt or a night out at a restaurant or concert, can mean the difference between life and death to more than one person somewhere in the world – and overseas aid agencies like Oxfam overcome the problem of acting at a distance.[13]

Singer does not claim that people owe each other the duty of rescue. But he maintains that individuals as well as political institutions are obliged to invest in the welfare of humankind. Indeed, all people are obliged to invest in rescuing individuals in desperate situations, doing the most good they can do.[14] According to the principles of an optimal return on investment and of marginal utility, everyone should give something to the very poor to enhance their living conditions.

Applied to climate ethics, Singer's parallel to the rescue scenario breaks down to the requirement to respond to urgent claims of climate victims and to support them. These duties are not restricted to local surroundings. Although restricted to cases in which they do not override the priority of personal well-being, rescue duties are universal and positive duties. This is not the place to relate the whole picture Singer presents in "One Atmosphere." Singer establishes a range of principles that together back up the claim that people should protect natural common-pool resource (NCPRs), including the atmosphere, from over-exploitation. He also asserts the historical responsibility of former polluters; in his view, those who spoiled the resource must also now fix it.[15]

Rescue scenarios have a high degree of plausibility in terms of duties among strangers. More specifically, the imbalance between a minor impact and a big effect on behalf of a victim contributes to the insight that although not generally obliged to make other people's lives better, the attenuation of severe and life-threatening harm stands on other grounds. Thorough distinctions of principles of beneficence that apply in a rescue scenario not only take into account the way individuals are interconnected, they also consider the degree of suffering of another human being as compared to the costs of rescue to the rescuer.[16]

Yet Singer appears to overlook one important trait of rescue scenarios in discussing them as applied to climate change. In the case of climate change, individuals are not in a situation to rescue climate victims or to prevent severe harm. Much more plausible is an application of the idea of rescue to a joint rescue case.[17] Even when not in a position to rescue climate victims by means of single action, acting together might contribute to effective help. The interpretation of the climate scenario as a joint rescue scenario will be discussed at length in Chapter 8. In the context of this chapter, the most important aspect of Singer's interpretation of the climate scenario is the theoretical option to establish duties of individuals, even without describing the causal responsibility for the rescue scenario.[18]

7.2 The human right not to suffer climate harm

Unlike rescue scenarios, climate duties have also been established in the context of human rights. Although it is difficult to argue that there is a human right to an intact atmosphere or to climate stability, it can be argued that individuals have a human right not to suffer arbitrary and extreme harm from climate change. Simon Caney has formulated this proposal.[19] In his writings on climate ethics, Simon Caney argues that climate duties are in line with a cosmopolitan no-harm principle.[20] People are considered to be endowed with human rights, the most important of which are summarized in the list of human rights as also accepted in 1948 as a normative backdrop for the world community.[21] More specifically, human rights are not tied to any specific institutional framework, as, for instance, the constitution of a nation-state. Commenting on climate duties, Caney does not support the view that a "right to the environment" is part of the list of human rights. Instead, he explains that climate rights as well as environmental rights need to be framed differently. I shall first depict his core argument and then comment on it.

The argument begins with the assumption that

(P1) A person has a right to X when X is a fundamental interest that is weighty enough to generate obligations on others.

[...]

The next step in the argument maintains that

(P2) Persons have fundamental interests in not suffering from: (a) drought and crop failure; (b) heatstroke; (c) infectious diseases (such as malaria, cholera and dengue); (d) flooding and the destruction of homes and infrastructure; (e) enforced relocation; and (f) rapid, unpredictable and dramatic changes to their natural, social and economic world ...

(C) Persons have the human right not to suffer from the disadvantages generated by global climate change.[22]

This argument is based on an interpretation of human rights according to which fundamental interests count as a justificatory background for the claim that a person holds a human right. More specifically, this argument does not lead to the conclusion that every single person needs to account for the human right of every other person not to suffer from disadvantages caused by other people. Instead, to formulate duties that relate to justified claims in terms of human rights, Caney offers a *hybrid approach* that combines a set of different principles according to which the responsibility of individual actors is allocated.[23] In line with Singer's proposal, Caney also presents an account that "does not necessarily rest on the assumption that climate change is human-induced."[24] Indeed, it is not the causation of the harmful events that supports the view of climate obligations. Instead, Caney is very careful in arguing that advantages that result from high emission rates need to be considered when determining obligations related to climate goals. In addition, to cohere with distributive justice, a fair quota of emission budgets needs to be negotiated; and they need to cohere with the ability to pay of those who are already better off.[25] Overall, Caney proposes "to balance a persons' interests in engaging in activities that involve the emission of GHGs, on the one hand, with a persons' interests in not suffering the harms listed in (P2), on the other. We also need to employ a distributive principle."[26] One of the outcomes of this procedure is the view that the least advantaged have a right to emit higher GHG emissions and that the wealthy have to shoulder the costs for mitigation and adaptation to a higher degree.[27]

Basically, Simon Caney's arguments demonstrate that obligations to assist climate victims and duties to contribute a fair share of burdens to achieve mitigation do exist. By appeal to the moral power of human rights claims, individuals' justified interests in not suffering serious harm due to incidents of climate change are taken into account.[28] This still leaves much space for interpreting climate duties in detail and for discussing the issue of whether or not the duties Caney explores are moral duties in the strict sense.[29] More specifically, Caney groups individuals together as "rich" or "poor" and also relates climate obligations to the fact of belonging to one or another group. This is a common procedure, but nevertheless a questionable one.

Although climate duties can be established, they still differ from other types of moral obligation. Besides the fact that they cannot be established without also accounting for an interpretation of some effects as harmful to people and a general interpretation of what individuals are obliged to do, they cannot be established accordingly without also being cognizant that they are "conditional duties" in some respects. The justification of climate duties in its concrete shape depends not only on the prior claim that people have a human right not to suffer from climate effects or on the prior claim that people possess a right to be rescued from a miserable and life-threatening situation; they instead also depend on a distinct view of distributive justice. No one really claims that climate duties need to be fulfilled independently of further conditions. Instead, they need to be fulfilled once they rest on a just allocation of burdens.[30]

But even when these restrictions are taken into account, another issue is still on the table. It has been argued that climate duties do not belong to single actors for two reasons: first, single actors are not responsible for climate change; second, even when willing to bring about change, individuals cannot change anything on their own. When "ought" implies "can," it is better to attach climate obligations to collectives, rather than to individuals. The next section discusses this option.

7.3 Duty-holders in the climate case

In recent debates on climate ethics, a case for climate obligations has been made. Although the claims in climate ethics are contested, the discussion on climate ethics has recently developed into a debate on climate duties.[31] Yet neither the content of these duties nor the issue of who really is a duty-holder has so far been settled. The specific question of whether collectives count as holders of climate duties is a matter of vigorous debate. In order to prepare the discussion of climate duties, it is necessary to also take a look into the debate on duty-holders in the context of climate ethics.

A minimum requirement for ascribing a moral obligation to X is the capacity of X to act.[32] Although the concept of "action" is at the heart of sophisticated philosophical debates nowadays, a general idea can be framed thus: in order to count as an actor, an individual should possess, at a minimum, desires and beliefs and, in addition, the capacity to form a goal. As such, actions are undertaken by individuals. Yet in the literature on groups, it is also a common presupposition that groups, too, are able to act. The minimum requirements are derived from the desire/belief-model of agency. Stephanie Collins and Holly Lawford-Smith define agency as

having (1) something that plays the role of desires (e.g. wanted outcomes, goals, and preferences), which, in combination with (2) something that plays the role of beliefs about one's environment, move one to make (3) something that plays the role of decisions about how to act in that environment.[33]

These authors explain that – under certain circumstances – groups act in this interpretation of action. What the group needs is a "group-level decision-making procedure"[34] that is "operationally distinct from the procedures held respectively by the members."[35] In addition to a decision-making procedure that helps the group members develop a shared position on joint goals, the group also needs to have some internal organization.[36] According to these claims, structured collectives, that is groups that possess internal organization and a decision-making structure, count as collectives that can act. Whereas long-term groups such as states, firms, churches, and international organizations have the properties of agents,[37] the controversial cases are groups lacking such a high degree of organization. According to Peter French, "aggregates simply fail the tests for membership within the moral community."[38]

An approach that is helpful in the context of the exploration of climate agency rejects a general decision about the capacity of groups to act, but instead distinguishes various types of groups. At the high end are groups having an identity that is more than the identity of their parts and that persists over time, and possessing a decision-making structure and a conception of themselves as a unit. According to Toni Erskine, these are *"institutional moral agents."*[39] They can have duties and they can also be held accountable for flawed actions and for not living up to their responsibilities. In addition, we have a *range of structured collectives*, such as groups that possess some continuity over time, a decision-making structure, and some form of internal organization. They, too, count as agents. Finally, there are *random groups* and *conglomerates of people*. Such bodies do not possess a decision-making structure, nor do they necessarily have continuity in time or a self-understanding of the role of a group member.

So far, the discussion has been focused on the attribute "agency." Yet in the climate debate, the discussion of agency is primarily defined as a preliminary to the discussion of moral obligations and duties. In the case of a moral duty, the justification results from a moral framework that explains that the expectation that X should carry out a certain action is justified on moral grounds. More specifically, constraints regarding the duty-holder need to be considered. One important constraint is framed as a "capacity principle."[40] Whenever a duty is debated, the assumption is made that the actor really can discharge that duty by acting in a certain way.

The capacity principle raises two questions. First, there is the question of whether individuals really have the capacity to act, assuming the action is about a goal they cannot realistically reach on their own. Individuals may be able to reach climate goals together, but they certainly cannot do so on their own. If so, it might be the case that people do not have a duty, because they cannot fulfil the corresponding act.[41] Another fundamental question is whether the collectives that qualify as actors also qualify as "moral actors," holders of moral duties. The second question is whether individuals or groups count as subjects of duties. If so, it might also be possible to think about *obligatory climate action* either of individuals or of group actors. Indeed, philosophers have recently taken up this debate.

Instead of presenting a general picture of the current debate on responsibility of actors with respect to climate goals, I shall highlight two of its aspects. To begin with, I intend to demonstrate that the observation of a more generous interpretation of joint responsibility than the focus on either individuals or on groups can be defended. In the climate debate, it has been discussed that various types of groups hold climate duties, including spontaneous groups that together succeed in rescue operations. I think this presumption is right. Moreover, I intend to highlight a proposal according to which even random collectives can bear responsibility, assuming their action is goal oriented.

To tackle the first point, thinkers have handled the capacity of collectives to act with caution. Yet when addressing moral duties, the overall picture is shifting. It appears as if the conditions that need to be met to count as a "moral actor" are not directly related to the agency conditions. Indeed, even random collectives can count as morally responsible actors. In the fundamental contribution to this debate "Can a Random Collection of Individuals be Morally Responsible?,"[42] Virginia Held defends the view that random collectives without any decision-making structure can be held morally responsible. She does not argue on general grounds in favour of the responsibility of random collectives, but discusses a rescue case in which individuals together face a stranger in dire need of rescue and in which they can save his life. In a case like that, she claims that a random collective is responsible for acting.[43] In addition, she also claims that the individuals can be held responsible for failing to form into an organized group capable of acting in a situation like that.[44]

This proposal deviates from two other approaches in which collectives are considered to be responsible actors. Whereas Peter French claims a high degree of organization features as a precondition for responsibility,[45] Held argues in favour of the responsibility even of spontaneous groups – though under highly restricted circumstances. A *middle position* is represented by Larry May's approach.[46] May thinks it is flawed to reduce conglomerates to individual members. Instead, he claims that mobs and teams provide examples for groups that "are defined by reference to the solidarity that allows the members of the group to engage in joint purposive behaviour."[47] As a consequence, May does not lay emphasis on the organizational structure. Still, he regards groups as entities that can facilitate joint action and common interest, even when lacking organizational structures.

Whereas this already indicates that criteria for responsibilities of groups can be lowered, I now present a proposal that should be particularly helpful in the debate on climate obligations and that reflects on the second point regarding goal-directedness and obligations. It establishes that joint actors, as explained in the theories of joint agency presented by Tuomela and Bratman, do count as moral joint actors. According to Toni Erskine, two cases of collective agency need to be clearly distinguished. The first is a scenario in which random collectives hold responsibility, even without possessing a decision-making structure. These are cases in which random collectives "can be blamed for not performing given certain conditions."[48] In that case, limited cooperation suffices. Yet, unlike those cases, responsibility is also connected to collectives when performing "joint purposive action."[49] Erskine concedes that to be capable of that type of action, some degree of prior deliberation and special coordination suffices. She explains:

> To be able to exercise joint purposive action, the members of a collectivity must have: (1) compatible interests (although not common or even necessarily complementary motivations); (2) a concomitant willingness to cooperate (something which might be called "participatory intention"); and (3) the capacity to deliberate (however informally) to coordinate their actions (even imperfectly) in circumstances in which the required collective action is not obvious but, rather, open to disagreement.[50]

If all three conditions are being met, a conglomerate and a spontaneous group can bring about an action in the true sense. In addition, it can be held responsible for actions under specified circumstances.[51]

This insight is not only important in that it sets the discussion of responsible actors apart from the debate on agency of structured collectives. It is also remarkable that it highlights a specific feature of joint purposive action. "Joint purposive action is thereby distinct from action which is best described at the level of the organization as a whole, and which generates coherent accounts of moral responsibility which cannot be reduced to its individual constituents."[52] According to this interpretation of joint action, coalitions of the willing, as well as groups that have been at the heart of our analysis of joint action, count as responsible actors.[53] Yet the way responsibility is generated and is distributed to the individual participants differs from the simpler cases of either individual or cooperative agency.

In the remainder of this chapter, the discussion focuses on climate duties.[54] Here, the distinction between obligations of individuals and obligations of groups is not at the centre of discussion. Instead, individuals as well as joint and collective actors can figure as duty-holders. We therefore have an answer to our problem. But what is still not clear are the conditions under which the capacity principle applies to the case of climate change. More specifically, the backdrop of climate duties as duties that are described in a very specific scenario needs to be explained.

7.4 The role of schemes of appropriation of a resource

Although the approaches that have been described so far demonstrate that there are ways to define climate duties that are connected to individuals under certain presumptions, it is still necessary to qualify the content and the category of those duties. Obviously, duties in rescue scenarios and duties that result from human rights claims differ significantly from moral duties more generally. Even when focusing on climate victims, it is far from clear whether people are *morally* obliged to help individuals in desperate situations when victimhood is caused by mega-events in the atmosphere. Moreover, the relationship between claims of distributive justice on the one hand and claims of burden-sharing for assistance to climate victims is far from clear.[55]

In addition, the concepts of climate duties that have been discussed so far also relate to a very specific concept of climate justice. People who debate climate duties not only focus on the victims of climate change and on obligations for supporting them. Instead, they also share common ground in highlighting the urgency of mitigation and the need to support adaptation worldwide. Yet it is also far from clear whether people really are *morally* obliged to act in favour of one of these goals.

Instead of discussing moral obligations for rescuing climate victims and preventing further severe harm, I want to start the discussion of climate obligations from a different angle, congruent with the overarching theme of this book. Remember that the problem of climate change was framed as a "tragedy of the commons." Although taking the specific severity of this tragedy into account, I rejected the interpretation usually given in addressing that tragedy as a cause for climate change. Instead, the tragedy occurs because fair and successful schemes of appropriation of a scarce and shared natural resource have not been elaborated so far or have not been accomplished by beneficiaries. A fair scheme of appropriation needs to cohere with basic values for addressing shared natural resources. The "ethos" includes respect for the integrity of that resource, fairness in benefiting from the resource, and moderate eco-centrism.[56] More specifically, appropriation in terms of joint action assumes shared and fair goals. In addition, these schemes of appropriation cannot be established once and for all. Rather, they need to be worked out with respect to future goals of groups of people. A theoretical exploration can explain the prerequisites and the shape of climate goals, but the definite content needs to be negotiated among the users of the shared resource. Yet when responding to those conditions, fair schemes of appropriation present the best possible way to appropriate a shared resource together. I have established that much. I now add the claim that *in a situation like that, individuals as well as groups are obliged to support that scheme of appropriation.* In short, they are obliged to cooperate. The argument proceeds in several steps.

As a beginning, I want to bring up an argument presented by Bailor L. Johnson. He claims that moral duties have a special shape and a special content in situations in which "tragedies of the commons" threaten the sustainable use of a shared resource.[57] In accordance with arguments in this study, Johnson maintains that environmental duties cannot be established in terms of a duty to simply restrict the individual use of a natural good to a level that contributes to overall sustainability. This would generate dependence of the existence of an obligation to the condition "if accomplished by all possible users."[58] Instead, in a context like that, the fulfilment of a duty needs to be interpreted as an act that fulfils a goal that has been explicated as a moral goal in advance. It then can be established that every single person holds the duty to act accordingly.[59] Yet even then, the conditions for moral duties differ from usual cases. In order to protect the natural commons effectively, "one's moral obligation is to work for and adhere to a collective scheme to protect the commons."[60]

According to this proposal, individuals do not hold an individual duty to protect a natural commons. Instead, they have *the duty to cooperate in favour of a collective scheme that protects the common good effectively*. The underlying claims of this proposal are twofold: on the one hand, it is reasonable to claim respect for the constraints resulting from the sustainability conditions of goods. Johnson restricts this claim to conditions that also include realistic expectations of success.[61] I shall shortly offer a slightly different interpretation of constraining factors. Yet even more important, duties that correlate with benefits from an NCPR differ from "unilateral duties." Instead, they are tied to a context in which cooperation as a collective endeavour is obligatory to avoid undermining the currency of cooperation. On the other hand, there is a strong argument for regarding a duty to cooperate in the specified circumstances as moral obligation. As Johnson states, an underlying rationale to claim cooperation is that free-riding in these particular cases is morally false.[62]

At this point, Johnson leaves space for further arguments. Instead of relying on the subsistence rights of individuals as a moral backdrop or the dangers that come with over-exploitation and that can threaten the lives of many, my argument is that in a scenario in which *shared goals are right in a morally deep sense* and in which cooperation is goal-driven – whereas the *goal can only be achieved when all participants cooperate* – defection not only puts the achievement of the goals at risk, it is also morally false. One might argue that the defection of only one person might not threaten the goal. True, the scheme of appropriation is not fixed once and for all times, but it can only be accomplished by actions that instantiate the processes towards the shared goal. Remember the lake example. It is not only the case that all beneficiaries have to work together to achieve a shared goal. Moreover, one big emitter can spoil not only the efforts of all other beneficiaries from the shared water resource; he also jeopardizes the scheme of appropriation. In addition, one person who defects invites more defectors, and in the end the scheme of appropriation is undermined.

As for the appropriation of the atmosphere as a dumping ground, the same principle holds. One high-level emitter might not undermine a fair appropriation of the atmosphere as a dump by other beneficiaries, yet a handful of high-level emitters undermine cooperation in a much more fundamental way.[63] They drive a *rise to the bottom*, instead of supporting a dynamic that, if not perfect, at least supports the opposite, that is, a *crowding-in*. When group action is successful, individuals are attracted to that group, and – in the best case – they wish to enter the group.

In short, a scheme of appropriation that is justified and fair depends on continued support and cooperation in terms of voluntary compliance with that scheme by all potential beneficiaries. They have an obligation to cooperate, as long as the scheme of appropriation is fair. In sum, the protection of the natural commons and the individual contribution to that protection differs from other moral duties more generally. The obligations to cooperate are related to a fair scheme of appropriation of a shared natural resource.[64] More specifically, Johnson also introduces the distinction between individual duties and unilateral duties in that context.[65] Yet this idea needs to be discussed in detail. Here, only some general comments will be made. Arguments that reject the duty to reduce greenhouse gas emissions usually respond to the problem of ineffectiveness and also the moral inadequacy of unilateral duties – duties that are claimed without taking into account whether or not others are also complying with the same set of duties.[66] Yet when conceptualized as a duty to comply with a fair scheme of appropriation, things look different. It can then be said that non-compliance is voluntarily chosen and that it puts the goals at risk.

A similar point about the special quality of climate duties as non-unilateral duties has recently been made by Anton Leist.[67] He maintains that climate duties differ in a fundamental way from other types of duties by mandating the modification of a behaviour that cannot be conceived other than as already being integrated into a collective scheme. Each person needs to consume air and needs to dispose waste into the air as much as she needs to use water and other life-sustaining substances. As social beings, people are also already involved in cooperative schemes. Water is provided by institutions that also determine ways of cooperation. The same holds for the atmosphere. Yet the way patterns of cooperation are modelled and changed depends on each person's behaviour. In a context of mutual dependence, it is reasonable to claim that destructive single behaviour is morally flawed.[68] In addition, a duty to change behaviour also needs to be framed as claims for contributing to changing these patterns of cooperation.

Both arguments prepare grounds for an alternative defence of climate duties. Conceived as actions already happening in a cooperative context, climate action is a contribution to a distinct context. More specifically, distinct types of cooperation are accompanied by schemes of exchange and appropriation of a natural resource. These schemes are either sustainable or they contribute to the devastation of that resource. Johnson and Leist claim a duty to cooperate, based on a duty not to defect in these specific circumstances. Leist adds the duty to care for a good cooperative scheme. In order to summarize the outcome so far, I introduce two different types of climate duties in the next section.

7.5 Further arguments for duties to cooperate

Starting with the observation that natural goods are part of a web of "indispensable natural common goods" on which societies and individuals depend for their survival, the transformation of these goods is obviously also necessary. This transformation is implemented through schemes of appropriation with many different qualities. I have maintained that a fair scheme of appropriation needs to resonate with the ethos of the group of beneficiaries of a shared resource. More specifically, schemes of appropriation serve various goals. Concerning scarce and life-sustaining resources, it is wise to pay attention to their relatedness to "integrated environmental goals" in addition to goals that result from the urgent claim to repair a shared good or to remain within the space of a "safe operational space." More specifically, over-exploitation and flawed acquisition of the natural goods by means of destructive patterns of use have been proven as causally relevant for what Jared Diamond calls the "collapse" of societies.[69] As advanced as societies might be in economic and cultural terms, the protection of NCPRs remains a central need for their survival and development.

Systemic natural goods, such as water cycles and atmospheric patterns of exchange, can easily degrade by over-exploitation and by hazardous influences. But a cooperative approach to shared goals that takes the vulnerability of these goods into account and that serves as a protective institution can support the long-term integrity of that resource. In a scenario like that, a duty not to destroy successful patterns of cooperation – as enshrined in a wise management system – is justified. The duty not to undermine a successful scheme of cooperation, including the duty not to free-ride on the effort of the other people who comply with that scheme of cooperation, is not dependent on the cooperation of other beneficiaries. Instead, it results from a duty to comply with a generally acceptable scheme of appropriation that also serves shared goals of groups of people.

This proposal gets additional support from a perspective on ethics that differs from ethics as enshrined in moral obligations among individuals. That ethics implies a new general assessment of cooperation. Robert H. Myers has discussed an ethical theory that regards cooperation as central.[70] He maintains that moral claims and cooperation are tied together much more closely than usually proposed. In his view, moral activity is best framed as cooperation in a real engagement with others in the promotion of value.[71] More specifically, cooperation is not conceived as an instrument for achieving a certain goal whose fulfilment corresponds to a duty. Instead, moral obligations are framed as duties to cooperate in favour of justified goals. Assuming it is right to define climate goals in such a way that they not only enshrine values, but correspond to a reasonable environmental ethos – and assuming they are important goals, exemplifying long-term successful patterns of fair cooperation on shared natural resources – then climate goals are truly moral goals for cooperation. Cooperation in favour of accomplishing them does not depend on particular worldviews, but on insights about the deep morality of these goals.

In the context of this study, it is not necessary to reframe moral obligations as obligations to cooperate, broadly conceived. Instead, it suffices to assert the much more restricted case that people have a moral obligation to contribute to environmental cooperation on certain basic natural goods, including water, the biosphere, soil and – foremost – the atmosphere, when schemes of cooperation are fair and conducive to justified goals. These obligations are conditional. They depend on a framing of environmental goals, and climate goals in particular, as intrinsically justified goals. They also depend on a clear-cut cooperative scheme that exemplifies a fair expression of cooperation and its continuation.

This argument for the duty to cooperate and not to free-ride results from an assessment of the meaning of a distinct type of cooperation as related to natural resources. It also gets support from another perspective. Richard J. Arneson takes a fresh view of schemes of cooperation and the relevance of the possibility of really "opting out" as against cases in which this option does not exist.[72] Arneson explains that obligations to contribute actively to a cooperative scheme cannot be established in cases in which the possibility of "opting out" of the scheme is not a realistic expectation. This appears to be true of patterns of exchange that relate to the most indispensable natural goods. Yet Arneson also describes an exemption from that general case.

Here is the lengthy argument. The context is the debate on a principle of fairness that states that when a number of persons conduct any joint enterprise according to rules and thus restrict their liberty, those who have submitted to these restrictions when required have a right to a similar submission from those who have benefited by their submission. Although this is generally right, Arneson emphasizes the fact that it is a peculiarity of public-good scenarios that people cannot "opt out" of receiving benefits. Therefore, exchange in accordance with rules of exchange cannot be said to be "voluntary" in the strict sense. Furthermore: "One cannot voluntarily accept a good one cannot voluntarily reject."[73] This is why an obligation to contribute to that scheme does not exist.

Nevertheless – and in contrast to this assessment of scenarios in which cooperation is not voluntary in that benefits are connected to each potential beneficiary – Arneson also defends the view that "accepting or even simply receiving the benefits of a cooperative scheme can sometimes obligate an individual to contribute to the support of the scheme, even though the individual has not actually consented to it."[74] Whether this claim is valid depends on the specific circumstances regarding the scheme of appropriation. In Arneson's view, a minimal condition for a reasonable interpretation of obligations that derive from benefits from a collective good is that "exclusion of anybody from consumption is unfeasible, but individuals may choose whether to engage in consumption."[75] This is precisely the case regarding water systems, but also regarding the emissions budgets that, in the end, are causing climate change. Moreover, Arneson claims that a minimal obligation to go along with the scheme is only in place when the scheme that distributes burdens is fair.[76] The scheme's fairness includes – inter alia – that the payments requested from each beneficiary are fair and affordable; second, that the choices to contribute to the scheme supplying the collective good are mutually independent, and that the benefits are distributed equitably.[77]

As for joint environmental action related to natural common goods, this argument can be applied straightforwardly. No one is in a situation to reject one side of cooperative schemes in terms of participating in exchange patterns. People naturally benefit from the atmosphere as well as from water. Yet people nevertheless can accept existing patterns of exchange or reject and try to change them. The argument that people are perfectly free to choose a scheme of appropriation is not right. Yet the insistence on limited freedom also prompts the idea of another set of duties.

Assuming the appropriation of natural resources is mediated through institutions and the market, and assuming this also includes alternatives, the argument can be extended in the following way. Compliance with schemes of appropriation is channelled through institutions that sell commodities and that supply individuals with goods, including services derived from a shared resource. But whether individuals contribute to these schemes and – in the case of plurality of distributive institutions – whether individuals address changes in these schemes of appropriation is up to each individual. Obligations to pay attention to the means of appropriation result from this limited freedom of choice. More specifically, as long as the burdens that investments in joint environmental action produce are not distributed according to principles of fairness, people will not be obliged to comply with the modes of exchange enshrined in that pattern.[78] They are not only free to cooperate once a fair scheme of cooperation is available; they also should reject schemes of appropriation that are unfair and that destroy the shared resource.

In sum, climate duties are duties to cooperate once a successful and fair scheme of appropriation of a shared resource frames the cooperation. This is the duty to cooperate on fair schemes of appropriation, in short, an *environmental cooperation duty* or a *climate cooperation duty*. Whether it is restricted to individuals or also extends to collectives remains to be discussed. Overall, this duty is restricted to cases in which cooperative schemes of appropriation are available. In scenarios like that, continued participation in destructive schemes of appropriation is morally false, whereas cooperation is obligatory.[79]

7.6 Conditional duties

So far, the discussion of duties has been restricted to cases in which (a) a fair pattern of appropriation is in place, (b) actors are free to choose either to cooperate or to defect from that scheme of cooperation, and (c) alternatives are in place that are neither fair nor goal oriented. In scenarios like that, every single actor has a duty to comply with the fair scheme. Two remaining questions are whether this duty also holds when it comes to an exorbitant price – this is the question of *demandingness* of a moral claim; and whether individuals are also obliged to change existing patterns of appropriation on their behalf, or at least to try to do so. In short, do individuals or groups hold *climate-scheme-change duties?*[80]

As for the second point, the discussion can be abbreviated here. I define those duties at length in the next chapter, but they are restricted to actors who are capable of enforcing schemes of cooperation and schemes of appropriation of resource respectively; and they come with what Caney has termed "second-order responsibility," which, in our context, means the responsibility to initiate and to institutionalize responsible action.[81]

Yet it is still necessary to discuss the issue of demandingness. Is there always a duty to cooperate in all circumstances? Or is it restricted to situations in which individuals merely have to make small commitments and pay low prices in order to shift from an unfair pattern of appropriation to a fair one? Against the backdrop of an analysis of rescue scenarios, Schwenkenbecher proposes the following:

> If, due to one or more agents' defection, a random group does not achieve the joint outcome, the remaining individual members of that group are not accountable for that failure and not blameworthy if (i) they did contribute or took credible steps toward establishing joint action, and (ii) as a result they had good reason to believe that an insufficient number of others would not contribute.[82]

The situation Schwenkenbecher is discussing is a rescue scenario framed as joint action. It therefore differs in significant respects from the proposal to frame climate duties as duties to participate cooperatively in a distinct pattern of cooperation. Yet the initial idea – that the duty does not imply that each person is fully doing all she can do, but should do the best she can – is nevertheless important. Bill Wringe has offered a comparable proposal, though in a different framing, in the context of a debate on obligations of the world population to work against climate change:

> [T]he suggestion is that where there is a collective obligation but no collective agent the individuals who are the addressees of the obligation acquire obligations to do things which are appropriately related to the carrying out of the action whose performance would constitute fulfilment of the collective obligation.[83]

Wringe's argument is that the "addressability requirement" can be detached from the agency principle. As applied to schemes of cooperation, this insight supports again the view that individuals are not obliged to fully cooperate, but they can be asked to work in favour of a cooperative scheme to the best of their abilities. Anton Leist tackles the issue from another perspective. He thinks that climate duties cannot be claimed unless the effectiveness of the act that fulfils the duty is guaranteed.[84] Yet whether the individual contribution can ever be assessed as a real causal contribution is a matter of debate.

At a minimum, a joint climate action approach is in a much better position to specify what "reasonably demanding" actually means than are proposals to assist strangers in severe need more generally – because the goals are already framed as goals to be brought about by joint effort. Assuming environmental goals cannot be achieved unless people work together, the contribution of every single person can be broken down to smaller units. Since collaboration is indispensable for achieving a goal, a joint action approach helps to specify the share. According to a proposal of Garrett Cullity, in an accumulative approach "we should ask not about the size of the sacrifice I need to make to save each person, but about the size of the aggregative sacrifice that would be involved in helping them collectively."[85] Yet in order to determine the right degree of demandingness, it is also important to acknowledge that failures to reach climate goals will have severe consequences for humankind, and not only for some people. Moreover, small investments, if contributed by immense groups of people, will have important effects.[86]

Overall, the range of responsibility individuals bear regarding the institutions performing daily services for them is highly controversial. As for institutions that have an environmental impact, recurrent themes are the right to information and also the duty to provide information. Whether or not individuals should inform themselves regarding the environmental impact of institutions performing services is an important issue to discuss.[87] Some of these issues will be broached later in this book. It suffices to state that cooperation is obligatory, as long as it is not overly demanding. As for accumulative goals, I later develop a framework of principles that helps to bundle the burdens of cooperation accordingly. In this chapter, my main goal was to define a duty to cooperate once a fair scheme of cooperation is in place. Further debate is also needed in order to explain the responsibilities to enable that type of cooperation.

One final point is particularly important regarding the duty to cooperate. Unlike former approaches to climate duties that derive either from a commitment to make good on a historical harm or on a legacy of benefits, the duties established in this section are tied to specific goals. The climate goals that have been established rely on new ways to appropriate shared natural resources together. Basically, appropriation of resources is accomplished by schemes of appropriation. This chapter was not about the goals themselves; the responsibility for goals is "prospective responsibility."[88] Instead, the duties that were addressed in this chapter are duties to cooperate once opportunities for cooperation are available.

7.7 Universal climate duties

In establishing this version of climate duties, some of the most central reservations against climate duties as individual duties can be removed. My proposal relies on a new interpretation of cooperation in the context of the appropriation of a shared natural resource. Individuals do not bear the duty to achieve climate goals, but they have a duty to cooperate once a fair scheme of appropriation is in place. In addition, they should also take responsibility for initiating action in response to climate change when a situation is disastrous. Much more has to be said regarding a fair claim for various actors. In addition, more discussion is needed to adjust this general claim to the various types of goals as distinguished in Chapter 6.

In addition, a thorough exploration of constraints is needed. Even when equipped with conditions of effectiveness, with a condition of reasonable demandingness, and when implying a distinction between "single" and "unilateral" duties, some reservations remain. One reservation is that it is possibly unfair to state that climate duties are actually moral duties. That frame might be too demanding. I have tried to demonstrate that – although this reservation is important – it depends on the chosen interpretation of a moral framework. Assuming morality is not exclusively about mutual claims of respect and of rights, but also about duties to joint goals that qualify as moral goals, the claim to integrate climate duties into a moral framework is not that improbable.

Two other objections, however, might appear to be serious. Duties need to be assessed in the context of other duties. More specifically, this section has not established that duties to cooperate on successful and fair schemes of cooperation and to change others override rescue duties.

Furthermore, it appears that the capacity of various actors needs to be taken into account. Nor should action in favour of climate protection reduce capacities to act unfairly in other respects. Another important issue is the question of whether duties exist even when the individual who is willing to comply knows for sure that other people – perhaps all other people – are not complying. These problems will be analysed separately. Yet they point to another important issue. Assuming success is never assured in advance, can duties really be claimed when their fulfilment is uncertain?

This speaks in favour of taking up another dimension of the discussion on climate duties: the investigation of duties of groups and of collectives, including the recent debate on duties to collectivize. Criteria are also needed that transcend the scenario of moral obligations in terms of mutual respect for rights of individuals. Instead, the following chapters get back to the scenario of "joint action" and explore the content and meaning of obligations in the context of people working together for the realization of environmental goals.

Overall, the outcome up to this point is still remarkable. Climate duties not only exist, they can even be connected to every individual. They need to be adjusted according to reasonable demandingness, they are related to schemes of cooperation that promise to be effective, and they are not unilateral duties. In sum, every single person is obliged to cooperate at his or her best in order to contribute to joint environmental goals based on mutual dependence, especially when it comes to climate goals.

Notes

1 The main sources for establishing climate duties are currently human-rights approaches, an extension of rescue duties, and principles of beneficence that include rescue duties regarding climate victims. Proponents include Simon Caney, "Two Kinds of Climate Justice: Avoiding Harm and Sharing Burdens," *Journal of Political Philosophy* 22, no. 2 (2014): 125–149; Simon Caney, "Climate Change and the Duties of the Advantaged," *Critical Review of International Social and Political Philosophy* 13, no. 1 (March 2010): 203–228, https://doi.org/10.1080/13698230903326331; Simon Caney, "Human Rights, Climate Change, and Discounting," *Environmental Politics* 17, no. 4 (August 2008): 536–555, https://doi.org/10.1080/09644010802193401; Stephanie Collins, "Collectives' Duties and Collectivization Duties," *Australasian Journal of Philosophy* 91, no. 2 (2013): 231–248, https://doi.org/10.1080/00048402.2012.717533; Toni Erskine, "Coalitions of the Willing and Responsibilities to Protect: Informal Associations, Enhanced Capacities, and Shared Moral Burdens," *Ethics & International Affairs* 28, no. 1 (2014): 115–145, https://doi.org/10.1017/S0892679414000094; Bill Wringe, "Global Obligations and the Agency Objection," *Ratio* 23, no. 2 (2010): 217–231, https://doi.org/10.1111/j.1467-9329.2010.00462.x; Anne Schwenkenbecher, "Joint Moral Duties," *Midwest Studies in Philosophy* 38, no. 1 (2014): 58–74, https://doi.org/10.1111/misp.12016; Peter Singer, *The Most Good You Can Do: How Effective Altruism is Changing Ideas about Living Ethically* (New Haven, CT/London: Yale University Press, 2015); Tracey Skillington, *Climate Justice and Human Rights* (New York: Palgrave MacMillan, 2017); Bill Wringe, "From Global Collective Obligations to Institutional Obligations," *Midwest Studies in Philosophy* 38, no. 1 (2014): 171–186, https://doi.org/10.1111/misp.12022.

2 "Climate duties" is an abbreviation for a variety of more concrete duties, all framed as duties to contribute in one way or another to bring about the climate goals that were discussed and determined in Chapter 6.

3 See Peter Singer, "One Atmosphere," in *One World. The Ethics of Globalization* (New Haven, CT: Yale University Press, 2002), 14–50.

4 For a short version of the arguments, see Simon Caney, "Cosmopolitan Justice, Responsibility, and Global Climate Change," *Leiden Journal of International Law* 18 (2005): 762–773.

5 The fact that climate change causes human suffering and deaths is a generally accepted fact in climate ethics. But for different interpretations, see, e.g., Singer, "One Atmosphere," 118–119; John Broome, *Climate Matters. Ethics in a Warming World* (New York/London: W.W. Norton, 2012), 11–12; Simon Caney, "Justice and the Distribution of Greenhouse Gas Emissions," *Journal of Global Ethics* 5, no. 2 (2009): 768.

6 One way to circumvent the conclusion that responsibility for this harm needs to be shouldered by current generations is discussed in Derek Parfit's "non-identity problem." See Derek Parfit, *Reasons and Persons* (Oxford: Clarendon Press, 1984), 200ff. This problem has been at the centre of a vigorous debate in climate ethics since then. For a position that rightly characterizes the limited force of this argument for climate ethics more generally, see Edward Page, *Climate Change, Justice and Future Generations* (Cheltenham, UK/Northampton, MA: Edward Elgar, 2007), 131–610.

7 For different opinions regarding the importance of causal involvement for co-responsibility, see Björn Petersson, "Co-Responsibility and Causal Involvement," *Philosophia* 41, no. 3 (September 2013): 847–866, https://doi.org/10.1007/s11406-013-9413-x; Walter Sinnott-Armstrong, "It's Not My Fault: Global Warming and Individual Moral Obligations," *Advances in the Economics of Environmental Research* 5 (2005): 293–315.

8 The two most important possibilities are (a) individual rescue duties as argued by Singer and explained in this chapter and (b) joint rescue duties as argued by Schwenkenbecher; see Schwenkenbecher, "Joint Moral Duties." This proposal will be discussed in Chapter 8.

9 This view of Singer's approach to climate ethics already offers a distinct interpretation.

10 https://unfccc.int/process-and-meetings/the-paris-agreement/the-paris-agreement, accessed April 29, 2019.

11 Singer, "One Atmosphere," 42.

12 See ibid., 43.

13 Peter Singer, "The Drowning Child and the Expanding Circle," *New Internationalist* (April 1997).

14 Singer, *The Most Good You Can Do*.

15 Singer summarizes the discussion on the polluter-pays principle as follows: "If we believe that people should contribute to fixing something in proportion to their responsibility for breaking it, then the developed nations owe it to the rest of the world to fix the problem with the atmosphere" (Singer, "One Atmosphere," 33–34).

16 For a thorough discussion of this relationship in various rescue scenarios, see Elizabeth Cripps, *Climate Change and the Moral Agent: Individual Duties in an Interdependent World* (Oxford: Oxford University Press, 2013), 48–57.

17 Schwenkenbecher, "Joint Moral Duties."

18 Note that the fact that there is no direct causal link between the harm that a climate victim suffers and the causation by a distinct person is regarded as an argument against rescue duties. For this, see Sinnott-Armstrong, 297–302.

19 Simon Caney has written extensively on climate justice. His most important contributions include Caney, "Two Kinds of Climate Justice"; Caney, "Justice and the Distribution of Greenhouse Gas Emissions"; Caney, "Human Rights, Climate Change, and Discounting"; Caney, "Cosmopolitan Justice," 2005.

20 For the "hybrid account," which includes a cosmopolitan framework and principles of justice, see Simon Caney, "Just Emissions," *Philosophy and Public Affairs* 40, no. 4 (2012): 255–300; Caney, "Cosmopolitan Justice," 2005.

21 United Nations, "Universal Declaration of Human Rights," 1948, accessed January 10, 2019, www.un.org/en/universal-declaration-human-rights/.

22 Caney, "Cosmopolitan Justice" (2005), 767–768.

23 See Caney, "Cosmopolitan Justice" (2005); Caney, "Just Emissions."

24 Caney, "Cosmopolitan Justice" (2005), 768.

25 Ibid., 769.

26 Ibid., 770.

27 Ibid.

28 For an approach that also focuses on a more specific set of human rights, see Skillington. She specifically discusses the rights of self-determination over natural resources (Skillington, 129–135) and the rights of the climate-displaced (Skillington, 151–175).

29 For a discussion of individual obligations as related to climate change, see Cripps; Marion Hourdequin, "Climate, Collective Action and Individual Ethical Obligation," *Environmental Values* 19, no. 4 (2010): 443–464. For a rejection of individual duties as related to climate change policies, see Sinnott-Armstrong.

30 David A. Weisbach offers a fundamental critique of distributive justice in climate ethics: "Distributive justice is doing no work ... [A]s long as there is inequality, distributive justice demands that we deviate from the cost-effective solution to help the poor. That is, suppose that pure efficiency concerns imply that developed countries go first and do more than developing countries. When we add distributive concerns, developed countries would have to do even more still because the distributive gains would be worth the efficiency losses." Stephen M.

Gardiner and David A. Weisbach, *Debating Climate Ethics* (New York: Oxford University Press, 2016), 205–206. Although it sounds as if efficiency is the basic argument, another line of thought is also presented by Weisbach, who argues that there is a "blinders argument" that rests on the assumption that an unequal sharing of scarce resources impels claims of correcting that injustice. Yet he correctly asserts that scarce resources simply cannot and should not be shared according to distributive principles. See Gardiner and Weisbach, 219.

31 See Stephanie Collins, "Duties of Group Agents and Group Members," *Journal of Social Philosophy* 48, no. 1 (2017): 38–57, https://doi.org/10.1111/josp.12181; Collins, "Collectives' Duties and Collectivization Duties"; Erskine, "Coalitions of the Willing and Responsibilities to Protect"; Toni Erskine, "Assigning Responsibilities to Institutional Moral Agents: The Case of States and Quasi-States," *Ethics & International Affairs* 15, no. 2 (2011): 67–85, https://doi.org/10.1111/j.1747-7093.2001.tb00359.x; Schwenkenbecher, "Joint Moral Duties"; Anne Schwenkenbecher, "Joint Duties and Global Moral Obligations," *Ratio* 26, no. 3 (2013): 310–328, https://doi.org/10.1111/rati.12010; Wringe, "From Global Collective Obligations to Institutional Obligations."

32 For this assumption in the climate debate, see the general emphasis on the "agency principle" in Schwenkenbecher, "Joint Duties and Global Moral Obligations," 318.

33 Stephanie Collins and Holly Lawford-Smith, "Collectives' and Individuals' Obligations: A Parity Argument," *Canadian Journal of Philosophy* 46, no. 1 (January 2, 2016): 41, https://doi.org/10.1080/00455091.2015.1116350. In discussing these "agency principles," the authors rely on a model proposed by List and Pettit. See Christian List and Philip Pettit, *Group Agency: The Possibility, Design, and Status of Corporate Agents* (Oxford/New York: Oxford University Press, 2011), 20.

34 Collins and Lawford-Smith, 41.

35 Ibid.

36 For the combination of decision-making procedures and an additional internal organization needed for groups to count as responsible actors, see Peter A. French, *Collective and Corporate Responsibility* (New York: Columbia University Press, 1984), 13.

37 Collins and Lawford-Smith, 42.

38 French, *Collective and Corporate Responsibility*, 44. Erskine summarizes French's position thus: "A collectivity is a candidate for moral agency if it has the following: an identity that is more than the sum of the identities of its constitutive parts and, therefore, does not rely on a determinate membership; a decision-making structure, an identity over time, and a conception of itself as a unit" (Erskine, "Coalitions of the Willing and Responsibilities to Protect," 119).

39 Erskine, "Coalitions of the Willing and Responsibilities to Protect," 119.

40 For different formulations of the "capacity principles" see Collins, "Collectives' Duties and Collectivization Duties," 240; Schwenkenbecher, "Joint Duties and Global Moral Obligations," 318.

41 The issue of whether they instead have "joint duties" that can only be fulfilled by acting together is discussed by Schwenkenbecher. See Schwenkenbecher, "Joint Moral Duties." See the discussion of joint duties in Chapter 8.

42 Virginia Held, "Can a Random Collection of Individuals be Morally Responsible?" *Journal of Philosophy* 67, no. 14 (1970): 476, https://doi.org/10.2307/2024108.

43 Held, 476. I do not discuss it here; I cover it in Chapter 8 in the context of a discussion of collective rescue duties.

44 Held, 476.

45 French, *Collective and Corporate Responsibility*.

46 See Larry May, *Sharing Responsibility* (Chicago: University of Chicago Press, 1996).

47 Larry May, *The Morality of Groups: Collective Responsibility, Group-Based Harm, and Corporate Rights* (Notre Dame, IN: University of Notre Dame Press, 1987), 22.

48 Erskine, "Coalitions of the Willing and Responsibilities to Protect," 134.

49 Ibid.

50 Ibid.

51 Note that this proposal differs from literature on responsibility of joint actors that highlights responsibility for inaction, for failures, and for non-achievements. See the arguments in May, *Sharing Responsibility*; Christopher Kutz, *Complicity: Ethics and Law for a Collective Age* (Cambridge: Cambridge University Press, 2000).

52 Erskine, "Coalitions of the Willing and Responsibilities to Protect," 134.

53 This proposal resonates with a subclass of groups introduced as such by Tracy Isaacs: "Goal-oriented collectives are more loosely structured than organizations and arise out of shared understandings and a sense of common purpose." Tracy Isaacs, "Collective Responsibility and Collective Obligation," *Midwest Studies in Philosophy* 38, no. 1 (September 2014): 42, https://doi.org/10.1111/misp.12015. She also notes that blame- and praiseworthiness needs to be distinguished from duties and obligations. Specifically, she claims "that to be prospectively responsible is simply to have a responsibility to perform some action in the future" (Isaacs, 43). This type of responsibility is at the heart of climate duties according to the interpretation given in this study.

54 Note that this proposal does not yet include the justification of concrete duties. Regarding the "sources of duties," recurrent themes in the climate debate are the correlation between responsibilities and the duty to either make good on previous flawed behaviour in terms of remedy for a situation. See Collins, "Distributing States' Duties," 346. Another theme is prior-committed "joint wrongs." See Stephanie Collins, "Filling Collective Duty Gaps," *Journal of Philosophy* 114, no. 11 (2017): 580ff, https://doi.org/10.5840/jphil20171141141.

55 For a fundamental critique, see Weisbach's points in Gardiner and Weisbach, *Debating Climate Ethics*, 203–244.

56 Gardiner explicitly supports the view that a moderate eco-centric view is right. In outlining the basic ecological storm as a new element in the interpretation of the "perfect moral storm," he states: "A third view is straightforwardly non-anthropocentric: at least some nonhuman entities, relationships, and processes have value beyond their contribution to human projects. Such positions are often treated as marginal, as well as too controversial to be taken seriously in both public policy and mainstream political philosophy. Still, I have come to believe that this is a mistake. In fact, most of us are nonanthropocentrists, to at least some degree" (Gardiner and Weisbach, 34).

57 Baylor Johnson, "Ethical Obligations in a Tragedy of the Commons," *Environmental Values* 12, no. 3 (2003): 271–287.

58 Ibid., 272.

59 In arguing in a similar way, Johnson overlooks that it is difficult to parallel "sustainability" with other morally binding goals. I shall discuss this point later in more detail.

60 Johnson, "Ethical Obligations in a Tragedy of the Commons," 272.

61 Ibid., 277.

62 Ibid., 272.

63 More on the logic of defection will be said in Chapter 10.

64 Hourdequin objects to this reasoning and states that the approach to moral duties fails to take into account the virtue of integrity. Hourdequin, "Climate, Collective Action and Individual Ethical Obligation," 448. Moreover, she rejects an approach to duties that regards the individual as primary bearer of duty; instead, she argues that the self needs to be defined in a relational way (Hourdequin,

"Climate, Collective Action and Individual Ethical Obligation," 453). In my view, this argument falls short of addressing relational duties. Hourdequin conflates individual duties that individuals bear with collective duties that collective actors bear. Moreover, she needs to distinguish more thoroughly between unilateral action and individual action, as does Johnson, too, with a self-critical undertone regarding his own arguments in Baylor Johnson, "The Possibility of a Joint Communiqué: My Response to Hourdequin," *Environmental Values* 20, no. 2 (2011): 148. Even after having introduced this distinction, Johnson still holds his opinion that, whereas unilateral action, if not in a realistic assessment ever successful in abating climate change, cannot be claimed, individual action, in particular individual action that contributes to cooperation and collective solutions, can be claimed. Johnson, "The Possibility of a Joint Communiqué."

65 See also Johnson, "The Possibility of a Joint Communiqué."

66 An elaborate approach on joint duties is also presented in Cripps. Unlike the theoretical steps proposed in this chapter, Cripps chooses as point of departure a group that is not aware of being a group; she then proposes should-be groups in cases comparable to rescue scenarios. Cripps specifically focuses on the notion of harm and the duties of beneficence. See Cripps, 59–81.

67 Bernward Gesang, "Gibt es politische Pflichten zum individuellen Klimaschutz?," in *Klimagerechtigkeit und Klimaethik*, ed. Angela Kallhoff (Berlin/Boston: De Gruyter, 2015), 135–142.

68 Ibid.

69 See Jared M. Diamond and Christopher Murney, *Collapse: How Societies Choose to Fail or Succeed* (New York: Penguin Audio, 2004).

70 Robert H. Myers, "Cooperating to Promote the Good," *Analyse & Kritik* 33, no. 1 (2011): 123–139.

71 Myers argues that "acting morally is a substantive way of engaging with other people in the promotion of value" (Myers, 124). Specifically, his claim is restricted to a situation "where the concept of cooperation denotes not a formal ideal to be given content through reasoning but a substantive way of engaging with others" (Myers, 124). The arguments for supporting cooperation do not result from utilitarian approaches to the best outcome, but from the value of engagement with the other person for a promoted value. Since Myers is interested in a general moral theory of cooperative obligations, he makes a much wider case than my study intends to. He introduces the figure of an ideal judge who is in a situation to judge the fair shares of individuals (Myers, 134). In that respect, his approach comes close to virtue ethics (ibid.). Another difference between Myers' proposal and the one of this study results from the fact that he works with a general notion of cooperation, whereas I have explained cooperation in the context of a model of joint action. Due to a loose notion of cooperation, Myers needs to explore the distinction between agent-relative reasons to promote the good and agent-neutral reasons to establish a foundation (Myers, 136–138).

72 Richard J. Arneson, "The Principle of Fairness and Free-Rider Problems," *Ethics* 92 (July 1982): 616–633.

73 Ibid., 619.

74 Ibid., 623.

75 Ibid., 621.

76 Ibid.

77 Arneson.

78 Note that the arguments in this section differ from the argument for complicity in cooperative schemes as put forward by Kutz in *Complicity*. Kutz's proposal is particularly helpful in a context in which structural harm is caused by a group of people. This proposal is discussed in the context of environmental claims in terms of a duty to collectivize in order to fill a "collective duty gap" (Collins,

"Filling Collective Duty Gaps"). This gap exists when "a group caused (or will cause) harm that requires remedying but no member did harm serious enough to impose a remedial duty on them" (Collins, "Filling Collective Duty Gaps," 574). Joint wrongs are events that harm other people and that are generated by many people together. Following Kutz's complicity principle, people have a duty not to contribute to joint harm. This presupposes that every single person is to some degree responsible for that harm: "I am accountable for what others do when I intentionally participate in the wrong they do or the harm they cause ... I am accountable for the harm or wrong we do together, independently of the actual difference we make" (Kutz, 122).

79 Note that this argument also differs from Larry May's argument about collective responsibility. His main claim is that incidents of "collective inaction," when causing dramatic outcomes, need to be addressed as moral shortcomings, even when undertaken by spontaneous groups. For the full argument, see May, *Sharing Responsibility*, 105–124. In my understanding of climate duties, individuals as well as groups are positively responsible for bringing about certain goals.

80 These duties correspond to one type of co-agency as argued in Section 6.10.

81 For Caney on "second-order responsibility," see Caney, "Two Kinds of Climate Justice," 141.

82 Schwenkenbecher, "Joint Moral Duties," 72.

83 Wringe, "Global Obligations and the Agency Objection," 227.

84 Anton Leist, "Klima auf Gegenseitigkeit," *Jahrbuch für Wissenschaft und Ethik* 16, no. 1 (2012): 159–178.

85 Garrett Cullity, "Asking Too Much," *The Monist* 86, no. 3 (2003): 413. Cullity defends the view that it is "asking too much" to save lives even by means of small spending, since this is an overly altruistic demand. Instead, he proposes to support agencies so that they can save a life, yet this is also under restricted conditions. Specifically, the spending needs to be balanced against alternatives. For the intricate argumentation, see Cullity, 415.

86 A principle that helps to determine a fair sacrifice is possibly the proportionality criterion. Unlike a utilitarian calculus, it does not balance outcomes and costs, but argues that instruments that support a major good need to be balanced against the urgency and importance of the goal. For an explication of this principle in the context of war ethics as a non-utilitarian principle, see Thomas Hurka, "Proportionality in the Morality of War," *Philosophy and Public Affairs* 33, no. 1 (2005): 34–66.

87 As for the reduction of this question to the freedoms of consumers, I have argued that the freedom of consumers is indeed limited. See Angela Kallhoff, "The Normative Limits of Consumer Citizenship," *Journal of Agricultural and Environmental Ethics* 29, no. 1 (2016): 23–34.

88 For the distinction between retrospective responsibility and prospective responsibility in the climate debate, see Schwenkenbecher, "Joint Moral Duties," 311ff. Unlike Schwenkenbecher, I do not think that it suffices to highlight the importance and urgency of climate goals; instead, these goals are justified as moral goals. This is the reason why individuals have to contribute their fair share to the realization of these goals.

8   Climate duties as joint action duties

Bert: I do now understand that it was good to work through another chapter. There is not just one single argument to work in favour of climate goals – there are many different ways to address obligations. I'm particularly convinced by the idea of rescue duties. What do you think?

Cindy: I like the idea that nobody really should suffer climate harms. But honestly, both approaches are not very helpful to really detect what each single person has to do.

Bert: You're right. And it's also once again kind of an accusation to address harm and victimhood. Even though both are important, it's not particularly well suited to motivate people to act, right?

Cindy: Yes, I think so. The alternative are duties to cooperate in favour of climate goals. I also like the argument: assuming climate goals can be achieved by a concerted effort, and assuming joint climate action is the best suited model to explain it, don't you also think that there are also obligations to cooperate?

Bert: Yes, but the argument is not that people should cooperate because it works. Instead, the argument is that in a scenario in which each person depends on the achievement of the goal and the goal cannot be achieved unless concerted action takes place, the one who spoils that cooperative option is doing a moral wrong. Got it?

Cindy: Yes, you're right. That is the argument. But it doesn't work for each possible goal. I think it's important to understand that the achievement of climate goals really is life-saving. And it also depends on the framing of climate goals. As presented in the earlier chapters, climate goals are very urgent goals, because failures to achieve are so costly and so detrimental, perhaps even life-threatening.

Bert: You're right. Is the debate now finished?

Cindy: Sorry, I think there is something more to clarify. Even when it's reasonable to claim that each single person has climate duties, two more loopholes need to be closed. The first loophole is something like shifting the burdens to someone else. And people do so for several reasons. One could argue that in cases in which single persons cannot succeed, the conclusion is not that all persons have to work together; instead, one could argue that no one has a duty to act. This is debated as the "collective rescue case." And there are many more problems like that. Philosophers call them "duty gaps." They think that a duty has to be rendered very concrete in order to be allocated to single persons. But in cases of collective action, you can never be sure what other persons do. And unless you're sure that others do their part, it is unfair to claim that you should do it. Got it?

Bert: Wow, difficult.

Cindy: Not difficult enough for philosophers. They not only discuss how to allocate duties within groups of people who together could fulfil a rescue duty, they also ask whether the world community has a duty to act.

Bert: And what do you think? Does the world community have to act?

Cindy: I think the answer depends pretty much on what you mean by "duties of the world community." I'm not convinced that the world community is able to act. I have never experienced the world community to really act.

Bert: Right.

Cindy: I think the question is wrong. It shouldn't be: does the world community have to act in favour of climate goals? The better question is: who is obliged to do what in order to achieve climate goals? The theory we've learned so far gives a good background for thinking about collective action as goal-driven and concerted action. I think we could possibly skip this chapter and look into the problem of allocation of burdens – what do you think?

Bert: Good idea.

Climate duties can be established under specified circumstances. This is the most important outcome of the preceding chapter. Such duties, however, are tied to a specific scenario, including the following: individuals are necessarily involved in a scheme of appropriation of a shared natural resource. Their choice is restricted to either cooperating with a proposed fair scheme of appropriation or defecting. When defection has severe, possibly even disastrous, effects with respect to the overall scheme of cooperation, each individual is obliged to cooperate. As for climate duties, the scheme of appropriation is either one of eco-services performed by a stable climate system, or it can be applied to the utilization of the atmosphere as a waste dump. Whereas the latter has been at the centre of debates on distributive climate justice, the focus in this study is the former.

It is important to note that this proposal for climate duties deviates significantly from previous arguments in climate ethics in asserting that single actors are obliged to support climate goals. Moral arguments for duties to either protect climate victims or even to rescue them prevail. In those interpretations, moral arguments in favour of climate duties do not rest exclusively on the presumption that climate victims suffer severe harm from climate change; instead, they identify a moral failure not to support goals that are framed as ethical goals. Above all, arguments in favour of climate duties do not rest on an emergency ethics.[1] Climate duties have not been established as an answer to the imminent threat to people in general or even the world population. Rather, the arguments rest on the assumption that environmental action relates to goals people have good reasons to share. The goals resonate with an ethos that includes fairness in appropriating a resource, protection of the integrity of that resource, and conservation of the resource as a space for living beings.

Against this backdrop, the relationship between climate goals and joint action has to be discussed once again. The argument that climate duties are linked to individuals does not undermine the claim that climate duties are *also* linked to collective actors. On the contrary: supposing individuals have already a duty to act in favour of climate duties, would it not also be reasonable to explore the duties of actors who are stronger than individuals?

This chapter brings the discussion back to joint action and thereby also to the baseline of the discussion of climate change in this book. The inquiry focuses on two questions: can joint actors be held responsible for the achievement of climate goals? And, if so, what are the theoretical prerequisites to support this claim? Overall, at least two theoretical steps need to be added to the former analysis: a re-examination of rescue scenarios in the context of joint action; and the claim that, besides climate duties of individuals and groups of people, the duty to collectivization also needs to be taken into account.

The chapter is organized in eight sections. Section 8.1 surveys the theoretical landscape with respect to joint climate duties. Important distinctions are noted concerning the category of "actors" as well as the issue of whether the debate starts with previously constituted joint actors or with individuals who collectivize. Section 8.2 summarizes the interpretation of joint actors as unstructured collectives. Section 8.3 discusses rescue scenarios as situations in which not an individual, but a group of people, can accomplish a rescue action. One particularly important observation stems once again from rescue scenarios. Assuming individuals can rescue a victim and that the way to achieve this is the spontaneous formation of a group of helpers, each individual can then also be held responsible for acting. This then creates another problem, as explained in Section 8.4. Since the accomplishment of the duty to rescue depends on the success of many, it has to be asked whether what is at stake is the duty of the group or the duty of individuals as members of the group. After having discussed the issue of a specific allocation of duties, another set of problems will be raised. Section 8.5 is dedicated to "duties of collectivization." It explores the view that individuals are obliged to form and to establish collectives, as long as those collectives can achieve rescue goals. Although this proposal helps to settle the issue of duty-holders, it does not help to fill each type of duty gap. Section 8.6 discusses remaining duty gaps. Overall, the sections help to support the claim that duties to collectivize cannot be defined narrowly. Section 8.7 raises an issue that philosophers have also posed in the context of collectivization duties. Can the world population be held responsible as a collective actor in order to bring about climate goals? Section 8.8 highlights important outcomes of this chapter.

8.1 The theoretical landscape

Groups that need to be considered in the climate debate fulfil three claims.[2] To begin with, groups exist of single actors who together form a group in order to bring about a joint goal. There is no need to claim that this joint actor also has the properties of single actors as independent entities. There is, for example, no memory of that entity besides the memory of the single actors involved in the group; there is also no joint intention besides the many intentions of the various actors. Furthermore, group action is based on a group ethos that binds the group members together and gives them reasons to act in favour of a shared goal. Individuals are able to act in favour of a shared goal that is willed as something that will be brought into existence together. Individuals who do so, are taking a "We-stance," as opposed to the "I-stance," which is their usual mode of action. The glue holding the group together and providing the shift to that "We" is a shared ethos.

Finally, reaching a shared goal is a process that includes further sub-goals. As for environmental action, an example of this fact is the existence of accumulative goals that are realized by means of the addition of partial contributions. Another example is transformation goals that are sub-goals of the overarching, integrated developmental goals. Bratman also explains that people not only come to work together; they are also involved in a process of mutual response to sub-goals of other participants in the joint process of achieving the goal together.[3] It is also important to assume that the people involved in joint action are not only aware of the shared goal; instead, they know that it can only be brought into existence as a joint goal.

In order to discuss climate duties possibly linked to groups that in turn qualify as joint actors, it is necessary to distinguish various elements. The claim that unstructured groups have duties is intricate and several questions need to be distinguished. The first one is whether joint actors are capable at all of bearing responsibility. Although it is not difficult to demonstrate that structured collectives qualify as actors and that responsibilities belong to those actors, the groups that are now discussed are usually *unstructured* and possibly also won't last very long. The individuals who form an unstructured collective might also be goal-oriented; still, participation can be random. Notably, there is no decision-making structure that has been implanted by the group for the group.[4]

The second question is how duties are linked to an unstructured collective.[5] Is it right to claim that every single actor has a duty as a member of a group bearing a duty to act? Or is it the group that has responsibility, notwithstanding the problem that it might happen that duties somehow can't be pinned down anywhere? This is the question of how duties can best be allocated on a group level – assuming random joint actors have duties to act.

Third, philosophers have also asked whether there is any duty of individuals or of existing groups to collectivize.[6] This question has two components. It once again addresses the conditions that need to be met in order to hold a group responsible for an action. It also includes the question of under which conditions a duty to collectivize can be justified. It has been claimed that – under certain circumstances – people can be asked not only to join existing groups, but also to actually form groups. One of these particular cases appears to be the case of climate change.[7] A background assumption of many publications in this field of research is that climate goals can *only* be achieved by means of joint action. Whether or not this is a good argument in favour of collective duties must be examined.[8] We have to discuss whether this is the right conclusion.

Fourth, another issue also has to be raised in the context of obligations of joint actors regarding climate goals. We have to discuss whether individuals are obliged to prioritize joint action in favour of climate goals over private endeavours.

In addition to this already complex picture, I also intend to raise another issue that has – to some degree – been unduly neglected in the debate on joint climate duties. This is the question of whether pre-existing groups should be obliged to take responsibility for climate goals. Imagine a skating group that meets each Friday in the park; let's assume this group is composed of people who sometimes show up once in a while, and others who join the event more regularly. Moreover, there is no decision-making structure at all when it comes to conflicts. It really is an unstructured group. The question that a discussion of climate duties includes is: should this group be obligated to support climate goals? Instead of claiming collectivization of individuals as a means of achieving demanding goals together, the question of whether groups that already exist in one form or another should now act responsibly regarding climate goals also needs to be raised. I call this issue the question of the "environmentalization" of groups.

8.2 Unstructured groups as joint actors and emergency cases

In order to outline the theoretical prerequisites for a discussion of joint duties in the case of climate change, I want to start with an example on a small scale. Imagine the following: citizens in a city are convinced that greening their city is an important goal for coping with climate change. It will help to cool the city in summer. Growing more plants in the city will also help to support the water cycles. In a meeting at city hall, the mayor claims that each citizen is obliged to contribute to the goal of "greening the city." However, some people question whether these obligations really are justified. Primarily, they think that groups of people cannot be held responsible for acting on environmental goals, but institutions such as the government of the city need to shoulder responsibility. The mayor suggests instead that individuals should take responsibility and start greening the city. He says that each individual is obliged either to contribute to greening the city on her own behalf or to ask neighbours and others to plant trees. They should collectivize in order to fulfil their duties regarding climate goals.

This example includes some elements of the debate on single duties versus duties of collectives with respect to environmental goals. In discussing obligations to act in favour of a joint goal, the question of whether groups can also be held responsible for acting in favour of legitimate goals is an important question. I shall outline some key questions first and then discuss the claim in more detail. The first question with regard to whether random collectives can be held responsible for acting in favour of a joint goal is whether those groups qualify as actors in the first place. Note that this discussion is not about *group agents* who already qualify as actors and that have – as many believe – the capacity to act.[9] Instead, the discussion is about obligations that may appear to be justified, but whose allocation is unclear. Notably, the duties relate to goals individuals cannot achieve on their own.

Assuming individuals who do things together form unstructured or random groups, this group still differs from structured groups in that they do not have an established decision-making procedure, nor do they have an identity that is independent of the identity of every single member. Above all, a joint actor exists not only when several people pursue a goal that each individual subscribes to as a shared goal; instead, each individual must also be convinced that other people also subscribe to the very same goal. Whether this includes the conviction that others will also contribute their share towards the joint goal is another issue to discuss.

The scenario of acting together as an *unstructured collective* has been framed in a variety of ways. In some ways, the collectives that this discussion addresses can be compared to "goal-oriented collectives." For Tracy Isaacs, they share the following features:

> Goal-oriented collectives are more loosely structured than organizations and arise out of shared understandings and a sense of common purpose. They also have collective intentions, that is, intentions which are distinct from the intentions of individual members and which propel the collective's actions towards the achievement of its goals.[10]

Principally, as unstructured collectives, goal-oriented collectives are also distinguished from social groups. Social groups consist of people who "are grouped together because of some feature that they share or are assumed to share."[11] At a minimum, the people who are involved in the joint action of an *unstructured collective* as discussed in the debate on climate duties do not contribute to a goal ignorantly or accidentally. Instead, the joint action is from the viewpoint of individuals as "something they team up to perform."[12]

In order to support the claim that such an actor exists, it is necessary to spell out the criteria that render this group an actor. This cannot be done without also going over the minimum requirements for qualifying an event as an action. As for a group of people meeting accidentally in a café and deciding to pledge money together for "greening the city," it does not suffice that all the individuals in the café pledge money. Instead, they undertake a joint action when they pledge the money together intentionally. Notably, they do not just put money together, but do so in order to achieve a shared goal together. An example that is much easier to interpret as an unstructured collective undertaking a joint action is the situation where green activists meet and talk about goal-oriented activities. They decide to undertake activities to contribute to "greening the city." This is a goal-oriented collective, and they want to take responsibility.[13]

In cases in which unstructured collectives act in a goal-directed way, the question of whether they really can be held responsible as actors for a certain act is based on two principles. In order to be able not only to act, but also to qualify as an actor who can bear responsibility, the minimum requirements are the fulfilment of a *capacity principle* and an *agency principle*.[14] Put briefly, the agency principle claims that "only agents can hold moral duties"; the capacity principles states: "An agent can only hold a moral duty if the agent is capable of discharging that duty."[15] The latter has already been suggested for groups in the discussion of joint action. It defines the minimum criteria for a collective to count as an actor.[16] As for the capacity principle, much of the discussion in the remainder of this chapter is about the relationship between the capacity to fulfil a goal and the obligation to fulfil the goal. But an oversimplified view of that principle will be explained and made clear right at the beginning. The assumption that individuals can only reach a goal together, and that they really can bring the goal about, is not a sufficient argument for concluding that they also have the duty to bring this goal about together.

8.3 The collective rescue case

Let's assume that a group exists that is capable of acting in favour of a joint goal. The question that can now be raised is whether such a group can be held morally responsible for fulfilling an obligation related to a distinct goal. An example supporting the view that such a duty exists has been discussed by Anne Schwenkenbecher.[17] Although parts of Schwenkenbecher's discussion were mentioned in the previous chapter (Section 7.6), her argument must now be presented in full.

She chooses a scenario in which a spontaneous group of random individuals who pass by jointly assist the victim of a traffic accident. This is a rescue scenario. It resonates with an idea that Peter Singer spells out when debating rescue cases as an explication of two characteristics of the case of climate change: we have climate victims who suffer life-threatening harm, and we have rescuers who could by their actions rescue the climate victims.[18] Notwithstanding the causal link to the existing harm, the moral obligation results from the ability of rescuers to save a life and the duty to assist victims in a situation of severe harm. Unlike Singer's discussion, rescue cases are also being debated as a situation in which not an individual, but several individuals, can by joint effort rescue a victim who is about to suffer severe harm.

Schwenkenbecher discusses an example in order to illustrate the claim that unstructured collectives can be held responsible for acting together in order to rescue a person:

> [A] motorcyclist is trapped underneath a burning car after a collision. Passersby approach the car realizing that in order to retrieve him, the car must be lifted. After a couple of unsuccessful attempts with insufficient people, eight people eventually manage to jointly lift the car off the injured man, saving his life. Would we say that the passersby were under an obligation to team up and lift the car to save him?[19]

Schwenkenbecher maintains that this obligation holds and that people are obliged to rescue the victim together. In order to discuss this scenario in moral theory, Schwenkenbecher relies on a presumption suggested by Larry May. He states that in a scenario in which shared responsibility of group action is given,[20] collective inaction can be as wrong as intentional wrong-doing.[21] Furthermore, it is important to note that duties are restricted by an interpretation of the "capacity principle" that states: "An agent can only hold a moral duty if the agent is capable of discharging that duty."[22] In the rescue scenario, single individuals are free either to join the group and together rescue the victim or to discharge the duty by not acting together.

In the philosophical debate on duties in the collective-rescue case, two interpretations need to be distinguished. The first is an *individualistic interpretation* of the moral duties involved. An individualistic interpretation states that each individual has the duty to perform a single action in order to save the victim. According to this reading,

> the individual members of the group (passersby or bystanders) hold no duty to ... lift the car. Such a duty simply does not exist because there is no agent who could hold that duty given that there is no agent who could discharge the duty. The individuals, however, have a moral duty to perform individual actions toward a joint goal.[23]

This argument is based on the principle that "ought" needs to imply "can" if it is to work as a sound moral argument. Since no member of the potential group can solve the problem on her own, she does not have a duty to solve the problem. According to the capacity principle, each person can be asked to fulfil her share of whatever is needed to resolve the problem. The group of people can resolve the problem; if unstructured collectives can be held responsible for their action, the consequence is the moral claim that it should rescue the victim.

As applied to the case of climate change, it can be claimed that supposing people can achieve the goal of rescuing a single person who would die from the effects of climate change by joining efforts, each individual is also obliged to do so. Obviously, complications arise from the fact that we do not know with certainty whether a climate victim can be rescued by group action. By readjusting the goals – for instance, by introducing transformation goals that, once fulfilled, will suffice to reach an integrated development goal – the causal chain between acts of group actors and the fulfilment of goals can be re-established. Assuming the achievements – in terms of integrated development goals – do rescue people from severe harm, a case comparable to the car-lifting example can be suggested. The reason for establishing obligations in rescue cases results not only from the capacity to help a person, but to really rescue her. In addition, individuals have duties, not to fully reach the goal, but to contribute their share in order to help bring about the goal. Still, there is a different interpretation of the unstructured-collective case of joint agency.

In a *non-individualistic interpretation*, obligations are allocated in a different manner. In addition to single actions, individuals also can and ought to perform a joint action as an unstructured collective. When they willingly and knowingly assist in lifting the car, they fulfil the duty of the group actor to rescue the victim. According to this interpretation, they also jointly have a duty to lift the car in order to save the person.[24] The duty is allocated to a group of people who together fulfil the claim to rescue the person from severe harm, possibly even death.

This distinction between an individualistic and a non-individualistic interpretation of the joint rescue case is not just an interesting philosophical issue, but also has consequences. According to the first interpretation, individuals should obey the duty to contribute to a rescue scenario. According to the second interpretation, the randomly generated group has the duty to lift the car. Individuals have duties only in a derivative sense. Schwenkenbecher maintains that the first interpretation might fit better with moral intuitions in comparable scenarios.[25] This becomes apparent in the negative case. Suppose one person defects, and in the end the person under the car really dies. We are likely to accuse the single bystander not only of not having done her share in rescuing the individual, but also for being to some degree responsible for the death of the victim of the accident.[26] One the other hand, the non-individualistic interpretation might also create, once again, the no-one-is-really-responsible problem.[27] Without allocating a duty correctly, it is difficult to allocate it at all.

Another reservation also needs to be addressed. Both interpretations appear to be restricted to an immediate, clear-cut emergency case. Although the second version might be easier to understand – Wringe calls it the "primitive obligation to cooperate account"[28] – its plausibility depends in a particular way on the rescue situation. In the description of the event, individuals are already marked as bystanders. Only as bystanders are they able to bring about visible and immediate change. And they are not asked to do anything particularly costly or exceptional.[29] Whether acting with other individuals or not, they just fulfil a duty that is also inscribed in many legal frameworks: *not* assisting a person in life-threatening situations is the *failure* to assist a person in danger. Non-assistance in comparable cases is not only morally wrong, but can also be punished. In short, the issue of whether individuals act together is not the main problem here. Instead, the more pressing question is whether the interpretation of rescue cases can rely on moral exceptions that are accepted as such. Principally, a non-individualistic interpretation works against the generally accepted claim that failure to assist a person in danger is morally wrong.

Two additional problems need to be discussed. The interpretation of a scenario as a rescue scenario depends on a moral framework that needs to be outlined in detail. Obviously, each person living in a big city walks past people who desperately need help on a daily basis. Moreover, some of them also need rescue action, even immediate action. But claiming action in favour of individuals in desperate situations is not what philosophers evoke when drawing on rescue scenarios. Instead, the situation is dramatic; the victim is demonstrating her need to get rescued; she screams or behaves in a wild manner; and only immediate help can bring about change. In short, rescue scenarios work with distinct intuitions about the dramaturgy of a situation in which individuals need urgent help. As for the issue of climate goals, it is debatable whether the overall situation really is sufficiently comparable to draw the same moral conclusions. Although the life-threatening nature of climate change is real and has been acknowledged, the victims are usually not close by, nor does immediate action suffice to rescue them.

In addition to this problem, the assessment of moral duties in rescue cases also depends on a defence of those goals as either themselves qualifying as moral goals or as legitimate goals in a moral framework. Without the qualifications as moral goals, it is not easy to address duties of joint actors regarding goals that do not have that stamp as part of already morally conceived scenarios. If accepted as such, rescue cases in the context of climate goals only provide a small part of the overall space of goals. Supposing it is possible to define duties of collectives for bringing about climate goals, it is possibly not enough to only discuss the most dramatic cases. Nevertheless, before coming to a conclusive assessment, further theoretical challenges and further theoretical options to address climate duties as duties of either members of collectives or of collectives themselves can be addressed.

8.4 The problem of a distinct allocation of responsibilities

The current debate on duties of joint collectives is portrayed in a more general observation by Virginia Held about joint action.[30] She is not primarily interested in acts undertaken by random collectives; instead, her focus is on a scenario in which individuals fail to act together and thereby cause severe harm to others:

> I shall confine my discussion to the nonperformance of an action, and argue that when the action called for in a given situation is obvious to the reasonable person and when the expected outcome of the action is clearly favorable, a random collection of individuals may be held responsible for not taking a collective action.[31]

Specifically, Held points to a distinct problem rescue scenarios face when addressing unstructured groups. She maintains that linking action to unstructured groups is – at least on pragmatic grounds – reasonable. In scenarios of rescue, a reduction of an omission to the failed action of individuals appears to be irritating. She mentions statements like "The corporation manufactures X" and "State W provides higher welfare payments than state Y"[32] as statements that bolster the claim that a non-reductionist approach to group action is acceptable. Since 1970, the exploration of premises that contribute to agency in cases like that has been refined in many respects. The claim that it is philosophically sound to attribute action to collectives has been formulated in detail. Still, with regard to the allocation of responsibility to collectives, many questions are still unanswered. One reason it is not that easy to ascribe responsibility to collectives is their reliance on two assumptions.

To begin with, the person who takes over a moral responsibility for an action needs to be aware of "the moral nature" of the action in question.[33] This is broadly conceived as an awareness of the moral consequences and the moral value of the act. This does not amount to strong positive conditions, "[b]ut they preclude holding a person responsible for the thoroughly unascertainable aspects of an action he performs."[34] Moreover, knowledge about the moral character of an action is still not enough, either. Instead, the person also needs to be able to decide which action to take. And this second aspect is particularly difficult to attribute to individuals who are part of a random group.[35] Whereas an organized group that has an implemented decision-making structure can allocate distinct parts of responsibility of a joint action to individuals, a random collection of people lacks this trait. When confronted with a variety of different actions, people who are all in a group of observers or bystanders do not necessarily choose distinct action, but might be unsure what to do. As a consequence, it can be stated that the organized group should have done at least one of the actions proposed; the decision procedures help figure out which one will be chosen. Held states:

> The judgment that "An organized group in these circumstances ought to have done action₁ or action₂ or action₃" can be considered valid. But which action to take was not obvious to the reasonable member of the random collection; so, again, an essential requirement for holding a

random collection of individuals responsible for the nonperformance of an action was missing.[36]

The decision on a distinct course of action is impeded for individuals as members of an unstructured collective.

Conclusions from this assessment need to be drawn with caution. Held does not claim that a random group of passersby in an emergency situation is free from responsibility. Instead, she contends that decision-making procedures are important to avoid a situation in which duties cannot be defended because of a lack of unanimity.[37] When a group of people lacks the means to define an appropriate action, there is still a failure to act. Furthermore, another duty now surfaces. This is the duty of "collectivization," which, according to Held, is the transformation of a random group into a structured group equipped with a decision-making structure. In addition, and depending on the particular scenario, there is also a duty to take action that falls on the individual. Held again:

> The judgment, then, that "Random collection R is morally responsible for not constituting itself into a group capable of deciding upon an action" is sometimes valid when it is obvious to the reasonable person that action rather than inaction by the collection is called for.[38]

The problem Held highlights has many different facets, some of which have also been reconsidered in the recent debate on joint climate duties. First, "duties to collectivize" is particularly interesting when discussing climate duties. Assuming individuals cannot achieve climate goals on their own, and assuming one of the biggest shortcomings regarding effective climate action is the non-existence of structured groups capable of achieving climate goals,[39] the question is then whether actors are obliged to form collectives that can achieve climate goals together. Possibly, this is not only a duty of every single actor, but a duty of people chosen and elected as representatives of people. But still is it important to discuss whether individuals or representatives of people are morally obliged to collectivize in order to realize climate goals.

A second issue is that, supposing there are duties to collectivize in certain cases, another problem surfaces. This is the problem of an infinite regress, as Held puts it: "Can there be responsibility for not deciding upon a method to decide upon a method to decide upon a method … to act?"[40] Even if a duty to collectivize can be established, the responsibilities for forming collectives also need to be linked reasonably. This presumes an answer to the question of who is responsible for the responsibility to form effective collectives.

Held's discussion also highlights a third problem, that of incoherent, even possibly dilemmatic structures within groups. As long as climate goals are accumulative goals, distributive principles can help to identify the bit each participant has to do to fulfil a specific duty. Besides, the overall situation is much more complicated. In that case, it is not easy to determine every single contributor's share of a joint goal. Some of the most important climate goals are accumulative goals. They cannot be achieved without the effort of many different actors; it may even be necessary to coordinate action and thereby to form a joint actor. However, as long as this group is not structured and does not possess a decision-making procedure, it is difficult to explain how the question of individual contributions can be settled.

A fourth issue is that the analysis of the responsibility of joint actors also allows for exemptions. Rescue cases are framed as being closely focused on an incident that the group of individuals who shares the experience as bystanders has to address. Still, joint action of an unstructured collective is not only related to that incident, but also reduced to that group of individuals. For climate goals specifically, it has to be clarified whether there is an argument to exempt anyone from complying with climate duties. Assuming the argument that fair schemes of appropriation of a shared natural resource need to be all-encompassing if they are to prevent defection – at least in the case of a global natural good such as the atmosphere – there is no reason why any individual should be exempt from contributing to climate goals. Here again, the question of moral obligation needs to be distinguished thoroughly from the question of whether goals can be accomplished without such a joint effort.

In order to decide whether these problems can be resolved, it is necessary to explore duties of collectivization in more detail.

8.5 Duties of collectivization

In Held's proposal, individuals cannot only fail to fulfil their prospective responsibilities, but they can also fail in a moral duty to form a group to take the demanded action. Larry May has made an argument to bolster this latter claim.[41] He affirms that a putative group – a group that could be constituted and is even likely to be constituted, but has not yet been constituted – can act collectively by forming itself as an organized group. Such groups thus not only have a distinct decision-making procedure, but also enough leadership potential to organize themselves in order to perform a collective action successfully. This means that the realistic prospect of acting as a group might depend on the transformation of an unstructured collective into an organized group. Although it might be possible for groups to establish themselves as organized groups, the more far-reaching issue is not the question of whether it is empirically possible to do so. Rather, it is whether it is legitimate to require this transformation into an organized group.

The standard scenario in which duties of collectivization have been debated has specific characteristics. First, it is presumed that there is not only some existing problem, but a problem in need of urgent resolution. Climate change figures as such a problem, because not addressing it successfully may lead to catastrophic events. Second, it is presumed that this problem can certainly not be resolved unless people act together to resolve it. The only remedy to such a dramatic event is a form of collective action. Third, according to Stephanie Collins, scenarios in which collectivization can be demanded also presume that responsive actions are insufficient:

> In some such cases [in which responsive actions are insufficient], each individual incurs a duty to perform responsive actions with a view to there being a collective which can reliably address the circumstance. These are individual duties to collectivize.[42]

In a situation like that, it has to be disputed "when and why individuals might severally bear duties to take steps towards creating a collective agent."[43] In short, if they exist at all, duties to collectivize relate to specifically dramatic scenarios, which cannot be addressed successfully by existing group agents. Instead, success depends on the proactive activity of forming a collective capable of addressing that urgent problem successfully.

Collectivization duties differ significantly both from single duties of individuals and from duties of structured or unstructured collectives. The duties of collectivization are, above all, not identical with the pre-existing duties of individuals who are about to form a collective. This is an important observation because otherwise duties of collectivization could be reduced to pre-existing duties. In fact, duties of collectivization form a distinct body of duties. They therefore also need to be established on separate grounds. Another important fact is that unless a person takes responsive steps towards forming a "collective-that-can-φ" there is no body in existence that bears a duty to undertake φ.[44] The joint duties at stake when discussing duties of collectivization are strictly counterfactual and depend on the existence of a joint actor. This joint actor is not in place when duties of collectivization are in place. All these traits call for a philosophical exploration.

In order to render the moral implications explicit, it is helpful to discuss a specific case first. It is presumed that the duty to collectivize fills a "collective duty gap."[45] This gap exists when "a group caused (or will cause) harm that requires remedying but no member did harm serious enough to impose a remedial duty on them."[46] Joint wrongs are events that harm other people and that are generated by many people together. In such cases, it appears particularly reasonable to claim collectivization in order to take over a duty that needs to be allocated correctly to the collective. Individuals are regarded as accountable for remedying harm, yet only when regarded as a group actor is the harm serious enough to justify the remedial claim. This example resonates with the complicity principle as defined by Christopher Kutz.[47] According to Kutz's complicity principle, accountability can be attributed to a collective when individuals participate intentionally in the wrong that other individuals do. He states: "I am accountable for what others do when I intentionally participate in the wrong they do or the harm they cause ... I am accountable for the harm or wrong we do together, independently of the actual difference we make."[48] In other words, a collective is held responsible for a harmful action even though individual members of that group cannot be held responsible for that action.

The proposal to fill a "collective duty gap" has some plausibility in scenarios of complicity. But the proposal also has problematic implications. These implications apply even in a positive scenario, as long as individuals cannot remedy climate change in terms of not being able to change the course of events in a way that the most significant harms resulting from accelerating climate change can be prevented. Besides, integrated climate goals that also prevent significant harm to a local population can be achieved by means of collectivization. Isn't it plausible to think that people should collectivize in order to fulfil the duty to remedy climate change, even though no single individual has this duty on her own?

Appealing as this thought might be, it nevertheless has severe short-comings. One problem is that the step between the act of forming a collective and collectively aiming to accomplish a goal presumes that a change of the individual's contribution does not have an impact on the claim. When individuals succeed in establishing a collective, this is not enough to pin down the "collective duty gap" bridging the duties of the then-existing collective and the not-yet-existing duties of the individuals. According to Schwenkenbecher, as members of the collective, the group members no longer have the power to really determine the course of action:[49]

> Group agents such as states or business corporations act in ways that some of its members do not approve of all the time. Individual members of group agents have – more often than not – very limited opportunity for determining the group's actions. It is simply not under their control what the group decides and does.[50]

Although this is an empirical problem, it also requires that the individuals acquire two duties: first, the duty to form a collective, and second, a duty to perform according to the goals of the collective. As for duties that climate action might imply, Schwenkenbecher concludes:

> The more plausible argument in favour of global moral obligations would be one which merely establishes a joint duty for individuals to form a collective agent with a view to remedying the situation and a subsequent duty for the group agent to address the dire situation.[51]

This proposal differs from the proposal to fill a collective duty gap in which it already includes a duty of individuals to form a collective with the distinct goal of fulfilling climate duties as a collective. This proposal can also be reduced to the proposal to conceive of climate duties as non-unilateral, as discussed in Chapter 7.

Schwenkenbecher offers a clear analysis of the complex scenario. However, it still rests on the assumption that collectivization is demanded as a remedy to a scenario with a nearly catastrophic outcome. Presuming the world community has a duty to achieve certain environmental goals, and presuming this duty results from the implications of these goals as the only possible means of providing fair and forward-looking solutions to a situation in which some benefit from acts that harm and destroy life on earth, whereas others suffer from the effects of this destructive behaviour, which is defection on an otherwise reasonably fair scheme of appropriation, does this overall situation suffice in order to maintain that people should collectivize in order to fulfil their individual, but not unilateral, duty to work together towards environmental goals?

In my view, there are more assumptions at work that are not declared as such. It is not only assumed that those goals really answer legitimate claims of justice. Moreover, it is also presumed that overarching goals, such as "greening the city," "sustainable practices of water appropriation," and "protection of wild forests" can *only* and *exclusively* be achieved by group actors who do not yet exist. The arguments in favour of collectivization do not take into account the plausible claim that many of the joint duties can be fulfilled by already-existing collectives. Cities do have city governments – would it not be fair to ask them first to perform action in favour of green cities? In short, the collectivization debate overlooks an obvious case, which is a case in which already-existing collectives are asked to fulfil their duties with respect to environmental goals.

In another respect, however, the application to the case of climate change is straightforward. Suppose individuals cannot remedy a dramatic situation in favour of rescuing a victim, but a group of people could. It has been maintained that in situations in which groups could remedy the situation, it is obligatory to form a group: "The idea is that in situations which require remedy by a group agent, individuals in putative groups acquire moral duties to form a group agent which then can remedy the situation in question."[52]

Yet at that point the discourse shifts in a different direction. This discussion is not about duties to collectivize in order to fulfil duties that only groups can hold. Instead, this is a discussion about collectives and the institutions guaranteeing the continued existence of groups that, once established and continued, could be successful in achieving climate goals. This is a discussion about structures and organizations that are needed but have not been created as yet. This debate is important for the case of climate change. In my view, this latter twist is useful and right for discussing the role of individuals and of institutions anew. I take up these issues in Chapter 10. It is different from the question of whether individuals should build up groups and "collectivize" in order to be in a position to achieve environmental goals. Before discussing that issue, however, more philosophical problems with duties to collectivize need to be addressed.

8.6 Proposals to fill duty gaps

In the model of joint action outlined in this book, climate goals can be broken down into various types. Obviously, not all actors can fulfil every type of aim. When focusing on duties, the problems no longer relate to empirical facts. Instead, the underlying moral assumptions are at the centre of concern. In general, duties are not linked to individuals arbitrarily. Instead, duties are legitimized by means of moral argument, which in turn highlights the need to give moral reasons for certain claims. Specifically, the principles of agency and capacity – which explain why actors are classified only as such when the assumption of these entities really acting and being able to respond successfully to claims of action can be made – do not suffice. Above all, moral arguments are needed in order to allocate duties of various types.

In the context of collective duties, one problem that figures as an additional problem has distinct importance for climate change. This is the problem of *slack-taking duties*, which are duties to take on others' shares of collective burdens when those others fail to comply by taking on their share.[53] This problem is particularly severe when addressing collective duties that either do not have a concise decision-making structure or that lack institutions that could enforce distinct duties. Stephanie Collins discusses a case in which

> global leaders in the 1980s unfairly reneged on an emission-reducing duty, which was owed, at least in part, to present-day polluters (because it would make present-day polluters better off by removing the duty-gap conundrum). These non-compliers have unfairly reneged on (a) their duty to fellow actors. When and why should this intra-group unfairness have any bearing at all on (b) what those present-day actors owe third parties?[54]

She then argues convincingly that three aspects need to be distinguished in answering these questions, namely "the severity of the intra-group unfairness, as against the severity of the harm to third parties, as against the cost of remedying that harm."[55] This problem needs to be addressed. But actually, it consists of a range of different problems, each of which needs a thorough investigation. The next paragraph tries to address this problem with respect to insights into a theory of joint environmental action.

With regard to collective action, it is part of the project of this book not to put too much emphasis on compensatory justice, nor to analyse climate change in terms of reactive responsibility. Instead, the proposed frame starts with the current situation and defines future goals. The single duties do of course vary. However, they do not have to fill gaps, but instead the scenario is modelled in such a way that duties and types of duties change in relation to both the type of groups and actors and the role of individuals in these groups. Moreover, when it comes to accumulative goals, duties are best allocated to single actors and to groups of actors by means of principles of fairness, as outlined in the next chapter. Slack-taking duties can be avoided or at least reduced when schemes of appropriation are defined in such a way that single duties are allocated accordingly. Instead of slack-taking, the failures of doing one's fair share can then be classified as incidents of non-compliance with a well-reasoned set of duties to fulfil a distinct share in the collective undertaking. Actors who do not act are defectors, and defection is wrong.

Another problem that surfaces at several instances when addressing joint climate duties is the proper allocation of duties. In cases in which individuals have duties as members of groups, discharging these duties does not exclusively depend on the individual's activities. Instead, in collective action scenarios, discharging a single duty is conditional upon the actions of other group members, especially in unstructured groups.[56] Even worse, the individual's contribution is also dependent on what she believes regarding the future action of other group members.

One aspect of this problem thus relates to a standard problem in public-good scenarios. The outcome of the studies shows that there is this sort of dependence. Besides, the empirical findings also highlight a positive side. As long as people can be convinced that their contribution is critical to the sustenance or the establishment of a public good, whose enjoyment benefits everyone, the willingness to contribute will grow.[57] If the overall goal is a threshold goal, this effect is even greater. As for climate goals, the fact that dependence on the actions of other actors is a fact does not necessarily lead to individual non-compliance.

A second severe problem can possibly even be remedied. Goodin[58] and Lawford-Smith[59] both point out that mutual release is another problem. The more individual contributors to a joint goal are convinced that other possible contributors will not discharge their contribution, the more likely is the outcome that none of them follows through on their commitments. Everyone exempts herself from the duty because the duty is conditional, and the condition does not hold. Goodin also argues against this circle of reasoning; each person could instead say to the other: "I will if you will and I will (if you will if I will)."[60] Although this is a smart solution, for climate goals this problem needs to be framed in a different way. If framed properly, climate duties are not conditional. Conditionality results either from an unrealistic framing of climate goals – this causes the effectiveness condition regulating individuals, but not unilateral duties, not to hold – or the concept of defection is not taken seriously. Once the question is settled, whether people are really involved in a moral way, compliance with appropriation schemes and defiance can also be established accordingly.

In addition, it has to be said that many of the problems discussed so far do not result from mutual reactiveness in cooperative schemes. Instead, they also result from the fact that in unstructured collectives, decision-making procedures are lacking or imperfect. One way to think about alternatives also draws on another type of collectivization. For a rescue case, instead of discussing spontaneous groups that form by means of mutual engagement, it is much more reasonable to claim that (a) individual, but not unilateral, duties to contribute to climate goals exist – notwithstanding the fact that some people might defect – and (b) that climate goals need to be broken down to various types of goals, some of them definitively part of the responsibility that already-existing collective actors have. The next chapter is dedicated to taking a closer look at this claim. Before getting there, I address the issue of whether the world population has a duty to act on climate goals – and what this claim includes.

8.7 Duties of the world community?

The discussion of this chapter has bracketed one more peculiarity of joint duties. People can have duties together without being or forming an agent. Supposing this is right, it is then also reasonable to ask whether the world population, which is all living people together, holds climate duties. According to a proposal made by Bill Wringe, it is helpful to distinguish between an addressee of moral duties and the subject of a moral duty.[61] With regard to climate goals, it is reasonable that the world's population is the addressee of global moral obligations, whereas the subject of those obligations is groups or responsible agents who should care about climate goals. According to Wringe, the need to address the world community rests on certain premises:

1   There is a situation in need of remedy,
2   That situation can only be remedied by a collective agent, and
3   There is no collective agent to remedy that situation, and
4   There is a collective that is not an agent.

Then the obligation to remedy the situation falls on a collective which is not an agent, and the individual members of that (unstructured) collective acquire obligations to see to it that the obligation to remedy the situation is discharged.[62]

This proposal resonates with the insight that the world population needs to take action in order to prevent catastrophic climate outcomes. However, one central problem of the proposal to define the world population as a duty addressee in the case of climate change has already been analysed thoroughly by Schwenkenbecher. Wringe's proposal relies not only on the assumption that "ought" implies "can," but also necessarily presumes the possibility that "ought" in some cases also implies "someone else can."[63] Indeed, Wringe asserts that

there is nothing unusual about situations in which the addressee of an obligation acquires an obligation to do something which does not have the same content as the individual on whom the obligation falls (when the addressee and the individual on whom the obligation falls are different).[64]

The example Wringe chooses is one of children in a theatre who should be quiet, but whose parents ought to act on the obligation. Schwenkenbecher criticizes this proposal: "While I can have a moral duty to do X, I simply cannot have a moral duty that someone else do X."[65] Schwenkenbecher also thinks that Wringe conflates two different duties. This is the duty to remedy a morally problematic situation as opposed to group members' duty to contribute to achieving that goal.[66] It is not possible to (a) allow for a situation in which the subject of a duty and an addressee fall apart, which also has the consequence that two different duties are at stake, and (b) simultaneously treat both duties as if they were the same. Actually, the first duty is the duty to remedy a problematic situation, whereas the second duty is to make a contribution in order to realize the goal.[67]

In sum, this proposal gives further distinctions with respect to climate duties, but it falls short of providing a convincing approach. As a matter of fact, there is no actor as huge as the world community. And the reason is that joint actors need structures and institutions that work in accordance with the prerequisites of joint agency. At a minimum, they include a shared vision of goals, actors who really are capable of taking responsibility and of acting accordingly, and structures that help to readjust actions at various points. Another reason for criticism is the problem of scale. Not only in terms of incentives to cooperate, but also in terms of realistic options to develop cooperative schemes and to further them, do individuals as well as collective actors need a level of action that they can understand and in which cooperation can be framed successfully.[68]

8.8 From duties to collectivize to duties of collectives

Some important observations in this debate should be established at this point. One important way to think about climate change and the need to work together to achieve climate goals is that the concept of a moral duty gives space for a variety of different types of allocation to collectives. Above all, duties should not be conflated with moral obligations of individuals or of collectives. Instead, it is also important to discuss whether individuals should form collectives that are then able to fulfil rescue duties. Strange as this proposal might sound at first glance, it presents a range of further difficult aspects of the case of climate change. Overall, researchers cannot make a convincing case that people should form groups that are capable actors; chiefly, the gap between a collective bearing a duty and the individual equipped with possibly different duties (but not with the same duties as the collective or putative collective) is difficult to grasp. Nevertheless, three aspects of the discussion are particularly helpful when thinking through climate duties.

First, the argument that collectives that could fulfil climate duties are not (yet) in existence is *not* an argument against the existence of collective duties. Instead, distinctions between addressees and subjects of duties – as well as the observation that duties of collectives that exist, although the collective is not yet mobilized – are important observations for the climate debate. Primarily, in situations in which structural injustice is so significant that it causes the deaths of human beings, it can be maintained that this situation needs to be remedied. I am not so sure that the remedy is a call for collectivization: how can you ensure that the prospective collective will do any better? At a minimum, success needs to be guaranteed.

Second, slack-taking and other problems of duty gaps highlight the need to not only develop a theory of cooperation that forestalls these problems. It also demonstrates that even when individuals wilfully join a group and when they share a goal, the danger of not complying is still very real. In my view, thinking about these problems in terms of moral duties is not the best way to approach them. Instead, I propose to reframe the problems as *defection*, which can only be framed as such when the cooperative scheme is precise. Unless a "fair share" of actors is defined by means of principles of fairness, the danger of non-compliance is far greater. A scheme of cooperation that rests on voluntary engagement is a realistic assumption regarding schemes of cooperation among enthusiastic individuals who choose that goal as a personal concern. As for environmental goals, those groups exist. Nevertheless, regarding climate goals, the whole range of different goals needs to be considered.

Third, it is important to discuss joint action as a model of action that takes the possibility of forming group actors seriously. Instead of highlighting problems and dilemmas, it is also important to note that groups of people – including unstructured collectives – can together discharge duties that individuals cannot. After all, the duties to collectivize might also offset the efforts of voluntary groups. Much more important than generating duties to collectivize is that political bodies should not sabotage, but rather support, voluntary processes of collectivization.

Notes

1 Michael Walzer, *Arguing about War* (New Haven, CT/London: Yale University Press, 2004), 40.
2 These claims cohere with the ontology of groups in the context of joint action, as proposed in Chapter 9.
3 Michael E. Bratman, *Shared Agency. A Planning Theory of Acting Together* (New York: Oxford University Press, 2014), 84.
4 These presumptions cohere with the definition of "unstructured collectives" in Stephanie Collins, "Duties of Group Agents and Group Members," *Journal of Social Philosophy* 48, no. 1 (2017): 47, https://doi.org/10.1111/josp.12181; Anne Schwenkenbecher, "Joint Duties and Global Moral Obligations," *Ratio* 26, no. 3 (2013): 313, https://doi.org/10.1111/rati.12010.
5 The discussion of duties of unstructured collectives includes a reformulation of the problem of whether "random collectives" can bear responsibility as discussed in Virginia Held, "Can a Random Collection of Individuals be Morally Responsible?," *Journal of Philosophy* 67, no. 14 (1970): 471, https://doi.org/10.2307/2024108.

6 Stephanie Collins, "Collectives' Duties and Collectivization Duties," *Australasian Journal of Philosophy* 91, no. 2 (2013): 231–248, https://doi.org/10.1080/00048402. 2012.717533; Bill Wringe, "Collective Obligations: Their Existence, Their Explanatory Power, and Their Supervenience on the Obligations of Individuals," *European Journal of Philosophy* 24, no. 2 (2016): 472–497, https://doi.org/10.1111/ejop. 12076.

7 Schwenkenbecher, "Joint Duties and Global Moral Obligations," 212–214.

8 Note that this is not my argument, but a presupposition that others have put forward. For this argument regarding climate goals as obligations of the world community, see the summary in Bill Wringe, "From Global Collective Obligations to Institutional Obligations," *Midwest Studies in Philosophy* 38, no. 1 (2014): 172–174, https://doi.org/10.1111/misp.12022.

9 Peter A. French, *Collective and Corporate Responsibility* (New York: Columbia University Press, 1984); Christian List and Philip Pettit, *Group Agency: The Possibility, Design, and Status of Corporate Agents* (Oxford/New York: Oxford University Press, 2011).

10 Tracy Isaacs, "Collective Responsibility and Collective Obligation," *Midwest Studies in Philosophy* 38, no. 1 (September 2014): 42, https://doi.org/10.1111/misp. 12015.

11 Ibid.

12 Philip Pettit and David Schweikard, "Joint Actions and Group Agents," *Philosophy of the Social Sciences* 36, no. 1 (March 2006): 19, https://doi.org/10.1177/ 0048393105284169.

13 Whether this type of responsibility is the same as in a single-action case needs to be discussed on separate grounds. For this discussion, see Isaacs, 43–44.

14 Note that the discussion is restricted to one sub-type of the general notion of responsibility, which is "prospective responsibility," which, according to Schwenkenbecher, is linked to actors in relation to future-oriented moral imperatives (Schwenkenbecher, "Joint Duties and Global Moral Obligations," 311). Another term for the same concept is Tracy Isaacs's concept of "prospective responsibility": "I claim that to be prospectively responsible is simply to have a responsibility to perform some action in the future" (Isaacs, 43). A different interpretation of the scenario needs to be given when "responsibility" is not goal-oriented responsibility, but a notion of "being held responsible" for something that has happened. See the discussion in Held.

15 Schwenkenbecher, "Joint Duties and Global Moral Obligations," 61.

16 For a detailed account of the minimum requirements of group actors to count as actors, see Pettit and Schweikard, 23–24.

17 Anne Schwenkenbecher, "Joint Moral Duties," *Midwest Studies in Philosophy* 38, no. 1 (2014): 58–74, https://doi.org/10.1111/misp.12016.

18 Peter Singer, *One World: The Ethics of Globalization* (New Haven, CT: Yale University Press, 2002), 35.

19 Schwenkenbecher, "Joint Moral Duties," 60.

20 Larry May, *Sharing Responsibility* (Chicago: University of Chicago Press, 1996), 105–106.

21 Schwenkenbecher, "Joint Moral Duties," 60.

22 Ibid., 61.

23 Ibid., 62.

24 Ibid., 63.

25 Ibid.

26 Ibid.

27 This problem was discussed in Chapter 7 as a key reservation against individual duties.

28 Wringe, "Collective Obligations," 12.

29 Schwenkenbecher reflects on all these aspects; she sees limitations of the duties in terms of "demandingness," "competing duties"; she also proposes that the approach is best suited to "situational simplicity" and – among others – "epistemic simplicity" (Schwenkenbecher, "Joint Moral Duties," 68).
30 Held.
31 Ibid., 476.
32 Ibid., 471–472.
33 Held, 472.
34 Ibid.
35 Ibid., 479.
36 Ibid.
37 Ibid.
38 Ibid.
39 This ties back to the arguments of Chapters 2 and 3.
40 Held, 481.
41 May, *Sharing Responsibility*, 105ff.
42 Collins, "Collectives' Duties and Collectivization Duties," 233.
43 Ibid.
44 See Collins, "Collectives' Duties and Collectivization Duties."
45 Stephanie Collins, "Filling Collective Duty Gaps," *Journal of Philosophy* 114, no. 11 (2017): 573–591, https://doi.org/10.5840/jphil20171141141.
46 Ibid., 574.
47 See Christopher Kutz, *Complicity: Ethics and Law for a Collective Age* (Cambridge: Cambridge University Press, 2000).
48 Kutz, 122.
49 Schwenkenbecher, "Joint Duties and Global Moral Obligations," 322.
50 Ibid.
51 Ibid.
52 Ibid., 317.
53 Collins, "Filling Collective Duty Gaps," 581.
54 Ibid.
55 Ibid.
56 Holly Lawford-Smith, "The Feasibility of Collectives' Actions," *Australasian Journal of Philosophy* 90, no. 3 (2012): 453–467, https://doi.org/10.1080/00048402.2011.594446.
57 For a discussion of these findings, see Angela Kallhoff, *Why Democracy Needs Public Goods* (Lanham, MD: Lexington Books, 2011).
58 Robert E. Goodin, "Excused by the Unwillingness of Others?," *Analysis* 72, no. 1 (2012): 18–24, https://doi.org/10.1093/analys/anr128.
59 Lawford-Smith, "The Feasibility of Collectives' Actions," 462.
60 Goodin, 24.
61 Bill Wringe, "Global Obligations and the Agency Objection," *Ratio* 23, no. 2 (2010): 225, https://doi.org/10.1111/j.1467-9329.2010.00462.x.
62 Ibid., 227.
63 Schwenkenbecher, "Joint Duties and Global Moral Obligations," 320.
64 Wringe, "Global Obligations and the Agency Objection," 228.
65 Schwenkenbecher, "Joint Duties and Global Moral Obligations," 320.
66 Ibid.
67 Ibid.
68 This claim and its implications will be argued in Chapter 10.

9 A fair share in accumulative goals

Cindy: Actually, I looked into the last chapter. It was interesting to see that philosophers even think about "duties of collectivization." They think that might be an obligation to create groups or to join groups, assuming these groups are important for the achievement of climate goals.

Bert: At least a better idea than enforcing collectivization. But honestly, that sounds like the next step.

Cindy: I'm not sure. I think it is also possible to go in another direction. I don't think that people have to join groups or to build up groups in order to achieve climate goals. Instead, there are already so many organizations, groups, and other sorts of collectives that could make a difference. Why not ask nation-states, industries, cities, or even neighbourhoods to contribute something to climate goals? Wouldn't that be fair?

Bert: I think it would be fair. But I think it would be very unfair to ask poor people to shoulder heavy burdens of mitigation. Poor people do not contribute significantly to climate change, they have a very small ecological footprint. And I think efficiency also somehow matters – even when addressing collective actors and not single actors.

Cindy: I think the debate has now shifted to another issue. It has been argued convincingly that concerted action is obligatory; we now have to find out how the burdens of climate goals should best be distributed. And "best" means according to principles of fairness.

Bert: Yes. And philosophers will certainly have an idea of principles that help to allocate burdens in a fair and suitable way.

Cindy: Let's see how they do it.

Climate goals have a range of very specific characteristics. They cannot be achieved by individuals, but need to be supported by groups of actors; they cannot be reduced to a single goal, but rather include adaptation, mitigation, and technological solutions, as well as also sub-goals such as "greening the city," "reducing air pollution," or "conservation of a water cycle as an important local cooling mechanism." So far, the rationale of collectivization and of collectives caring for climate goals has been discussed. One more specific feature, however, remains to be explored. Some of the most important climate goals are not literal joint action goals, but instead accumulative goals. They can be achieved by many single contributions, which together reach the climate goal. In order to distinguish those goals from literal joint action goals, I shall call them "accumulative climate goals."

This chapter starts with a proposal to discuss accumulative climate goals with respect to a "fair share of burdens." Instead of equalizing this share with an equal effort of every single actor, this proposal rests on the assumption that a fair share of burdens should be allocated to various actors according to principles relating to the properties of actors, such as wealth, ability to contribute, and liability. This reassessment of the debate on distributive climate justice is also related to a critique of the interpretation of climate justice as distributive justice. Framed as an issue of a fair distribution of emissions budgets that should apply to actors, the leading paradigm is still the "cake" metaphor with respect to the atmosphere as a dumping ground for greenhouse gases. More generally, natural common-pool resources (NCPRs) are interpreted as dumping grounds with limited space; according to principles of fairness, the remaining space should be allocated to actors according to principles of fairness. As Gernot Wagner and Martin L. Weitzman correctly observe, the idea of a cake that should be divided according to principles of justice is flawed. Actually, there is no cake left over for current and for future generations.

Climate theorists have not only discussed a fair share of the atmosphere as a dumping ground, but also a fair share of secondary goods and burdens. The goods and burdens are related to greenhouse-gas emissions, in that beneficiaries of high emissions also enjoy a high living standard. Moreover, they also include costs of adaptation and costs for compensation of climate victims. Overall, the problem with the distributive picture is not only that there is almost no cake left, but also that unless the faucet is turned off, the flooding will become even worse. Instead, another serious problem is that concepts of fair burden-sharing do not automatically lead to joint action.[1] Even when enriched with further distributive dimensions, the leading paradigm is either an allocation of remaining emission budgets – which is difficult to reason – or the allocation of burdens and costs.

In three respects, the focus on distributive fairness needs to be criticized. First, fairness is not automatically supportive when it comes to cooperation among individuals who benefit from a shared resource. As already noted,[2] fairness in appropriating a shared resource needs to be negotiated among beneficiaries; it is not necessarily the same as distributive fairness. Second, a focus on distribution overlooks that current practices not only either add to justice or further undermine it; instead, current actions also determine the range of available eco-services in the future. The most desirable "safe operational space"[3] can be jeopardized by current practices – even when perfectly fair according to principles of a fair distribution. Third, even when a fair scheme of appropriation is critical, this is only one side of environmental cooperation.

The other side is responsibility for acting in such a way that future productive use of the natural resource is not impeded, but in possibly new ways prepared and enabled.

These initial observations do not prevent a discussion of fairness in terms of distributive justice. Instead, they demonstrate that distributive justice is part of a larger picture that needs to be defined in terms of joint climate action. With respect to climate goals that are accumulative goals, a discussion of principles of fairness is helpful. But at the centre of concern is a fair share of burdens – burdens that result from the costs and efforts to achieve a shared goal. Theorists term this "proactive responsibility"[4] and think that burdens that relate to this type of responsibility can be defined accordingly. Although still conceptualized in a distributive frame, the object of distribution is *a fair share of costs and burdens* when accumulative goals are at stake. Moreover, fairness is determined as a fair share in the common responsibility to act in order to achieve a shared goal. This type of justice is correctly framed as "burden-sharing justice" and addresses responsible actors who should shoulder their part of a shared goal.[5]

Regarding climate goals, the shares of responsibility need to be defined against the background of observations about the climate duties that individuals as well as collectives hold. The "weight" of the burdens differs; it is not the same for all actors. It is not fair, for example, to claim the same contributions from poor actors as from rich actors; simultaneously it is also not fair to address the most influential actors last. Instead, burden-sharing justice builds on criteria that help to assign a fair share of burdens in achieving climate goals according to various *properties of actors* that determine their capacities and obligations to act. What certainly counts is the ability to contribute to goals, but also a record of environmental performance that has been either helpful or detrimental. In addition, responsibility needs to be allocated according to concepts of role ethics. When political institutions are obliged to care for the common good, the engagement in climate goals is part of this duty.

Another distinction is critical for outlining a fair share of responsibility in addressing climate goals. Assuming collectives can be held responsible to contribute a fair share to the accomplishment of climate goals, it is then still possible to distinguish two different types of responsibility. First, actors have a responsibility to act in accordance with their fair share of burdens. This is *first-order responsibility*. Second, an actor might also be able to hold another type of responsibility, which is *second-order responsibility*.[6] An institution or an organization might be held responsible not only for acting on their behalf, but also for convincing other actors to do their share, possibly even for enforcing compliance with existing rules or to design institutions that are effective in addressing climate goals. Specifically, collective actors might also be under an obligation to prepare opportunities for individuals to contribute to climate goals.

With respect to both levels of responsibility, principles that address a fair share of burdens for various actors need to be defined. This chapter is restricted to some of the core principles. As for first-order responsibility, they include the *ability-to-pay principle*, the *liability principle*, the *principle of justified reductions*, and the *principle of special obligations*. Each of the principles can only be portrayed in a very simplified way – a thorough discussion would take much more room than this chapter offers. As for second-order responsibility, this chapter introduces only one of the definitive criteria, the power-capacity principle. Much more will be said in the next chapter about the obligations of institutions that are able to bear second-order responsibilities.

As for the goals that are at the centre of this chapter, three different types of goals need to be distinguished according to the proposals of Chapter 6. The first goal, mitigation, is an *accumulative goal*. Contributions to this goal do not have to be provided by co-agency in the sense outlined in theories of joint action. Instead, it suffices to break it down to shares; all shares together accomplish the goal. In another respect, however, mitigation is also a joint action goal. When interpreted with respect to the underlying activities, joint action can achieve mitigation by means of the protection of forests, by means of greening a city, or by means of a carbon-neutral infrastructure. Theories of distributive justice are best suited to explain how burdens that result from both types of achievements should be allocated to various actors. The second goal, *adaptation*, is an exigency goal. Accomplishing adaptation is especially costly, but the question of who should support which kind and which area of adaptation is another matter. In order to establish a fair share of contributions, a theory of justice is also able to explain why actors should help people who suffer from climate change. I shall not address adaptation in this chapter, because it is, in my view, a subclass of duties I defined in Chapter 7 as victim-related duties. A theory of joint action helps to support the view that *in situ* adaptation needs to be arranged according to principles of joint action. In any case, when claiming support for adaptation at a place far away from the current living location, theories about rescue cases need to be consulted. I do not claim that this is not important; quite to the contrary, a theory of justice for adaptation goals demands much more than this chapter can offer. The third group of goals is *integrated environmental goals* and *respective transformation goals*. The achieving of those goals causes burdens; these burdens also need to be distributed among various actors. The principles explained in this chapter also apply to them when they match the descriptions of an accumulative goal.

This chapter has eight sections. The first explains the shift towards an interpretation of fair burdens as part of a collectively held proactive responsibility to achieve climate goals. This section, like the following sections, builds on the general observation that climate duties are universal duties. It adds a detailed assessment of the share in taking that responsibility by various actors. This discussion also includes an explanation of how the current chapter is related to the broader discussion on climate justice and a critique of another interpretation of distributive justice in the context of climate justice. Section 9.2 discusses the first level of fairness in the framework of joint environmental action. A joint environmental action approach highlights proactive and goal-related burdens; they should be distributed according to principles of fairness. Section 9.3 provides another critical step in the debate on proactive burden-sharing among various actors. It justifies the claim that duties not only apply to individuals, but also to structured collectives. This includes organizations such as associations, churches, institutions of education, and governments. The actors involved differ from the formerly discussed "joint actors" in that they are true collective actors. The duties to act belong to them as collectives. Sections 9.4–9.7 discuss principles that help to assign a fair share of burdens to responsible actors: the ability-to-pay principle, the principle of liability, the principle of justified reductions, and the principle of special obligations. In sum, this discussion helps to defend a new view of responsibilities to shoulder the burdens for the reaching of climate goals. Primarily, I intend to present an alternative to (a) theories of justice, including historical and corrective justice, and (b) theories that rely primarily on a role ethics. It is not right to claim responsibility of governments, but to exempt the high-level emitters, including big oil companies and an ill-directed agricultural business, from responsibility. Instead, my proposal includes an allocation of burdens according to properties of actors such as ability to pay, liability because of a history of high emissions, and justified reductions for poor actors and for actors who have no chance to cushion from side effects of rapid cuts in emissions. Section 9.8 gives a summary of this proposal and presents a chart.

Note also that this chapter still works on the systematic grounds prepared in the preceding chapters. It is presumed that climate duties regarding the goals that have been portrayed as climate goals are not only justified, but that their acknowledgement and acceptance also generate duties. Notably, no single actor is exempt from the duty to contribute to the achievement of those goals – the issue I am dealing with in this chapter is merely the question of the "fair amount" of the contribution that should be made by individual and collective actors.

9.1 A fair share in proactive responsibility

It is the main goal of this chapter to change this narrative of distributive justice in terms of a fair share of benefits from the atmosphere as a dumping ground and in terms of a fair share of costs of mitigation and adaptation. Instead, I want to define a fair share of proactive responsibility for accumulative goals. According to the arguments already presented, single actors are obliged to support climate goals in terms of joint action goals under certain conditions. Individuals should co-act in favour of climate goals, assuming appropriate schemes of cooperation are available and the shift towards those schemes is not extraordinarily costly. This view of climate goals leaves some questions unanswered. What about collectives that already exist, but that do not yet engage for climate goals? Do they also have duties to contribute to climate goals? And what are the obligations of which type of actors?

The first thing to recognize is that justice has two sides: first, and somewhat in line with theories of distributive justice, principles of justice help to adjust a fair share of burdens to various actors. Unlike alternative approaches, the burdens are now framed as costs and efforts that result from the engagement in joint action goals. They are discussed as elements in "burden-sharing justice."[7] Second, principles of justice are also important for the engagement of individuals in favour of joint environmental goals. I have suggested that this side of justice can best be interpreted as part of an ethos that helps individuals to get engaged in joint goals. Although a concrete scheme of a fair appropriation of eco-services has not been established with respect to climate goals, the expectation that the achievement of goals also results in fair access to eco-services is still critical.

This chapter defends two claims in terms of distributive justice. On the one hand, the *allocation of burdens* correlated with the achievement of fair environmental goals needs to be guided by principles that allocate burdens to reach the goals in a fair way. This not only assumes that climate goals are legitimate goals, but that, when framed as joint environmental goals, they should be supported by every single individual as well as by every single collective actor. Still, the burdens that result from the duties to support these goals need to be adjusted according to principles of a fair share. In other words, distributive justice is a necessary means for adjusting the duties that actors hold when it comes to environmental goals and climate goals more specifically.

On the other hand, principles of fairness in allocating burdens are not restricted to individuals, but also apply to actors who qualify as collective actors. Indeed, collectives also have a duty to facilitate climate action. In order to develop this argument, it is once again necessary not only to carve out the interpretation of collectives, which is important for that claim, but also the reasons why they are responsible actors with respect to the specific climate goals. Unlike the earlier discussion of collectives in this book, this discussion does not focus on spontaneous groups such as rescue groups or on groups that begin as peer groups wanting to achieve a shared environmental goal. Mainly, the focus is now on collective actors who are already in place. This includes groups, but also organizations and institutions that are structured collectives and that serve purposes different from climate goals. The question that I raise is not whether they also have climate duties: since actors have climate duties, collective actors will also hold them. The question is, rather, which actors count as collective actors and how can duties be allocated to them according to principles of fairness.

One more proviso is necessary. Obviously, the concepts of "burdens" and of "shares" serve as metaphors for the "weight" of responsibilities. Since not all goals are accumulative goals, it is also necessary to explain what the concept of burdens means. Stephanie Collins has presented an intuitively clear proposal. She explains that once duties can be tracked to sources of duty-holders, it is also not far-fetched that costs can be tracked to these sources too.[8] Allocating burdens is a practice to allocate costs to sources that result from discharging specific duties. Collins also highlights the need to track the costs down to further levels when collectives are addressed.[9] The concept of "tracking" responsibilities and correlative costs and the application of this idea help to avoid a naive view according to which governments should pay for climate goals; even when they do so, it is not their money that is paid, but the taxes of citizens or other collectively owned revenues.[10] They are the sources of payments that help to achieve costly goals.

In addition, in some circumstances, *principles of fairness that guarantee a fair share in climate gains* also need to be taken into account. Even when climate goals are not about the global resource of "climate stability," but instead include goals of a healthy microclimate or healthy water cycles, the gains from the achievement of environmental goals do not necessarily apply to all contributors to a shared goal. Instead, when talking about micro-goals, a guarantee of a fair share of benefits according to well-established principles needs to be outlined. So far, I have claimed that the weight of duties to cooperate, which also includes the real costs of climate cooperation and restitution of already injured natural goods, needs to be allocated to actors according to principles of fairness. Whereas each actor has a duty to cooperate, the weight of the duty varies. Whereas nation-states and political organizations have special obligations, these duties are not derivative of the duties of single individuals.

In sum, this interpretation of climate justice provides an alternative to interpretations of climate justice in terms of a fair share of the atmosphere as a dumping ground for greenhouse gases. Although the debate on climate ethics is advanced, one persistent basic assumption is still that climate justice is about a fair share of emission budgets that need to be allocated according to principles of fairness. However, the debate has also turned to so-called secondary burdens, which are burdens that result from mitigation and adaptation. And this also includes discussion about principles of fairness that recognise the fact that the currently living generation did not literally cause climate change, but that parts of the now-living population benefit to a considerable degree from an economic performance that drove climate change and now accounts for its wealth and high standard of living. As a consequence, it has also been suggested that it is not only the polluters who have to pay the price of climate change, but also the beneficiaries. The relationship between those who benefit from high emissions and climate victims is defined in terms of "restitution," even "rectification" of harmful incidents as another instance of climate justice.[11]

It also has to be said that the initial restriction of climate justice to fairness in distributing emission budgets and costs of climate change goals has already received a fundamental critique.[12] This critique has also been formulated towards proposals of a climate contract that does not fully account for climate justice.[13] That discussion has also contributed to the insight that new ideas about a goal-directed approaches are needed.[14] My point here is not that concepts and principles of climate justice have become obsolete.[15] On the contrary, it is obligatory still to discuss the gap between rich and poor and between economically advanced and less advanced countries. In any case, in order to design distributive rules that account for those who benefit from high emissions as well as climate victims, the prior concern is not an adequate contract, but principles that establish a fair share in proactive responsibility and related burdens. It is therefore not possible to rely exclusively on economic proposals and solutions.[16] Against the background of joint environmental action, this dichotomy is flawed. Instead, principles need to be set up that account for the special set of goals that figure as climate goals and to whose achievement every single actor is to some degree obliged to contribute.

9.2 Fairness in allocating proactive burdens

A first step in the process of establishing principles for allocating goal-related burdens according to claims of distributive fairness is a recapitulation of the various burdens that are at stake in the context of a joint environmental action approach. Note that this discussion is already focused on environmental goals and the related duties of actors to bring about the goals formulated in earlier chapters of this study. Three types of burdens, which correlate with four types of goals, need to be distinguished.

First, burdens result from a reduction of impact on a common-pool resource, as with, for instance, mitigating greenhouse gases in order to protect the atmosphere and the climate from environmental hazards. When a collective decides to reduce waste, this is usually costly. The installation of new technologies and alternative methods of waste-disposal are examples of costly transformations. Costs often result from the change itself; they are also termed "opportunity costs." Second, there are burdens that result from the reparation and restoration of natural resources. When societies or groups of people decide that they cannot live with certain types of hazards or that they should be repaired, these costs also need to be shouldered. Integrated environmental goals often include these costs. Third, burdens that result from conservation or reparation of a shared natural resource also need to be addressed. Mitigation is a means of stopping further hazardous practices in order to protect the climate from further damage. In that respect, mitigation is not only an accumulative goal, but also a goal that results from the consensus on conservation practices with respect to the atmosphere.

In addition to distinguishing three types of burdens according to their relationship to the types of goals that will be accomplished, burdens also vary with respect to their specific relationship to the goals of actors. According to the distinction between goals that are primarily *reactive* in that they relate to former incidents of harm, and goals that are *proactive* in that they represent desirable goals that help to bring about a good future design of schemes of appropriation, I distinguish the related burdens as (a) *reactive burdens* and (b) *proactive burdens*. Much of what has been said in the debate on climate justice is part of the package of *reactive burdens*. These burdens result from costs that are created when wilfully reacting to environmental degradation and to the effects of climate change in particular. These burdens are different from the burdens that will be addressed in this chapter.

With regard to joint environmental action, burdens are related to goals that groups of individuals envision in order to design goals with respect to natural resources and according to normative principles. *Proactive burdens* are at the heart of a joint environmental action approach. Note that both types of burdens resonate with different normative claims. Notably, as Simon Caney thoroughly explains, the focus on victims and potential victims, which he calls "harm avoidance justice," sometimes but not always coincides with the broader "burden-sharing justice" approach.[17] At a minimum, burden-sharing justice applies when "(1) there is an important goal the realization of which involves burden; (2) the equitable sharing of the burden requires a certain distribution of burdens."[18] Specifically, Caney also notes that the distinction between harm-avoidance justice and burden-sharing justice cannot be paralleled with the distinction of backward-looking versus forward-looking approaches to justice.[19]

The main claim of this chapter is that in order to achieve climate goals, proactive burdens need to be allocated according to principles of fairness. In order to give a clear view of this claim, the principles of fairness established in this chapter also need to be distinguished from principles of distributive fairness that account for access to a fair share in a joint natural good. Instead, the principles are based on the more fundamental claim that all actors are responsible for bringing about climate goals.

The main claims of this chapter can thus be summarized: principles of justice help to allocate burdens that are already established as "climate duties." And they resonate with various capacities of actors to bring about change and according to special obligations resulting from the special role those actors hold. In addition, burdens should also be appropriately reduced regarding vulnerable groups. In contrast, actors whose current activities are directly linked to former hazards should – under certain circumstances – also be held liable for former environmental harm. In the remainder of this chapter, I address the principles in turn.

9.3 Extension of climate duties to structured collectives

Principles of justice apply to single actors as well as to collective actors. In order to fully back this claim, it must still be asked under which conditions already-established collective actors qualify as duty-holders. This section addresses this problem.

In order to determine which kind of collectives qualify as responsible agents, several proposals have been made in the literature on collective responsibility. Much of the current literature refers to an account presented by Peter A. French.[20] According to his proposal, the identity of an organization as a "conglomerate collectivity" is not exhausted by the identities of the individuals who make up that organization.[21] Examples of conglomerate collectivities include political parties, corporations, and associations.[22] According to that proposal, the criteria for accepting the claim that a conglomerate collectivity can be held responsible for acting include (a) an identity that is more than the sum of the identities of its constitutive parts and does not rely on determinate membership; (b) a decision-making structure; (c) an identity over time; and (d) a conception of itself as a unit.[23] But it has to be noted that French's proposal is just one approach among a range of further proposals to allocate the ability to hold responsibility to collectives.

A different account, which regards collective actors as "real agents" – and therefore also as responsible for the actions it accomplishes – has been proposed by Christin List and Philip Pettit.[24] They maintain that a group as a collective actor can be regarded as holding responsibility when the group agent (a) faces a normatively significant choice, (b) has judgement capacity, which means that the agent understands and has access to evidence required for making normative judgements about the options available, and (c) has the necessary capability, which is the capability required for choosing between the options.[25] In their view, the most problematic point is the third condition of responsible agency. Indeed, the collective agent itself needs to be regarded as capable of control over its deeds. Primarily, the control condition needs to apply to the collective agent itself.[26] Although this proposal has received much attention, it needs to be regarded as a complex proposal with respect to its metaphysical prerequisites. Obviously, this proposal is not only non-reductive with respect to the constituencies of a group agent, it also includes a range of distinct conditions with respect to the responsibility of a collective agent. This also includes the judgement capacity of a group actor. Even when procedures to make decisions as a group are carried out, it is still difficult to demonstrate that this group judgement is as valid as the judgement of single actors.

An account that covers middle ground with respect to metaphysical prerequisites and that also thinks through the responsibility of a group actor has been advanced by Collins and Lawford-Smith.[27] They state that the first step in explaining group obligations is a clarification of the accounts of obligation and of responsibility at work here. In accordance with this proposal, they do not start with an investigation of preconditions for calling an actor a true collective actor. Instead, they utilize a belief-desire model of agency in a functionalist interpretation of action in order to explain the prerequisites for responsibility. Notably, they defend the view that it is necessary to assume three elements for calling an event an action of a group actor. It needs something that can be labelled *desires*; something that can be labelled *beliefs*; and something that can be labelled *ideas* about how to act in the environment.[28] They then explain that decision-making procedures allow agents to move from desires to beliefs to decisions. This procedure needs to be "operationally distinct from the procedures held respectively by members."[29] In sum, a group of people who are organized in such a way that institutions are in place that help to integrate desires and beliefs on a group level, and that also adapt the possible act to the existing environment, is defined as acting as a group in a responsible way.

This ontologically minimal account of group responsibility supports the view that groups are able to actually hold responsibility for their actions. In addition, researchers demonstrate that group actors are also capable of bearing obligations. Notably, they deviate from the usual procedure of discussing the general presumptions that are to suffice as conditions for responsible action by a group actor. Instead, they discuss six more or less concrete kinds of obligations: to keep promises, not to cause harm, to end injustice, to be partial to those one is closest with, to assist those in need, and to rectify past harm.[30] The authors defend the view that each of these obligations enshrines a moral duty. Each duty can be attributed to a group actor. Their main argument is a negative one: differences between individuals and groups regarding the duties of both entities are not necessarily qualitative differences; rather, they rest on some former distinction or they are not real differences, but only quantitative differences.[31]

One way to apply this outcome to the issue debated in this chapter is that groups can be held responsible for the accomplishment of climate goals, presuming there is a general obligation to contribute to those goals. Obviously, this claim also assumes that climate duties immediately relate to the aforementioned duties that can be attributed to individuals as well as to group actors. In short, it would be necessary to demonstrate that climate duties also relate to duties to keep promises, not to cause harm, to end injustice, to be partial to those one is closest with, to assist those in need, and to rectify past harm. Even without a detailed discussion, but with respect to the discussion in Chapter 8 on climate duties, this is certainly the case. Not to accomplish climate goals causes harm and injustice; although it is reasonable to be partial with people who are somehow close to oneself, climate duties include the duty to act in favour of climate victims in remote areas, to rectify past harm, and to end injustice in a very fundamental sense.

One more difficulty needs to be discussed when it comes to moral obligations of groups. A particularly difficult issue to decide is whether groups can be held responsible to act in favour of future goals, since the identity of groups over time cannot be presumed. One strategy for arguing in favour of a group identity over time is to restrict a group responsibility account to groups that have a clear, long-standing identity, such as nation-states. However, it also suffices to tie the obligations to a group as long as it exists in a physically substantial way. This resonates with a proposal of Collins and Lawford-Smith in a discussion of historical responsibility:

> There is a minimal continuity when there are not abrupt and complete changes in membership, such that not one individual is a member at

both t_1 and t_2. *Substantial* continuity exists when a substantial number of members remain members between t_1 and t_2, even while new members join (or are born) and old members leave (or die). That means a group can entirely change its membership over time, without becoming a different group.[32]

Another problematic property of groups is the possibility of fusion and of fission. However, supposing there is still a relationship between the former constellation and the latter, responsibility can at least be tracked; moreover, a new allocation of responsibility can be discussed with respect to how the group has changed. That one group vanishes in fusing with another does not necessarily imply that responsibility also vanishes. And when a group splits into two or more groups, the responsibility needs to be straightened out.

Finally, one further distinction is helpful for thinking through the responsibility of structured collectives. Tracy Isaacs distinguishes between two kinds of collective moral agents, which are organizations and goal-oriented collectives:

> Organizations are more structured than goal-oriented collectives and have clear role definition, decision procedures, and mechanisms for acting in the world that might be outlined in terms of corporate structures and policies, the specification of corporate interests and the like.[33]

Organizations include corporations, universities, governments, clubs and non-profit groups, and the military. These collectives are considered to be capable of acting – in the aforementioned sense of acting – when a decision-making procedure is implemented. It also has to be observed that the issue of whether membership is voluntary does not bear on the question of whether collectives bear responsibility. The particular obligations collectives hold regarding climate goals and environmental goals do not result from the obligations of the members of these groups. Instead, in as much as structured collectives are regarded as responsible agents, the obligations also apply to that entity and cannot be reduced to obligations of the members of that organization. Theorists who defend the responsibility of structured collectives conceive of these entities as equipped with mechanisms that account for the actions of that entity and the decisions of that entity by means of a decision-making procedure. Although the outcome of decisions relates to the will of the individuals, it is not the same.

In sum, the insights into the ability of structured collectives to act and to hold responsibility support another additional way to discuss responsibility for climate goals than the already-debated climate duties approach. Instead of focusing exclusively on duties of individuals or of duties of unstructured collectives, it is also possible to discuss the responsibility of already-existing organizations with respect to climate goals. Notably, the differences between individuals as duty-holders and structured collectives as duty-holders are not substantial, assuming the collective is rightly classified as an actor.[34] Essentially, the duties to cooperate that have been established for single actors in the case of climate change also hold for structured collectives.[35] If reasoned accordingly, even acts of omission – that is, of not initiating activities that support the effective achievement of climate goals – need to be judged as morally flawed. Although unlike the single-person case, the fact that group actors have highly different capacities to act on climate goals and that they might also not all bear the same responsibility drives the debate about a fair allocation of climate duties with respect to group actors. The remainder of this chapter is dedicated to three important principles in that respect: the ability-to-pay principle; the principle of liability; and the principle of justified reduction of burdens.

9.4 The ability-to-pay principle

The ability-to-pay principle is not a free-standing principle establishing duties to cooperate and to invest in climate goals. Instead, supposing an actor is obliged to contribute to a joint goal, this principle helps to define the "fair share" for that single actor in accordance with the ability to contribute.[36] Simon Caney supports this claim. He states that the ability-to-pay principle is not a free-standing principle, but that it instead fills the gap of a fair allocation of a certain amount of burdens when forward-looking claims are determined:[37]

> Stated formally, this approach states that the duty to address some problem (in this case, bearing the burdens of climate change) should be borne by the wealthy, and moreover, that the duty should increase in line with an agent's wealth.[38]

Moreover, unlike duties to rectify former harm, the ability-to-pay principle is explicitly forward-looking.[39]

The discussion in Caney's text also raises a difficulty. Why should the wealthy make good on something that they possibly did not cause? He discusses the historical dimension that comes into play when judging wealth as something that people have earned or received.[40] Still, these questions are not crucial in the context of our analysis. Instead, in the context of joint environmental action, the principle relates rather to the already established principle that individuals should do the best they can to abate climate change. One measurement for the best possible an actor can do is wealth, assuming the reaching of climate goals costs money.

Another backdrop of the ability-to-pay principle is the general idea that when allocating duty-related burdens that the claims should be set within a frame that actors are able to accomplish. It is unfair to overburden individuals and single actors with duties that they cannot achieve. As for the allocation of the weight of a duty to cooperate, it is fair to claim that actors who are particularly effective in reaching an environmental goal due to their economic power should also contribute accordingly. The underlying idea is that the greater the capacity to act, the more weight an actor is able to carry. Since climate change is a real threat and perhaps even a special type of very large risk, it is fair to ask powerful actors to do more than usual actors. Particularly powerful actors include nation-states, international institutions, and large industrial corporations. In short, the ability-to-pay principle supports a fair allocation of proactive burdens to influential actors.

As established in Chapter 7, every single actor is responsible for cooperating when defection means undermining not only a fair scheme of cooperation, but also the achievement of shared and exigent goals. The ability-to-pay principle respects the observation that responsibility grows with power. This argument is comparable to the idea that property is not only a good, but also comes with responsibility for the common wealth.[41] It can also be maintained that burden-sharing responsibility grows proportionally with the strength of various actors. This observation could be implemented as a progressive rate of payments. This claim is also supported by Henry Shue, who states:

> Among a number of parties, all of whom are bound to contribute to some common endeavour, the parties who have the most resources normally should contribute the most to the endeavour. This principle of paying in accordance with ability to pay, if stated strictly, would specify what is often called a progressive rate of payment: insofar as a party's assets are greater, the rate at which the party should contribute to the enterprise in question also becomes greater.[42]

Furthermore, it is important to acknowledge that additional principles can be formulated to heighten the efficiency of the baseline in the ability-to-pay principle. According to Peter Singer's observations in "Effective Altruism,"[43] it is fair to set spending along lines of a utilitarian principle of most effective utilization: the more a person possesses, the less weight a certain amount of spending has; coupled with huge impacts on the lives of poor people, this principle combines effectiveness with efficiency. This is not the rationale underlying the ability-to-pay principle, but it can help to allocate scarce resources according to principles of a maximal useful spending.

9.5 The principle of liability

When actors cause environmental harm, when they do so wilfully, and when an incident of harm hinders the achievement of fair climate goals, they should possibly be held responsible for those harmful acts. This general observation has been framed in various ways.[44] At the centre is the claim that former polluters should now be held responsible and compensate victims of their harmful behaviour for the danger they created.[45] Unlike that interpretation of responsibility for former hazardous actions, acts that foster environmental hazards can also be framed as an incident of incautious, possibly even risky action. An actor who pollutes an area does not necessarily intend to pollute it, nor does he necessarily cause immediate environmental harm. Often incidents of environmental hazard cannot be traced to the sources of harm in a causally convincing way. Instead of taking this as an argument for releasing actors from responsibility, it is more apposite to frame their action as risky behaviour. In not contributing to climate goals, but also in spoiling the environment, they put the health and good life of people at risk.

An example of the debate on liability is provided in the example of car drivers. Whether car drivers can and should be held accountable for the harm – possibly even the death – of future people, has long been discussed.[46] So far, it has been very difficult to give reliable evidence of the causal chains that lead to climate victims. But it is not that difficult to argue that motor-vehicle drivers who spoil the environment heighten the risk of illness for passersby and for residents.

The appropriate moral framing for these observations is a *liability principle for environmental harm*. The question that needs to be raised in determining the fair share of the burden of a collective actor is the question of whether single collectives are liable either for environmental hazards or for the non-achievement of climate goals. Actors who heighten the risk of not achieving climate goals ultimately also put the good life of other people at risk. They deserve a particular share of the burdens of proactive responsibility because they can be held responsible for harmful action.[47] There has been a long discussion of whether actors who do not cause climate change, and whose causal connectedness with that mega-event is marginal, can be held responsible for their acts or for the compensation of climate victims.[48] Note that the line of thought with respect to liability is different. What counts is some causal involvement in a scenario in which effects are known to be risky – with respect to the health of other people.

In short, liability does not focus on a duty to compensate for wilful and destructive behaviour that causes harm to victims of these acts. Nor does it take for granted full knowledge of the causes and effects of an action. Instead, it suffices that an actor knows about the fact that his acts put the good life of other people at risk in order to regard him as liable for that risk.[49] This precondition is applicable to collective actors when defined along the lines of Section 9.3. When organizations have decision-making structures, it can also be assumed that they make decisions on future acts and on their activities. Today, organizations know that not complying with climate goals puts climate stability and the life of vulnerable groups on Planet Earth at risk. Unlike complete responsibility, liability does not have to rely on full intent of the effect, not even on full knowledge of the effects of the action. As a consequence, although possibly not intending to extinguish a shared and life-sustaining resource, and although not fully aware of the effects of their actions, actors can be regarded as liable for the damage they caused – even damage resulting from omissions, which count as failures with respect to climate duties and which enhance the risk of not achieving a fair climate goal.

Above all, the category of liability is best explained as responsibility, which applies to an actor who not only creates risks, but *changes the distribution of risks in an already risky situation.* [50] Unlike general responsibility for former harm, liability is a category that helps link duties to compensation according to the change of risk. As Myles Allen illustrates, the causal chains can be calculated even when causal histories are not obvious. Allen notes:

> If, at a given confidence level, past greenhouse-gas emissions have increased the risk of a flood tenfold, and that flood occurs, then we can attribute, at the confidence level, 90% of any damage to those past emissions ... So, if courts can accept the concept of averaging over possibilities to create an equitable distribution of liability, in theory, one day people driving up the local hill in their SUVs might be contributing to the cost of replacing the floors in Vicarage Road. [51]

In order to cope with complexities in causal histories, Allen also proposes agreements about amnesties, as for instance for pre-1990 emissions, without really affecting the final outcome. [52] Whether individuals and collectives can be sued for contributing significantly to putting the climate at risk is an issue that needs to be decided on other grounds. In the context of the proposal for joint climate action, liability could provide arguments to pose an extra load of burdens on the shoulders of high emitters.

When it is right that the world of actors is divided into supporters of climate cooperation and defectors, another rationale also helps to support the claim of additional rectification of former harms. Every single defector contributes to harm, whereas every single cooperator helps to overcome the dynamics of climate change. Whereas the causal impact of one single defector might not put the climate at risk, as part of a dynamic of either defecting or cooperating, the situation looks different. She might not generate the effect of actually reaching a tipping point, but she is part of a dynamic that either enhances the chance of achieving a climate goal or heightens the risks of not achieving urgent goals. And thus actors are also liable, possibly not for their particular impact on the dire situation today, but because they bear additional responsibility to rectify this situation now. Those who put the planet further at risk, who do so knowingly, and who have alternatives – above all the alternative of reducing risks – should shoulder the additional burdens of rectifying the drama of the commons through coordinated, future-oriented efforts.

9.6 Justified reductions of proactive burdens

Finally, a group of principles of a fair allocation of proactive burdens results from the observation that certain actors, even when responsible, should be exempt from the full claims of responsibility or at least should bear a smaller amount of responsibility. The arguments about *justified reductions of burdens* have several different observations working in the background. One is that, in the context of climate change policies, burdens cannot be shouldered by poor nations and poor actors in the same way. Whereas this claim is already covered by the ability-to-pay principle, this section maintains that a reduction of burdens does not result from poverty or an inability to contribute, but rather from a calculation of proportionality.

Nation-states that already suffer to a much higher degree from climate change than other nation-states should be exempt from bearing the full portion of proactive burdens. Moreover, nation-states that already invest in environmental goals, for instance in terms of guardianship duties – protecting forests as sinks, or protecting goods as natural heritage – should also be exempt from paying the whole range of costs for global mitigation goals. Another issue is their ability to cope with environmental threats, due to a lack of capacity to "cushion" or protect themselves from hazards. Natural goods are not only particularly unevenly distributed on Planet Earth. In addition, societies differ regarding their capacities to protect themselves against the impact of environmental hazards. Actors who suffer much more from worsened climate conditions because, for example, local agriculture happens to be the foundation of their economies, should also be exempt from paying the whole amount of the burdens for accumulative goals. Notably, Tracey Skillington is right in highlighting the fact that climate change relates to many particularly dramatic scenarios of human rights violations. Climate change forces people to migrate, causes the disappearance of nation-states, and creates climate refugees;[53] climate change undermines rights to self-determination over national resources;[54] and it threatens to undermine related environmental rights, such as the right to water.[55] People who already suffer these severe consequences should not be exempt from climate duties – this would hurt them. But they should be able to choose their environmental goals and their rescue goals according to their priorities.

9.7 Principles of special obligations in role ethics

A second group of principles for allocating burdens according to an idea of fairness results from a *role ethics*. It supports the idea of actors having special obligations according to their role in public life and in international relations. I have stated that joint climate goals should be supported by all actors. Different actors have different degrees of responsibility for the common good, though. Martha Nussbaum maintains that the idea that governments need to serve the common good, which dates back to Aristotle, has been one of the most persistent ideas in political philosophy.[56] One example is given by John Broome. In establishing a climate ethics, he distinguishes between a private and a public morality.[57] One important element of his proposal is the idea that governments have climate duties that differ from private duties, because they have to care for justice. Broome makes this distinction in the context of climate duties:

> It is partly because, if you as a private individual aim to promote good in the world, reducing your emissions of greenhouse gas is a poor way to do it. It does indeed bring real benefits, but you could do much more good by using whatever resources you have in other ways. Nevertheless, you ought to reduce emissions for the simple reason of justice that you are not morally permitted to harm other people even in order to bring about more good overall. Governments have a stronger moral mandate than individuals to make things better. It is one of their principal duties to make things better for their own citizens, and they should cooperate to make things better for everyone.[58]

In order to fulfil this duty towards climate goals, governments need to determine the most efficient and value-based options for enhancing climate goals. The issues Broome explores at length are the new roles of uncertainty in decision theory as well as the evaluation of population – with the not far-fetched expectation that populations might become extinct from the consequences of climate change.[59] In this approach, the role of governments is not only to promote the well-being of the population, but also includes the duty to engage in thorough evaluations that can best be prepared by scientists, but need to be put on the public table by governments.

Two different aspects of role ethics need to be distinguished here. In accordance with the distinction of first-order responsibility and second-order responsibility as presented in the introduction of this chapter,[60] role ethics for political actors also has two sides. On the one hand, it addresses the obligation to shoulder a certain share of burdens in fulfilling joint climate goals. These are first-order responsibilities. In addition to this proposal, role ethics also claims that actors who as legitimate political authorities also should protect individuals from harm and work in favour of the common good of the people also hold second-order responsibilities. Furthermore, a joint action approach also claims that they have the duty to arrange opportunities for joint action in favour of climate goals. They have to care about green infrastructure, the energy sector, and other public goods in a way that best honours climate goals.[61]

Two objections against a special obligation of governments to accomplish climate duties also need to be addressed. Governments have played an extraordinary role in the literature on climate justice, but global climate goals have yet to be achieved in a meaningful way. Representatives of nation-states sign contracts, including the Paris Agreement. And they can be held accountable when forfeiting the chances of future generations to lead a decent life. But it has to be noted that nation-states, although heavily responsible, are not exclusively responsible. Though they hold authorized power, they also need to be regarded as dependent on social readiness for climate protection and for a multitude of interests, many of which counter-balance strong climate politics.

Furthermore, governments are not passive elements in chains of events that they have not initiated, but should modify and channel them according to principles of justice and fairness. In other words, a political role ethics is welcome, and it is necessary to remind political institutions of their role, as many political philosophers do. But even political institutions will not be able to channel causal events that are created by many different actors. In addition, depending on the interpretation of the private sector, collectives in that field of performance should also be regarded as restricted not only by law, but also by codes of conduct that support the view that the enhancement of welfare is not the single goal of the activities.

9.8 Criteria for allocating a fair share of proactive burdens

At the outset, the debate on climate justice was a debate on principles to guide the allocation of the remaining space for greenhouse gases in the atmosphere in an appropriate way. It then developed into a discussion about a fair share of burdens, primarily since the claim to support climate victims as well as the call for collective action for adaptation became central. Overall, much ink has been spilled in determining fair distributions of the costs that result from mitigation, adaptation, and other types of preventive action.

In the context of this study, I have not followed this path. Instead, I have offered a joint action approach that includes group action on a local level. From this paradigmatic case of joint environmental action, the theoretical approach has been developed in various directions. One critical step was the proposal to construct a model of voluntary action within an obligatory action approach. Climate duties were formulated accordingly. In this chapter, the discussion was brought to another level. The model of climate duties was applied to collectives that count as responsible actors. In accordance with proposals to apply proactive burdens according to principles of a fair share in responsible action, the theoretical proposal focused on principles of fairness in allocating proactive burdens to collective actors.

The main outcome of the discussion of this chapter's debate is that distinct principles of fairness help to allocate proactive burdens in a justified way. The proactive burdens illustrate the responsibility to support the effort of actors to achieve environmental goals together. Above all, the principles legitimate the allocation of burdens without relying on general assumptions about the responsibilities of particular actors. Instead, they focus on properties that collective actors possess. This means that the fair share does not depend on whether a collective is a nation-state or a company. Instead, the fair share resonates with other properties of collectives, including their economic power, their capacity to enforce compliance, and the dedication to goals of common interest and the well-being of citizens.

Two more issues have not yet been discussed. They need to be settled in agreements among the actors who will work together. The first question is the one of off-setting. Should actors be allowed to buy out their contributions to climate goals by supporting other environmental goals?[62] In my view, this issue needs to be negotiated, since the impact on the goals and the probability of achieving them cannot be calculated unless the issues are rendered more precise. The other issue is how burdens should be handled if not related to one and the same goal, but to various goals. Is it justified to outweigh goals against each other? Or do some goals have absolute priority? These questions need to be carefully addressed.

At first glance, it appears the better strategy is to try to accomplish various important environmental goals simultaneously. Notably, it is helpful to regard the goals not as being exclusive, but rather as mutually supportive. Mitigation can help to reduce integrated environmental goals to aims that are much easier to achieve than goals without effective mitigation also in place. The same holds for adaptation as related to mitigation and to integrated environmental goals.

In addition to explaining each of the principles for adjusting the burden-sharing responsibilities of various actors, something more needs to be said regarding their relationship. Overall, these principles can provide guidelines for allocating proactive burdens. In any case, the definite outline of these principles also depends on the specific goal. For climate goals, the principles certainly help to frame the agreements on the fair share of costs of various nation-states in achieving the goals of the Paris Agreement. Regarding local environmental goals, the principles can only serve as a yardstick for negotiations among the actors involved in the various goal-setting processes.

Notes

1 This critique differs from the critique of Weisbach that theories of distributive justice work against feasible solutions, but also coheres with it at a distinct point. For Weisbach's critique of climate justice approaches, see Stephen M. Gardiner and David A. Weisbach, *Debating Climate Ethics* (New York: Oxford University Press, 2016), 202–236. I agree with Weisbach's concern that the whole idea of distributive justice, including the claim of corrective justice, is a "climate blinder" when framed as a distributive scheme that allocates a fair share of the atmosphere as a waste dump to various consumers. First, there is not enough space left for additional emissions; second, and more importantly, an institution that would actually account for the equal right of individuals to emission budgets would have to calculate payments according to existing unfair distributions. This would not only force high-emissions countries to make considerable payments, it would also force poor countries that are high emitters to make large payments. See Gardiner and Weisbach, 232–236. Overall, this proposal neither reduces emissions effectively nor supports joint action.
2 See Section 6.5 on negotiated fairness, according to Ostrom's example of the Thambesi River.
3 For this concept in the context of natural sciences, see Johan Rockström et al., "Planetary Boundaries: Exploring the Safe Operating Space for Humanity," *Ecology and Society* 14, no. 2 (2009).

4 For an explication of this term, see Simon Caney, "Two Kinds of Climate Justice: Avoiding Harm and Sharing Burdens," *Journal of Political Philosophy* 22, no. 2 (2014): 125–149.

5 Caney, "Two Kinds of Climate Justice," 127.

6 This distinction is introduced in a comparable way by Caney, "Two Kinds of Climate Justice," 134–135.

7 Caney, "Two Kinds of Climate Justice," 127.

8 Stephanie Collins, "Distributing States' Duties," *Journal of Political Philosophy* 24, no. 3 (2016): 347, https://doi.org/10.1111/jopp.12069.

9 See ibid., 352.

10 Note that the proposal also leaves room for a scenario in which duties are not over-determined, but assigned rather loosely to groups. See Collins, "Distributing States' Duties," 359.

11 One central rationale for defending claims of compensation is the polluter-pays principle, which – very simplified – says that actors who have caused environmental harm, should also compensate for it. See Simon Caney, "Cosmopolitan Justice, Responsibility, and Global Climate Change," *Leiden Journal of International Law* 18 (2005): 125–129; Stephen M. Gardiner, "Ethics and Global Climate Change," *Ethics* 114 (2004): 555–600; Robert E. Goodin, "Selling Environmental Indulgences," *Kyklos* 47 (1994): 243–244; Dale Jamieson, "Adaptation, Mitigation, and Justice," in *Perspectives on Climate Change*, ed. Richard Howarth (Amsterdam: Elsevier, 2005), 267; Anton Leist, "Klimagerechtigkeit," *Information Philosophie* 5 (2011): 8; Henry Shue, "Global Environment and International Inequality," *International Affairs* 75, no. 3 (1999): 103; Henry Shue, "Subsistence Emissions and Luxury Emissions," *Law and Policy* 15, no. 1 (1993): 39–59; Peter Singer, *One World: The Ethics of Globalization* (New Haven, CT: Yale University Press, 2002), 187ff.

12 Recently, some of the most important points have been cogently debated by Steve Gardiner and David A. Weisbach in *Debating Climate Ethics*.

13 For proposals for contracts on a global level that encourage action and that nevertheless integrate claims of fairness, without integrating the claim of distributive justice as also correcting historical injustice, see John Broome, *Climate Matters. Ethics in a Warming World* (New York/London: W.W. Norton, 2012); Eric A. Posner and David A. Weisbach, *Climate Change Justice* (Princeton, NJ: Princeton University Press, 2010). Both proposals have received a thorough critique from Gardiner. See Stephen M. Gardiner, "Climate Ethics in a Dark and Dangerous Time," *Ethics* 127, no. 2 (January 2017): 430–465, https://doi.org/10.1086/688746. Gardiner is right that a contract is unacceptable when it causes deep injustice, as for instance implied in the proposal not to correct historical injustice through exemptions for poorer countries.

14 See the proposals for goal-directedness and for a joint effort of all actors in Caney, "Two Kinds of Climate Justice"; Gardiner and Weisbach, 121–133; Henry Shue, "Climate Hope: Implementing the Exit Strategy," *Chicago Journal of International Law* 13, no. 2 (2013): 381–402.

15 This appears to be an implication of Weisbach's critique of climate change justice (Gardiner and Weisbach, 170–244). It is also reiterated in approaches that provide a win-win scenario by means of setting incentives in favour of climate goals, though with the price of overriding claims of justice.

16 Arguments for climate economics have been most prominently formulated in Nicholas Stern, *The Economics of Climate Change. The Stern Review* (Cambridge: Cambridge University Press, 2007). In the meantime, many more proposals have been made for designing a goal-driven climate politics that includes incentives for the achievement of climate goals. For a summary of these proposals, see Ottmar Edenhofer, Christian Flachsland, and Steffen Brunner, "Wer besitzt die Atmosphäre?:

Zur Politischen Ökonomie des Klimawandels," *Leviathan* 39, no. 2 (June 2011): 201–221, https://doi.org/10.1007/s11578-011-0115-0.

17 Caney, "Two Kinds of Climate Justice," 126.

18 Ibid., 127. Note that these are two conditions of a conclusion, in which Caney explains under which conditions both types of justice coincide. This is the case when harm-avoidance is part of the achievement conditions of the goal whose burdens will be distributed.

19 Caney, "Two Kinds of Climate Justice," 127.

20 Peter A. French, *Collective and Corporate Responsibility* (New York: Columbia University Press, 1984).

21 Ibid., 10–13.

22 Ibid., 13.

23 Toni Erskine, "Assigning Responsibilities to Institutional Moral Agents: The Case of States and Quasi-States," *Ethics & International Affairs* 15, no. 2 (2011): 72, https://doi.org/10.1111/j.1747-7093.2001.tb00359.x.

24 Christian List and Philip Pettit, *Group Agency: The Possibility, Design, and Status of Corporate Agents* (Oxford/New York: Oxford University Press, 2011); Philip Pettit and David Schweikard, "Joint Actions and Group Agents," *Philosophy of the Social Sciences* 36, no. 1 (March 2006): 18–39, https://doi.org/10.1177/0048393105284169.

25 This is a paraphrase of the account outlined in List and Pettit, *Group Agency*, 155.

26 List and Pettit see this problem. Their solution stems from the idea that in addition to a causally relevant reason for single actors to act, programming reasons apply to the group as an agent. This helps to control the agent's action, in addition to the remaining responsibility of group members to do their share in achieving joint responsibility. See List and Pettit, *Group Agency*, 159–163.

27 Stephanie Collins and Holly Lawford-Smith, "Collectives' and Individuals' Obligations: A Parity Argument," *Canadian Journal of Philosophy* 46, no. 1 (January 2, 2016): 38–58, https://doi.org/10.1080/00455091.2015.1116350.

28 Ibid., 41.

29 Ibid.

30 Ibid., 43.

31 Ibid., 55.

32 Ibid., 46. Note that t_1 and t_2 are arbitrarily chosen, consecutive moments of time.

33 Tracy Isaacs, "Collective Responsibility and Collective Obligation," *Midwest Studies in Philosophy* 38, no. 1 (September 2014): 42, https://doi.org/10.1111/misp.12015.

34 Note that this discussion differs from the question of whether obligations among members of a group need to be presumed. In the account of joint action that has been explored in Chapters 4 and 5 I have maintained that group members are tied to each other by a joint ethos and that they differ from actors in a market-style cooperation in that they are not predisposed to cheat. However, in order to undermine cheating in constellations of shared agency beyond single groups, it is helpful to have a decision-making structure and also to determine a scheme of cooperation that allocates burdens resulting from the shared goals to individuals according to principles of fairness.

35 Note that this claim is defended in, for example, Broome, 49–72; Martha C. Nussbaum, *Frontiers of Justice. Disability, Nationality, Species Membership* (Cambridge, MA: Harvard University Press, 2006).

36 This principle also differs in its pragmatic nature from the moral claim to adjust moral obligations of beneficence according to normative criteria; for a discussion of this relationship, see Elizabeth Cripps, *Climate Change and the Moral Agent: Individual Duties in an Interdependent World* (Oxford: Oxford University Press, 2013), 48–56.

37 Simon Caney, "Climate Change and the Duties of the Advantaged," *Critical Review of International Social and Political Philosophy* 13, no. 1 (March 2010): 213, https://doi.org/10.1080/13698230903326331.

38 Ibid.

39 Ibid. However, most of the discussion of the principle in this article is based on the assumption that ability to pay needs to be judged in the context of the causal history of prior emissions. Supposing the wealthy are already low-end emitters and their wealth is gained without any environmental harm, it might be unfair to claim additional burdens (Caney, "Climate Change and the Duties of the Advantaged," 213–218). Still, the link between former emissions and wealth does not have to play a role in discussing the validity of the principle.

40 Caney, "Climate Change and the Duties of the Advantaged," 213–218.

41 Similar arguments about payments in favour of the common good have been stated with respect to tax policies in Stephen Holmes and Cass R. Sunstein, *The Cost of Rights: Why Liberty Depends on Taxes* (New York: Norton, 1999).

42 Shue, "Global Environment and International Inequality," 537. An important objection is also raised by Shue. He discusses disincentive effects of progressive schemes; he also states that this is neither a sound objection against the principle, nor can motivational aspects be determined on general grounds, but need to be discussed case by case (Shue, "Global Environment and International Inequality," 539).

43 Peter Singer, *The Most Good You Can Do: How Effective Altruism is Changing Ideas about Living Ethically* (New Haven, CT/London: Yale University Press, 2015).

44 One strand of the debate is related to historical responsibility in the forms of restorative justice or of compensatory justice. For "historical responsibility," see the groundbreaking discussion in Lukas H. Meyer and Axel Gosseries, eds., *Intergenerational Justice* (Oxford/New York: Oxford University Press, 2009). For a critical discussion of the polluter-pays principle that many accept as the most fundamental principle in environmental ethics, including the main objections, see Caney, "Climate Change and the Duties of the Advantaged," 205–213. The introduction of the beneficiary-pays principles is based on the observation that it is difficult to hold current generations responsible for harm that former generations produced; what is instead arguable is a compensation according to today-gains from a causal history that (a) has a direct impact on the high-level activities of industries today, and that (b) was correlated with high emissions. For this idea, see Simon Caney, "Cosmopolitan Justice, Responsibility, and Global Climate Change," in *Climate Ethics. Essential Readings*, eds. Stephen M. Gardiner et al. (Oxford/New York: Oxford University Press, 2010), 128.

45 For a critical discussion of the polluter-pays principle that many accept as the most fundamental principle in environmental ethics, including the main objections, see Caney, "Climate Change and the Duties of the Advantaged," 205–213.

46 The question of whether motor-vehicle drivers are responsible for the suffering of climate victims has been part of an intense exchange between Gardiner and Jamieson, who hold different views. For a summary of the arguments through the eyes of Gardiner, see Gardiner, "Climate Ethics in a Dark and Dangerous Time," 437–440.

47 Arguments in favour of liability with respect to climate harm have immanent legal consequences. For an example, see Allen Myles, "Liability for Climate Change. Will it Ever be Possible to Sue Anyone for Damaging the Climate?," *Nature* 421 (February 27, 2003): 891–892.

48 For an overview of arguments along these lines, see Björn Petersson, "Co-Responsibility and Causal Involvement," *Philosophia* 41, no. 3 (September 2013): 847–866, https://doi.org/10.1007/s11406-013-9413-x.

49 An insightful discussion of the elements of knowledge, intention, causal connectedness, and culpability relates to the notion of liability as a central element in McMahan's approach to killing in war. See Saba Bazargan, "Complicitous Liability in War," *Philosophical Studies* 165, no. 1 (2013): 177–195; Jeff McMahan, "The Basis of Moral Liability to Defensive Killing," *Philosophical Issues* 15, no. 1 (2005): 386–405.

50 This is the underlying rationale for the liability criterion in war ethics. See Cécile Fabre, "Guns, Food, and Liability to Attack in War," *Ethics* 120, no. 1 (2009): 36–63, https://doi.org/10.1086/649218; Thomas Hurka, "Liability and Just Cause," *Ethics & International Affairs* 21, no. 2 (2017): 199–218.

51 Myles, 892.

52 Ibid.

53 Tracey Skillington, *Climate Justice and Human Rights* (New York: Palgrave Macmillan, 2017), 177–198.

54 Ibid., 126ff.

55 Ibid., 207–224.

56 See Martha C. Nussbaum, "Aristotelian Social Democracy," in *Liberalism and the Good*, eds. R. B. Douglass, G. M. Mara, and H. S. Richardson (New York: Routledge, 1990), 203–252; Nussbaum, *Frontiers of Justice. Disability, Nationality, Species Membership*.

57 See Broome, 49–96.

58 Ibid., 188.

59 See ibid., 178–183.

60 See Caney, "Two Kinds of Climate Justice," 134–135.

61 A comparable argument about political obligations has been made regarding single actors who as politicians hold responsibilities that differ from the responsibilities of citizens. Some also claim that a special obligation also results from the function of people as role models. Politicians as well as individuals in the media not only affect the climate by means of private actions, but also have an impact on how other people behave. They function as role models. Therefore, they have a special obligation to display praiseworthy behaviour. See Bernward Gesang, "Gibt es politische Pflichten zum individuellen Klimaschutz?," in *Klimagerechtigkeit und Klimaethik*, ed. Angela Kallhoff (Berlin/Boston: De Gruyter, 2015), 135–142. Although the logic of visible leadership is very important in joint action, and although it is highly desirable that public personalities help to prevent the climate catastrophe, it is difficult to make a moral claim based on that, for two reasons. First, the function of a role model is not an obligation, except in political public service, which comes with such an ethos. Second, I suspect that the effects of role models might be overstated when it comes to joint environmental action.

62 For the concept of "off-setting" and its moral implications, see Edward Page, *Climate Change, Justice and Future Generations* (Cheltenham, UK/Northampton, MA: Edward Elgar, 2007).

10 Some conclusions: Institutions and responsibilities

Bert: Wow, that's it. This is a perfect blueprint for claiming concerted action for climate goals.

Cindy: Yes, let's recall the most important outcomes. A theory of joint action explains how concerted action in favour of a shared goal works.

- This theory can be applied to climate goals, because climate goals can best be framed as "joint action goals." Climate goals include goals of mitigation, of adaptation, and of transformation.
- And climate goals can be adjusted to various actors. A condition for getting action done is precisely that: a framing of climate goals that works for distinct types of groups.

Bert: Does this mean that climate goals need to be reframed?

Cindy: Yes, I think this is an important lesson from the discussion of joint action. People do what they are convinced of. They need motivation; and groups are good at supporting the motivation of all. But this also has a shortcoming. What about goals that nobody wants to achieve, because they're costly, hard to get, and unattractive?

Bert: You mean something like "merit goods"? Merit goods are goods that each person should be in a situation to enjoy but that no one is willing to support. Traditionally, the welfare state cared for that type of goods. Do you really think climate goals are like health care etc.? This would make the problem even bigger.

Cindy: You're right; no – not merit goods. I think climate goals and other environmental goals are really more like future goals of groups that know of their influence on nature, but have some options to model their future behaviour. It's really more like "active co-designing with nature," as explained in Section 6.1. We, that is, humankind, have a deep influence on water resources, the climate etc. anyway. But "we" can decide what this influence looks like in the future. Climate scientists speak about possible "pathways" into the future. But a model of joint climate action is not interested in forecasting, but in goals that could be achieved by a concerted effort. It's more like thinking about goals that are important and scientifically

> proven to be important, like a certain amount of reduction in greenhouse
> gases or methods to achieve a good adaptation to rising sea levels. Actu-
> ally, I'm now convinced that climate ethics is not so much about what to do,
> but rather about how to work together.
>
> Bert: Right. And I think it's also fair to allocate responsibilities according
> to various principles – always claiming that governments should take
> responsibility is certainly not enough.
>
> Cindy: However, I'm still wondering what this theory includes for our real
> world.
>
> Bert: Yes, let's look into the conclusions.

Political philosophy and ethics are highly abstract subject areas. But climate
philosophy is different. Climate philosophers are driven by issues of high
social and moral significance. Climate ethics is also driven by the belief that
the achievement of climate goals is pressing and urgent. Nevertheless, from
the perspective of climate ethics, principles of fairness of burden-sharing and
principles of proactive responsibility need to be thought through first. They
are an essential element in practical proposals.

One way to interpret climate ethics corresponds with the distinction between
ideal theory and *non-ideal theory*. The distinction, which was introduced by John
Rawls in his philosophy of justice,[1] has since then been continually discussed.
Recently, it has been applied to climate ethics.[2] Since it might not be far-fetched
to frame a chapter on responsibility and institutional demands as the non-ideal
part of the joint climate action approach, it is necessary to first give a brief
comment on insights into "non-ideal theory."

Laura Valentini distinguishes three interpretations of "non-ideal theory" in
current philosophical discourse.[3] First, a non-ideal theory represents the shift
from the presumption of full compliance to partial compliance, whereas the
latter is the realistic assumption.[4] This interpretation comes closest to Rawls's
initial concept of "non-ideal theory."[5] Second, the distinction between ideal
and non-ideal can be equated with "utopian" or "idealistic" versus "realistic"
theory. A non-ideal theory considers that feasibility constraints not only matter
as such, but they also matter on the level of theory-building.[6] Third, the distinction also indicates the difference between an end-state theory and a transitional theory.[7] Whereas non-ideal in the first two meanings relates to a general
normative theory, the latter aims at societally possible change.[8]

According to this proposal regarding non-ideal theory, this chapter is part
of a non-ideal theory in its first meaning. It emphasizes the problem of non-
compliance with respect to climate duties; it also discusses various types of
defection that might occur in the context of already agreed-upon types of
joint climate action. But in another respect, the distinction between ideal and
non-ideal theory is misleading. In my interpretation, the second and the
third meanings apply to the theoretical approach to climate ethics as argued

in this book. The presumption underlying a theory of joint action coheres with assumptions in transitional theory, which propose how end-states (in terms of environmental goals) can best be achieved. In addition, it offers a realistic approach by also including facts about human nature, in particular, when it comes to facts about cooperation and about natural goods. It is not the goal of a theory of joint climate action to portray an ideal world or a utopia, but rather to develop proposals to get from an unsatisfactory status quo to a situation that is desirable and fair, even if still not ideal.

Nevertheless, this chapter also differs in certain ways from the rest of the book. It explicitly considers the possibility that (a) actors are not ready to invest in joint climate action, but instead prefer to defect, even when a joint approach to collective action is available and possibly even agreed upon; and that (b) today's political institutions are not yet ready for joint climate action, but instead are hindering the realization of the proposed terms of cooperation.

Although parts of a theory of joint climate action are about climate duties and although the questions of moral responsibility and moral duties do need to be answered, it is important to also explore the institutional frame supporting and enabling climate action. Joint climate action has been described as a voluntary endeavour that coheres with well-reasoned and fair goals of group actors. This chapter goes back to that paradigm and also explains what institutions – particularly political institutions – can do to support voluntary joint climate action. Still, the discussion of institutions does not end there. This chapter goes one step further in its exploration of an institutional frame for climate obligations. Although still tentative, this includes ideas about how institutions currently in place can reduce non-compliance with justified frameworks, including climate duties.

The chapter is organized in six sections. The first discusses the normative claim that is the backdrop for all the normative claims in this chapter: power is immediately related to responsibility. This means that powerful institutions, including political institutions, have to take responsibility for responding to climate change. However, this responsibility has several parts, and one of them is the duty to remove barriers to joint climate action. Section 10.2 explains this claim. Removing barriers presumes insight into the types of barriers, which will also be sketched in the section. In addition to real physical barriers, psychological as well as theoretical barriers are major obstacles to change. This section also addresses another often-overlooked problem, which is the invisibility of climate goals. In Section 10.3, I present another problem that is also part of the most basic assumptions in this study. This is the problem of non-compliance. Although multiple publications have already established that non-compliance should be taken seriously, this section addresses types of defection that immediately relate to inadequacies of collective action. Defection results not only from wilful deviation and from the unwillingness to cooperate, but also from an inability to cooperate or from strategic insights. A distinction of various types of defectors is obligatory for developing a clear view of appropriate instruments to stop defection or to ameliorate the means of cooperation. Section 10.3 describes various types of defectors. Section 10.4 introduces some initial thoughts about a better design of institutions in order to enhance joint action and to cope with barriers as well as with defection. Section 10.5 continues the debate on institutions and addresses institutions in the international arena. Section 10.6 presents some initial ideas for governmental duties and Section 10.7 concludes the chapter with a call for action. The whole chapter is limited to some initial thoughts of a theory of institutional design for collective climate action that is currently in preparation.

In particular, this chapter draws practical conclusions from the theoretical insights in this book. In particular, it presents ideas about necessary institutional change and of the need for new institutions for joint climate action. Although the chapter is too brief to address institutional change in depth, and too abstract to suggest practicable governance tools, I nonetheless hope it will help to deepen the ongoing discussion of applied climate ethics and of adequate social and institutional prerequisites for successful climate governance.[9]

It is also important to point out that this analysis of the responsibility of governments and political institutions is far from complete. It is not my aim here to develop an institutional theory. Instead, this chapter has the more modest intention of highlighting some of the consequences that need to be drawn from the interpretation of climate goals as joint action goals. In that interpretation, the first duty of governments is not the realization of climate goals – they are not even realistically able to do so. Instead, they should support, protect, and enable joint environmental action and joint actors respectively.[10] This chapter works through various layers of this claim. It starts with what might be called the "domestic responsibility" to support joint climate action, addressing the "empowering state" and the "enabling state." It then works through scenarios that are already framed in international environmental contracts such as the Paris Agreement. And it then also points to an even more comprehensive perspective: the world community.

10.1 Power and responsibility

This chapter starts with the now well-argued view that governments and legitimate political institutions more generally are not exclusive duty-holders towards climate change justice and the more encompassing climate goals. Instead, it has been argued that responsibility needs to be allocated to various actors; what counts for this allocation are features such as ability to pay, liability, and fairness.[11] In addition, this chapter starts with the thesis that Simon Caney calls the *power/responsibility principle*.[12] This principle states that "with power comes responsibility."[13] It maintains that powerful institutions hold responsibility, especially institutions that are obliged to protect the people. In Caney's interpretation, this principle has immediate relevance for climate justice: "It posits that those with the power to compel or induce or enable others to act in climate-friendly ways have a responsibility to do so."[14] Caney is, however, cautious in explaining the concrete content of the principle and its limitations. Government responsibilities regarding climate protection are not absolute, but can also be overridden by more urgent claims.[15] This insight is important in order to not drift into an emergency ethics and in order not to overstate the relevance of climate duties in concert with other obligations of responsible institutions.

To justify the link between power and responsibility, Caney points out the prospect of disastrous harms, the special role of governments, and their possible effectiveness.[16] Since arguments have already been provided for allocating responsibilities to structured collectives, and to governments in particular, we do not have to discuss Caney's argument at length. What is, instead, debatable is the *content* of second-order responsibilities.

Powerful institutions, and political institutions in particular, have many different responsibilities, even with respect to climate change. Yet one of the most important aspects is often overlooked. Governments should remove obstacles to joint climate action and to joint environmental action in general. This is a realistic expectation. The enabling state is not a recent invention – but it receives a new meaning when addressing it in the context of climate change. Governments are in a position to design law and to initiate governance practices; they are particularly in a position to enhance the conditions of climate-friendly transport systems, of a carbon-free energy supply, and of a food sector with climate-friendly practices. This is the role of government this chapter builds on – its baseline is care for the common good of the populace.

10.2 Barriers to joint climate action

One of the most important duties of governments is to support removing barriers in the way of collective climate action. The goal of removing barriers can, however, only be accomplished when the existing types of obstacles are known, studied, and interpreted in the right way. The attempt to reach demanding environmental goals and climate goals in particular is confronted with many different types of barriers: behavioural barriers, physical barriers, institutional barriers, and mental barriers. I shall discuss each of them in turn. It is particularly important to keep them apart in order to explore the logic of each one as a special type of obstacle.

Behavioral barriers. In their daily life, individuals act at least to some degree according to their own preferences.[17] More specifically, people do not rationalize every single act, but simply do things that accord with preset preferences. Still, not all preferences are fixed. Action that resonates with environmental goals, moreover, does not comprehensively depend on preferences, but on how social and environmental goals are being framed. More specifically, the incentive to cooperate depends on the cohesion among individual values and social goals. In that context, group identity also plays an important role.[18] Another important aspect that has an immediate effect on climate change behaviour is the discrepancy between the scientific interpretation of climate change and its threats, as compared to the public understanding of climate change.[19] Although the proposal to frame climate action as joint climate action is not primarily based on the need to make people understand that climate change poses grave risks, the lessons from psychological studies are important. Obviously, the gap between scientific analysis and behavioural capacities needs to be taken seriously.

Physical barriers. Unlike patterns of behaviour related to psychological facts, obstacles to collective responses to climate change also result from physical barriers. Physical barriers are materialized constraints that prevent joint action in favour of climate goals. Regarding schemes of exchange with nature, the option to reap benefits from natural common-pool resources (NCPRs) and from nature in terms of water supply and food as well as waste disposal are pre-structured by many different institutions. They process food and provide societies with water services, channel movement through infrastructure, and so on. In addition, when institutions of supply are part of a market mechanism, it is likely that exchange is not channelled in favour of joint environmental action. This is not the place to discuss economics. But a general assessment includes the observation that a market economy does not automatically contribute to joint environmental action, but instead contributes to physical barriers that hinder environmental action.

Institutional barriers. Nation-states have sovereignty, which means that they cannot be forced to share worldwide burdens of mitigation and of adaptation. This means that institutional barriers need to be taken into account when framing international climate policy. Yet even on the level of nation-states, institutional barriers exist. Among the many particularly problematic aspects of climate change with respect to institutional change that Dale Jamieson explores,[20] one often-overlooked insight is particularly important for our investigation of barriers. Jamieson explores a set of deep differences between science and policy-making.[21] He argues that in the United States the split between the community of politicians and the community of natural scientists and academia more generally created a barrier that thwarts collective action in the case of climate change. This has many consequences, including the failure of climate science to provide concepts that policy-makers can use. Overall, Jamieson concludes: "Climate science has been a success story. Whatever its failures to adequately inform policy, the larger failure has been that we have not enacted climate policy. Ultimately, the failure to take action on climate change rests with our institutions of decision making, not on our ways of knowing."[22]

Mental barriers. Joint environmental action needs a climate and a surrounding in which people and representatives of groups are not only able to regard environmental problems as part of their agendas, but also able to imagine a coordinated solution to environmental problems. Yet the history of failures in climate negotiations demonstrates that this imagination has been spoiled by many factors. As Sayer notes:

The project of stopping global warming through regulation runs up against not only the accumulation-based nature of capitalist economy and culture, but the enormous social-spatial inequalities it has generated, for they in themselves present a huge barrier to the development of collective responses.[23]

But it is not only this aspect that challenges collective responses. In addition, the resignation even of theorists who are among the most prominent authors in the field of climate ethics is a widespread phenomenon. Furthermore, some theorems that support the view that environmental action is useless have also become mental barriers.

An example in this case is, if not the "Green Paradox,"[24] at least its popular understanding. As Sinn states, from an economic perspective, emissions of carbon dioxide from fossil fuels, which are still the largest source of greenhouse gas emissions, may not go down in response to policies demanding reduction. Instead, he claims that it is possible for global emissions to increase in reaction to green politics. Although this claim might be true, an overly general appreciation of this theoretical insight leads to a "mental barrier." It then puts off climate-friendly energy politics with the argument that "others" will, with possibly more dirty technologies, take the carbon-intensive energy carriers in any case.[25]

At least in a common interpretation, this theorem states that the saving of an environmental resource is useless, since other parties will immediately turn to that resource and exploit it. This line of thought also serves as a mental barrier to environmental action. Yet defences of the green paradox overlook that the alleged rationality of such claims is due to over-simplification. It supposes that everything that has to be done – and every-thing that could possibly be done – is a quantitative sum game. Yet the over-exploitation of natural resources is due to a range of further factors, includ-ing a lack of conservationist strategies and water management strategies, and a lack of mindful care for a critical resource.

Misinformation or lack of information. In addition to theorems that negate the availability of cooperative solutions, it also has to be taken into account that mental barriers result from not being educated regarding climate change. In order to be willing to act, people first need knowledge. Yet instead of framing this knowledge as insights into what natural scientists are saying about climate change, much of what has been debated publicly amounts to "organized denial."[26] People need to know what they can actually do to support climate goals. Misinformation should not be tolerated. In addition, programmes are needed that really explain to people what they can do on a daily basis to support climate goals.[27]

Invisibility of climate goals. To enhance climate cooperation, another pro-blem needs to be addressed: the sense that environmental goals are "far away" from people's daily lives – at least until people begin directly suffering from environmental hazards. Even when environmental hazards are doing damage, environmental goals are still often invisible to the population.[28] In order to achieve a shared commitment to environmental goals, it is important to make environmental goals visible. They should be put onto the agendas of local governments. They should also be debated openly and shared with the population in the media.

It has often been observed that the achievability of joint goals also depends on the scale. Global goals are difficult to realize; local goals instead cohere with the presumption that small groups enhance the motivation to participate.[29] This study argues a second claim. Climate goals need to be liberated in some respect and to some degree from their negative image, to be re-interpreted and re-integrated into practices of group actors related to achievable goals within their reach. In comparison to active co-designing with nature in the theorem of integrated water management, climate goals need to be reinterpreted as goals that groups of people wish to realize and that cohere with a "life with nature" that is fair and good.

This list of failings and "barriers" could easily be extended. This brief account suffices to reinforce the claim that it is not always a lack of motivation that stands in the way of joint climate action. Instead, forces undermining the will to cooperate and the capacity to act jointly on chosen goals should not be underestimated.

As a consequence, an important first step towards institutions that support climate action is the awareness of barriers and the attempt to remove them. In my view of the responsibility of political institutions, this is also an important aspect of the second-order responsibilities of governments. This might not, however, be sufficient in the current situation. Instead, governments have the additional duty to enforce the schemes of cooperation I have proposed in the preceding chapters.

10.3 Types of defectors

Non-compliance with respect to moral obligations that are well founded and that also imply duties of political institutions has already been discussed in the context of theories of climate ethics.[30] Unlike non-compliance in general, *defection* addresses a scenario in which cooperation takes place and in which one or several people do not comply with the agreed-upon terms of cooperation. This definition includes scenarios in which rules are not explicitly reasoned, though the terms of cooperation have been fixed. In the case of climate action, this includes small incidents of cooperation as in the example in which a group of people decides to "green the city" and to work together in order to realize that goal. It also includes large-scale schemes of cooperation, as agreed among nation-states in the Paris Agreement. Defection can be either in-group defection within the group (internal) or defection among group actors (a whole group defecting).

Imagine that seven out of nine nation-states are cooperating in favour of a joint environmental goal – for example, for a fair scheme of appropriation of water resources from a river that runs through all nine nations – but two nations do not cooperate. Since nation-states are sovereign, the two defectors cannot be forced to cooperate. The way to enhance cooperation might presume a better understanding of why the two nation-states defect. In order to remedy defection or possibly also to hold defectors responsible, a first step is the inquiry into the types of defectors.

Five different types of defectors can be distinguished. These are:

- The nervous defector
- The offended defector
- The strategic defector
- The incapable defector
- The reckless defector

In order to explain the figure of the defector, I first recapitulate five standard cases, two of which Richard Arneson has presented in his argument on defection.[31] The interpretation of cooperation in his theoretical framing was also important for outlining climate duties.[32] In the context of this chapter, it suffices to focus on the interpretation of defection in that context. The analysis of cooperation as well as of defection rests on the interpretation of a cooperative scheme as related to a public good.[33] In his analysis of a cooperative scheme, Arneson is concerned with the problem of the free-rider. In a standard analysis, public goods invite people to join the behaviour of a free-rider. A free-rider is not willing to cooperate in terms of contributing to the costs of supply; instead, she benefits from a situation in which a public good will be provided regardless of her cooperation. She focuses on her self-interest in benefitting from the good. Arneson argues that there are two further and related rationales for not contributing to the supply of "mutual benefit schemes supplying collective benefits."[34]

Case one is the *nervous cooperator*: "He fears that other individuals will fail to contribute to the required extent, that the scheme will collapse, and that B [the collective good] will not be supplied regardless of his own contribution."[35] Unlike the free-rider, the motivation of the nervous defector is not simply reluctance to pay a fair share despite partaking in the benefits. Instead, the nervous co-operator "does not want to waste resources in support of a lost cause."[36]

Case two is the *reluctant cooperator*. He "desires to contribute his assigned fair share of the costs of supplying B, provided that all others (or almost all others) also contribute their fair share."[37] In other words, the reluctant cooperator "is unwilling to allow himself to be, as he thinks, exploited by free riders."[38] These two types of defectors can be grouped into one category of *nervous defectors*. They do not defect just on a whim, but they are instead sceptical with respect to the other cooperators' contributions, and therefore they do not wish to take the first step.

In addition to Arneson's analyses, three more figures need to be explored. The second type is the *offended defector*. The offended defector does not look into the future, but looks into the past. He has been offended, therefore his willingness to cooperate is limited. What can we do with this defector? Interestingly, he is in a good position when it comes to theories of climate justice. He is the figure who claims restorative justice. Nevertheless, his claim goes beyond compensation. Possibly, he has been excluded by unfair means in the past. Possibly, he has been offended by not having been taken seriously in previous negotiations or something similar. He wants retribution – not only in terms of getting what he feels he is owed, but also in terms of no longer tolerating offensive acts.

Third, there is a *strategic defector*. Actually, this is the most common role in social choice theory and in the context of an analysis of the drama of the commons. Unlike the first type of defection, the strategic defector benefits from the cooperative behaviour of the other actors. This is parasitic defection, because it functions best when others are already cooperating.[39]

Fourth, we also have to address the *incapable defector*. This is an actor who – lacking competence – is forced to defect, unwillingly. This one has also been an important figure in debating climate justice. Ethicists have proposed exempting poor countries from climate duties, and not only because they have not contributed significantly to the current situation. They also argue that it is unfair to force climate victims to pay twice: first, they suffer the harm from climate events, and due to their poverty are not able to protect themselves; and second, they must also pay a price for a causal history of climate emissions whose benefits have gone largely to the rich nations. According to this logic, the incapable defector is a poor nation that is not able to support climate cooperation because of a lack of capacities at this point. Yet Henry Shue argues convincingly that this conclusion is not the only possible one. Instead, poor nations should be supported to find a direct path into a green future; they should be supported in "implementing the exit strategy" from a carbon-based industry.[40] Instead of granting them a right to additional emissions, poor countries need to be supported to invest in green technologies and not to repeat the flawed path of industrialization that rests on a cheap carbon-based energy supply.

Finally, there is the *reckless defector*, who has no interest at all in cooperating and in contributing to save the planet. She looks at her benefit and her interests, but does not give any attention to the needs and desires of other people. Indeed, she does so intentionally. She knows that an attempt to invest in environmental cooperation is obligatory, but she wilfully rejects the option.

In sum, it is important to distinguish these types of defection from the social dilemmas that were discussed in Chapter 2. Social dilemmas result from a situation in which individuals are free to either choose or reject cooperation, but nevertheless benefit from a scheme of appropriation of a shared resource. A defector, as addressed in this chapter, is supposed to already have a restricted choice. She either contributes to a scheme of appropriation of a resource that resonates with climate goals or appropriate environmental goals or she not only *does not* contribute her fair share, but also *undermines* the scheme of cooperation by not doing her share. This is the scenario that provides the background for arguing climate duties. I am not claiming that some types of defection are right, but rather that in order to provide adequate institutional means of stopping defection, insights into the motives and reasons of defectors are helpful. A non-capable defector should not be punished, but rather should be supported and assisted; a strategic defector should be punished when it comes to climate duties; and a nervous defector needs support in enhancing trust in the cooperative scheme. The offended defector, instead, needs practices of reconciliation in order to enhance the willingness to cooperate. The reckless defector needs to be convinced and possibly even forced to cooperate by legal means.

Table 10.1 Table of types of defectors and proposed remedies

Type of Defector	Characterization	Proposed Remedy
The nervous defector	Is afraid of a situation in which the distributive scheme collapses; is afraid of being exploited by other defectors	Develop reliable schemes of cooperation, enhance trust
The offended defector	Has suffered from offensive and non-cooperative behaviour of potential group members	Explain why forward-looking approaches do not rule out restorative justice; explain participatory and inclusive strategies
The strategic defector	Focuses exclusively on self-interest and calculates his gains relative to his defection	Support strategies of crowding-in; make use of moral psychology
The incapable defector	Does not have the means to actively participate in climate cooperation	Support for getting the means for cooperation (international funds)
The reckless defector	Undermines climate cooperation wilfully, although being able to cooperate	Try to convince; ultimately also needs to be pursued with legal action

10.4 Institutional answers to defection

In order to discuss institutional answers to defection, two perspectives on issues of defection need to be discussed first. It is one thing to argue for ways to reduce defection in the context of the voluntary cooperation of a group of people, and it is another thing to discuss defection in the international arena as non-compliance with widely shared climate goals and the necessary instruments for achieving them. This is the issue of *scenarios of defection* in relation to situations that include various facts of cooperation, such as scale, voluntariness, existence of a contract, and so on. On the other hand, institutional responses also need *strategies against defection*. Strategies include legal responses, institutional prerequisites, and so on. This section starts with a proposal to distinguish scenarios and then discusses strategies to punish and restrict defection.[41]

With regard to the types of defection along the lines of different background situations, three scenarios need to be distinguished. The first is one in which a group of individuals act together to reach an environmental goal. This is a case of joint action in favour of one of the environmental goals qualified as accumulative goals, transformation goals, or goals of active co-designing with nature. In this scenario, defection can be carried out by individuals who first join the group and then decide to free-ride. I have explained how this scenario might happen. However, it is only of minor importance, since the goal-oriented group action was first seen as voluntary cooperation.

But when transformed into obligatory cooperation, it is necessary to design institutions to discourage this type of defection. Political institutions, in particular, need to develop frameworks to deter in-group defection. These frameworks are possible, although possibly not perfect. One might think of access barriers comparable to the access barriers of club goods; other examples are trading systems such as the one installed in the Kyoto Protocol.[42]

The second scenario is one in which environmental goals are not yet successfully realized, but joint actors who work in favour of a realization of goals already exist. In such cases, individuals are obliged to contribute their share to the realization of these goals. Individuals, too, have an obligation to do their share in joint climate goals that are being communicated as particularly urgent.[43] Governments need to take care of institutions that put those goals on the agenda. They need to offer and support climate-friendly infrastructure, green energy-supply options, and products with small ecological footprints.

Ultimately, governments also have to do what the "power/responsibility principle" claims: "[T]hose with the power to compel or induce or enable others to act in climate-friendly ways have a responsibility to do so."[44] It is not my intention to discuss whether incentives, penalties, or legal institutions are best suited to realize this claim. It is also important to acknowledge the fact that governments have many different duties, and politics is always a matter of prioritization. It is, however, a duty of governments to support climate-friendly activities of citizens.

An additional argument for really getting tough on non-compliance, in cases in which non-compliance undermines the achievement of a group of people, can be deduced from an argument from Schwenkenbecher. She discusses failures in the rescue case, in which a random group is in a position to rescue a victim.[45] She highlights the distinction between an action that is unsuccessful and direct non-compliance. Regarding the latter, she distinguishes three types:

> (a) One or more individual members of the random group bail on the rest and undermine the group's joint ability. (b) People embark on joint action but one or more individuals fail to perform their contributory action. (c) No one does anything. They all just stand there and watch the person being killed by the attacker or the motorcyclist being killed by the flames.[46]

Not achieving success may be caused by the presence either of any defectors at all or by too many defectors. Regarding environmental goals, there is not often a distinct number or level that can be determined in advance in order to be successful. Instead, even a small number of defectors might spoil the attempts of the other cooperators. In scenarios in which cooperation is obligatory, governments are also allowed to enforce all-encompassing schemes of cooperation in order to avoid the scenario in which some defectors spoil the efforts and the success of the group of cooperators.

The third scenario is the most difficult one. In an international arena, governments will not only contribute a differently sized share regarding the encompassing climate goals. Instead, those who do not come through also need to be regarded as "slackers," as soon as a proposal for joint commitment of the world community has been developed. It would be very important to have institutions that enforce cooperation in the international arena; yet currently we do not have them. In a situation like that, the best nation-states can do is to build coalitions of the willing.[47] Coalitions of the willing can achieve a lot as joint actors; they should not be underestimated. If combined with tactics of "crowding-in," possibly also with "punishments for defectors,"[48] coalitions of the willing are powerful instruments to undermine defection.

In addition to responses to various scenarios, *strategies* against defection also need to be explored. Some ways to address the various types of defection have already been established in the last section. Another important insight is that the world community needs new ways to cooperate and new channels to organize environmental cooperation. So far, environmental politics is interpreted as restrictive laws that are costly and that have a bad effect on economic performance. At best, negative externalities, which are side effects with a detrimental and hazardous effect on environmental goods, are regarded as costs that need to be internalized into economic practices.[49] Yet more proposals are available, in particular when climate obligations no longer count as luxury claims, but as sound moral obligations that are associated with collective actors and to governments especially. Basically, three strategies for working with defectors can be distinguished.

To begin with, defection can and should be prosecuted. Defectors deserve punishment when acting intentionally and when cooperative schemes are available and are easy to join. The backdrop for this proposal is part of what this book has been about: climate duties are moral obligations that actors, and collective actors in particular, must fulfil – in proportion to their capacities and their powers. Non-cooperation not only causes harm to future generations, it already causes severe harm today and deaths due to climate change.[50] Individuals, as well as collective actors, should be punished if they fail to cooperate – at least when cooperative schemes are available. Since this book is not about environmental law, I need to leave it to the experts to translate climate obligations into law. Yet obviously, in order to litigate, institutions must be erected that are in a position to fulfil litigation. Proposals that have already been presented for an international institutional system include a global constitutional assembly[51] and a new International Court of Climate Justice.[52] I comment on these proposals and add another one in the last section of this chapter.

It is also possible to use strategies to help convince defectors that cooperation is a better choice. Strategies are also available when "crowding-in" takes place. Crowding-in is the process in which more and more actors join an already-existing joint actor, and when action that single actors take engenders respect and appreciation. Henry Shue has proposed taking leadership responsibility in this way:

> What is needed to break the current stalemate is leadership. We need one state to break the paralysis by unilaterally (if necessary) taking action in the hope that the others will respond to its example – and to their own comprehension of the inherent importance of the problem. In this way, one can then say not "after you", lest I be treated unfairly, but "I went first, so now you", lest you treat me unfairly – still appealing to fairness. This is obviously not a universally effective tactic, either, but it is particularly appropriate when one already has a moral duty to act and it is abundantly clear that one's fair share of the burden is well in excess of anything one has yet contributed.[53]

Additional proposals for strengthening "coalitions of the willing" have been presented since this one.[54]

Finally, the industry responsible for the largest share of the emissions budget needs to be forced to change its behaviour. Oil and coal are the dirtiest businesses; this is common knowledge. Agriculture also has an enormous carbon footprint. The goal of enforcing cooperation is pointless as long as the big emitters are still free to spoil the environment. Actually, they must be regarded as reckless defectors; and as such they should be forced to change their behaviour.

10.5 The role of international institutions and international law

Climate duties are justified duties. In particular, they have this status when allocated to various actors according to principles of fairness. These are important outcomes of this study. Yet, the debate has not stopped there. I have also argued that excuses from cooperative duties accrue to offended, incapable, possibly also nervous defectors. But what still remains is the reckless defector. This actor undermines climate cooperation wilfully, even though able to cooperate. It is fair to claim that responsible actors who could take action against climate change and who even have to do so according to principles of a role ethics (Section 9.7) need to be pursued by legal means if they do not fulfil their duties. This claim opens the wider question of whether the current international institutions are strong enough to pursue the reckless actors.

As explained from the outset, this book is not dedicated to delivering an institutional framework. This has to be argued somewhere else. But at first glance, two claims appear to be obvious: first, the climate duties to cooperate in favour of mitigation, adaptation, and transformation need an institutional backing that obstructs reckless behaviour and thereby also encourages voluntary cooperation. Second, without a strong international regime that supports cooperation in favour of climate goals and sanctions defection, the climate goals that need to be achieved are put at risk. This brings us to another question: the role of the United Nations and of international law.

Much ink has already been spilled on the failures of both with respect to environmental protection. Among the many proposals to reform the UN in order to really function as a strong international institution, two insights relate immediately to the climate debate. On the one hand, the values of the international community that serve as a backdrop for the work of the UN, the Universal Declaration of Human Rights, is a strong normative background for also claiming climate duties. The protection of human dignity and freedom cannot be secured without also protecting the environmental integrity of the global natural commons. As argued in Sections 2.3 and 2.4, the list of the shortcomings of the material side of the tragedy of the commons is long. Without institutions that prevent unintended, chaotic, and invisible exploitation, the integrity of water resources and the atmosphere cannot be secured. And a strong international Environmental Agency comparable to the World Health Organization or the World Trade Organization is still missing.

On the other hand, an agency that really could enforce the liability criterion is also a necessary supplement to international law and is also definitively missing. Whether the world community needs an International High Court of Climate Justice is an open question. But what definitively is needed are institutions that could pursue reckless actors and that could prevent reckless behaviour.

10.6 On governmental duties and new institutions

According to the proposal for joint climate action, climate goals need to be interpreted as joint action goals. The responsibility of institutions authorized to support climate goals as particularly important goals includes the power to enhance their realization. Since many other goods are related to immediate support for climate goals, it is justified to prioritize them. Yet, most importantly, governments should empower actors to care about climate goals and to realize climate goals together with other actors. Instead of starting with policies that also infringe on the liberty of citizens, the view that this chapter takes is based on three claims.

First, since climate goals are legitimate goals and since they in the end correspond to the most fundamental interests of every single citizen, governments are allowed to transform climate duties into climate law. Since the goals are not restricted to single nation-states, supranational institutions are also in charge of designing an effective international climate law.[55]

Second, governments have the responsibility to care for cooperative schemes that support active co-designing with nature and goals related to this ideal. More specifically, governments should also ensure cooperation once these options are in place. Yet there are two ways to do so. The first way is support for institutions that enhance mitigation and support climate-related goals, such as a green energy sector, housing that reduces carbon-based energy usage, ecological infrastructure, and the like. Most of these goals can be framed as "mitigation" and have already been discussed. Yet there is a second way to support cooperation in favour of climate goals. Governments need to support and empower voluntary joint climate action. The enabling state helps to remove barriers standing in the way of climate cooperation. According to the analysis presented here, the duty to empower cooperators and the need to reduce defection are among the most important goals governments need to pursue. In addition, free-riding and slacking in the international arena also need to be stopped. The Paris Agreement is a first step, but further steps need to follow.

Third, institutions that support and enable international cooperation are obligatory. One of the easiest means of supporting mitigation in an efficacious way is an internationally agreed-upon carbon tax. The current situation of climate change is in part caused by the incapacity to activate systems with buy-in from all members of the international arena, including both private and public sectors. In the context of climate ethics, Steve Gardiner has recently presented a far-reaching proposal to design an appropriate institution of international governance with respect to climate ethics. He calls for a "constitutional convention" that should primarily address the intergenerational gap he regards as a central moral failure of the current generation.[56] The proposal for a global constitutional convention has two parts. The first component is procedural and encapsulates a "call for action." It is dedicated to "an attempt to engage a range of actors, based on a claim that they have or should take on a set of responsibilities, and a view about how to go about discharging those responsibilities."[57] The second component is "substantive," including key ideas for elements of a global convention. More specifically, Gardiner proposes to create something like a world government, yet without also implementing the dangers of such an institution. The means of avoiding this is called "selective mirroring":

Specifically, a convention would seek to develop a broader system of institutions and practices that reflected the desirable features of a powerful and highly centralized global authority but neutralized the standing threats posed by it.[58]

In my view, conventions that really focus on climate goals and the interests of future generations do not have to be restricted to the international arena. It makes perfect sense to build up conventions within nation-states. The scope on future goals is severely limited by short-term interests of voters and governments alike. Conventions, instead, could focus on important future goals; they could work as counsels for desperately needed institutions that support climate goals; and they could help to put climate goals into context.

10.7 Collective action now

As this study closes, I also wish to highlight the need to act immediately. Obviously, individual action and voluntary agreements do not suffice to tackle climate problems efficiently.[59] It is instead an important goal, as Steve Gardiner argues, to "establish new collective practices."[60] Yet, the autonomy of the institution needs to be balanced against mutual accountability to other major international institutions.[61] In particular, it should be suited to deliver on the task.

This aspect could be made stronger by including the strategies that have been explained in this chapter. In order to really counteract buck-passing behaviour and undermine free-riding, it is important to address defection seriously and to offer ways to develop a more constructive roadmap for cooperation. In addition to these features, Gardiner also discusses the aims and the composition.[62] "Regionalism" is of particular importance for joint climate action: "Representation should be designed to reflect large-scale ecological realities at a supranational level."[63] As for the coordination of joint action regarding climate goals, as well as environmental goals more broadly conceived, it is particularly important to design new institutions that bring the various goals and agendas together. Moreover, it is important to think about representative bodies that can supervise success and failures likewise, and also to admonish governments and to rectify prioritization when they are flawed.

Among the organizations capable of acting and obliged to take on climate duties are political institutions. Unlike other organizations, legitimate political institutions are obliged to protect the well-being of the citizenry. This includes the protection and support of a range of public goods.[64] More specifically, political institutions need to shoulder responsibilities for the achievement of climate goals. This claim is not controversial in the community of experts who defend some version of climate justice. Yet what is controversial is the concrete level of responsibility assigned to governments and what type of responsibility has also been discussed. In addition, it has been questioned whether, in addition to the nation-state, further organizations also need to bear responsibility, such as groups of nation-states or even "coalitions of the willing."[65]

A necessary supplement to newly generated institutions on a supranational level are systems of cooperation based on voluntary consent to schemes of cooperation that nevertheless can be enforced. Whereas philosophers are sometimes overly optimistic with respect to possible win-win scenarios in the case of climate change,[66] economists warn that free-riding is a dramatic and persistent problem in contracts based on voluntary contributions to climate goals.[67] Instead of relying on voluntary cooperation, institutions and systems –including incentives for climate-friendly strategies and politics of nation-states – need to be negotiated. Edenhofer et al. recommend building on strategies that diminish costs by means of technological innovations, efficient climate politics, and additional positive effects.[68] In addition, they propose connecting national climate politics with an international system for transfer payments and agreements on technology transfer, as well as the threat of sanctions; in addition, reputation is an enhancing factor.[69] But self-enforcing international agreements are certainly only a first step; they need to be empowered to achieve the removal of barriers and the enhancement of cooperation and to make it obligatory. More specifically, it is also important to acknowledge that environmental goals cannot be achieved by exclusively pledging money. Instead, cooperation is critical; and this cooperation also needs to be carried out by everyone.

Overall, the most important prerequisite for establishing new institutions is provided when cooperation is in some important instances regarded as obligatory. This study has argued that under circumstances I have outlined in detail, climate cooperation is a moral duty. As such, each actor – single and collective – is obliged to contribute their share to the achievement of climate goals. But institutions that enable and guard this process are at best rudimentary today. More specifically, according to the framing of climate goals as prioritizing the integrity of the resource in question, institutions that safeguard environmental cooperation also need to focus on the resource side.

As applied to environmental goods more generally, rivers, the seas, forests, and natural reserves, as well as the atmosphere need to be regarded not only as resources, but as valuable and as centres of joint environmental action. The goal that institutions, as guardians of these resources, need to fulfil is an active support of schemes of cooperation that satisfy a range of justified interests, but that are ultimately restricted to types of cooperation that prevent over-exploitation and that cohere with moderate eco-centrism. In short, collectives not only need to be invited to support joint environmental action but also need to frame their goals in accordance with joint environmental action.

Notes

1 See John Rawls, *A Theory of Justice* (Cambridge, MA: Belknap Press, 2005), 8, 215.
2 Eric Brandstedt, "Non-Ideal Climate Justice," *Critical Review of International Social and Political Philosophy* 22, no. 2 (February 23, 2019): 221–234, https://doi.org/10.

1080/13698230.2017.1334439; Alexandre Gajevic Sayegh, "Justice in a Non-Ideal World: The Case of Climate Change," *Critical Review of International Social and Political Philosophy* 21, no. 4 (July 4, 2018): 407–432, https://doi.org/10.1080/13698230.2016.1144367; Laura Valentini, "The Natural Duty of Justice in Non-Ideal Circumstances: On the Moral Demands of Institution Building and Reform," *European Journal of Political Theory*, November 28, 2017, https://doi.org/10.1177/1474885117742094; Laura Valentini, "Ideal vs. Non-Ideal Theory: A Conceptual Map," *Philosophy Compass* 7, no. 9 (September 2012): 654–664, https://doi.org/10.1111/j.1747-9991.2012.00500.x.

3 Valentini, "Ideal vs. Non-Ideal Theory," 654–655.

4 Ibid., 655–656.

5 See Rawls, *A Theory of Justice*, 8, 215.

6 Valentini, "Ideal vs. Non-Ideal Theory," 656–657.

7 Ibid., 657–658.

8 A compendium of proposals for climate governance and for governance instruments, which are also regarded as instruments for implementing moral claims, is provided in Jennifer Clare Heyward and Dominic Roser, eds., *Climate Justice in a Non-Ideal World* (Oxford: Oxford University Press, 2016).

9 For a comprehensive view on "society and climate change" and proposals for state practices and governance structures, see John S. Dryzek, Richard B. Norgaard, and David Schlosberg, eds., *Oxford Handbook of Climate Change and Society* (Oxford: Oxford University Press, 2013).

10 This proposal is also distinguished from the more modest idea that governments should act as "norm entrepreneurs" in order to define appropriate and inappropriate behaviour. See Cass R. Sunstein, "Social Norms and Social Roles," *Columbia Law Review* 96 (1996): 909.

11 For these claims, see Chapter 9.

12 Simon Caney, "Two Kinds of Climate Justice: Avoiding Harm and Sharing Burdens," *Journal of Political Philosophy* 22, no. 2 (2014): 141.

13 Ibid.

14 Ibid.

15 Ibid., 144.

16 Ibid., 142–143.

17 The idea that individuals act in complete accord with existing preferences is only a plausible claim when individuals' values and reasons are also reduced to preferences. I think this is not a plausible interpretation of action, but is far too reductive in terms of explanatory resources for action.

18 Although there are findings about how the various parameters, including social goals, environmental goals, and individual values, relate to each other, research still leaves much space for further inquiries and interpretation. For an approach that argues, based on case studies, for interdependence, see David J. Hardisty et al., "About Time: An Integrative Approach to Effective Environmental Policy," *Global Environmental Change* 22 (2012): 684–694.

19 Elke U. Weber et al. provide evidence that there is a huge gap between the scientific interpretation of climate change and the understanding prevailing in the broad public. They argue that in order to achieve behavioural change, a first important step is a framing of climate change that responds to insights into psychology about the capacity to frame risks and to consider environmental needs. See Elke U. Weber, "Experienced-Based and Description-Based Perceptions of Long-Term Risk: Why Global Warming Does Not Scare Us (Yet)," *Climatic Change* 77 (2006): 103–120; Elke U. Weber and Paul C. Stern, "Public Understanding of Climate Change in the United States," *American Psychologist* 66, no. 4 (2011): 315–328, https://doi.org/10.1037/a0023253.

20 Dale Jamieson, *Reason in a Dark Time. Why the Struggle Against Climate Change Failed – and What it Means for Our Future* (Oxford/New York: Oxford University Press, 2014), 61–104.

21 Ibid., 78–81.

22 Ibid., 81.

23 Andrew Sayer, "Geography and Global Warming: Can Capitalism be Greened?," *Area* 41, no. 3 (September 2009): 351.

24 For the lengthy argument and a shortened version, see Hans-Werner Sinn, "Public Policies Against Global Warming," *International Tax and Public Finance* 15, no. 4 (2008): 360–394.

25 For a cautious and well-reasoned application of the "green paradox" as advice to design environmental politics, see Edwin van der Werf and C. Di Maria, "Imperfect Environmental Policy and Polluting Emissions: The Green Paradox and Beyond," *International Review of Environmental and Resource Economics* 6, no. 2 (March 30, 2012): 153–194, https://doi.org/10.1561/101.00000050. Van der Werf and Di Maria propose, among other things, that the green paradox helps to identify four different types of what they call "imperfect policy"; they also discuss possible ways to enhance the politics, including the support of green energy technologies. They summarize their outcomes: "Unilateral emission reductions can induce non-abating countries to change their emissions in response. We have identified five channels through which this may occur. None of the papers discussed above combines all the five channels ... Within the large literature on carbon leakage, only two analytical papers and one AGE paper found that under specific assumptions a green paradox may occur" (van der Werf and Di Maria, 188).

26 This term is chosen by Dale Jamieson to provide a detailed account of how the "denial industry" has been built up in the United States by a variety of actors, including the media and other industries. See Jamieson, *Reason in a Dark Time*, 81–93.

27 For constructive proposals regarding climate news and the media, see Maxwell T. Boykoff, *Who Speaks for the Climate? Making Sense of Media Reporting on Climate Change* (Cambridge/New York: Cambridge University Press, 2011).

28 An additional problem is the gap between the existence of high risks from environmental change and the understanding of those risks by the public. For an empirical study regarding this gap with respect to flooding, see W. J. Wouter Botzen, Howard Kunreuther, and Erwann Michel-Kerjan, "Divergence between Individual Perceptions and Objective Indicators of Tail Risks: Evidence from Floodplain Residents in New York City," *Judgment and Decision Making* 10, no. 4 (July 2015): 365–385.

29 For a critical discussion of the problem of scale in public-good scenarios, see Angela Kallhoff, *Why Democracy Needs Public Goods* (Lanham, MD: Lexington Books, 2011), 29–40.

30 See Simon Caney, "Climate Change and Non-Ideal Theory: Six Ways of Responding to Non-Compliance," in *Climate Justice in a Non-Ideal World*, eds. Jennifer Clare Heyward and Dominic Roser (Oxford: Oxford University Press, 2016), 21–42. This chapter prefers the concept of "defector," because it is closer than Caney's analysis to scenarios of joint action; in particular, this concept indicates that defection is undertaken by actors – both individual actors and collective actors.

31 Richard J. Arneson, "The Principle of Fairness and Free-Rider Problems," *Ethics* 92 (July 1982): 616–633.

32 For this discussion, see also Chapter 3.

33 Arneson, 618. Note that the third condition is not restricted to strict egalitarianism. Later, Arneson corrects the third description slightly and says that "all

citizens within a given territory must consume pretty much the same amount of them" (Arneson, 619).

34 Arneson, 623.
35 Ibid.
36 Ibid.
37 Ibid.
38 Ibid.
39 This kind of defection was analysed in Chapter 3.
40 See Henry Shue, "Climate Hope: Implementing the Exit Strategy," *Chicago Journal of International Law* 13, no. 2 (2013): 381–402.
41 For a more comprehensive analysis of institutional responses to non-compliance, though against the backdrop of schemes of cooperation that have already been arranged, see Caney, "Climate Change and Non-Ideal Theory."
42 For an enlightening interpretation of the contractarian ideas about the Kyoto Protocol and the need for another Kyoto Protocol with respect to the non-identity problem, see Edward Page, *Climate Change, Justice and Future Generations* (Cheltenham, UK/Northampton, MA: Edward Elgar, 2007), 132–150.
43 This condition is comparable to Caney's framing of "emergency": humanity faces a prospect of disastrous harm if mitigation is not quickly done; and "the time left to prevent the onset of dangerous climate change is quickly running out" (Caney, "Two Kinds of Climate Justice," 142–143).
44 Caney, "Two Kinds of Climate Justice," 141.
45 For a discussion of this scenario, see the detailed analysis in Section 7.1.
46 Anne Schwenkenbecher, "Joint Moral Duties," *Midwest Studies in Philosophy* 38, no. 1 (2014): 71, https://doi.org/10.1111/misp.12016.
47 For this proposal, see also Toni Erskine, "Coalitions of the Willing and Responsibilities to Protect: Informal Associations, Enhanced Capacities, and Shared Moral Burdens," *Ethics & International Affairs* 28, no. 1 (2014): 115–145, https://doi.org/10.1017/S0892679414000094.
48 See, for example, Ernst Fehr and Simon Gächter, "Cooperation and Punishment in Public Goods Experiments," *The American Economic Review* 90, no. 4 (September 2000): 980–994.
49 For a critique of this proposal, see Goodin's groundbreaking article: Robert E. Goodin, "Selling Environmental Indulgences," *Kyklos* 47 (1994): 573–596.
50 For a calculation and explication of the effects of the average American's greenhouse gas emissions in 2011, see John Nolt, "How Harmful are the Average American's Greenhouse Gas Emissions?," *Ethics, Policy & Environment* 14, no. 1 (March 2011): 3–10, https://doi.org/10.1080/21550085.2011.561584.
51 See Stephen M. Gardiner, "A Call for a Global Constitutional Convention Focused on Future Generations," *Ethics & International Affairs* 28, no. 3 (2014): 299–315, https://doi.org/10.1017/S0892679414000379.
52 See Tracey Skillington, *Climate Justice and Human Rights* (New York: Palgrave MacMillan, 2017), 252. Skillington's final chapter includes a broad range of further proposals for a new international institutional order. See Skillington, 231–260.
53 Henry Shue, "Face Reality? After You! – A Call for Leadership on Climate Change," *Ethics & International Affairs* 25, no. 1 (March 2011): 23–24, https://doi.org/10.1017/S0892679410000055.
54 Erskine, "Coalitions of the Willing and Responsibilities to Protect."
55 I am aware that this is a claim that needs to be argued at length. It includes the issue of whether an infringement on the sovereignty of nation-states should be allowed, which type of norms can best enshrine climate laws, and which institutions this would include. For far-reaching proposals, see also Skillington, 231–260.

56 Stephen M. Gardiner, "A Call for a Global Constitutional Convention Focused on Future Generations," *Ethics & International Affairs* 28, no. 3 (2014): 299–315, https://doi.org/10.1017/S0892679414000379.

57 Ibid., 305.

58 Ibid., 307.

59 Ibid., 308. See also Gardiner's detailed account of duties to the future in Stephen M. Gardiner, "The Threat of Intergenerational Extortion: On the Temptation to Become the Climate Mafia, Masquerading as an Intergenerational Robin Hood," *Canadian Journal of Philosophy* 47, no. 2–3 (2017): 386, https://doi.org/10.1080/00455091.2017.1302249.

60 Gardiner, "The Threat of Intergenerational Extortion," 307.

61 Gardiner, "A Call for a Global Constitutional Convention Focused on Future Generations," 310.

62 Ibid., 311–312.

63 Ibid., 312.

64 See Kallhoff, *Why Democracy Needs Public Goods*; Angela Kallhoff, "Why Societies Need Public Goods," *Critical Review of International Social and Political Philosophy* 17, no. 6 (2014): 635–651.

65 See J. Ronald Engel, Laura Westra, and Klaus Bosselmann, *Democracy, Ecological Integrity and International Law* (Newcastle upon Tyne, UK: Cambridge Scholars Publishing, 2010); Erskine, "Coalitions of the Willing and Responsibilities to Protect."

66 Broome appears to think that a climate contract can be modelled at least as a cooperative scheme without sacrifices. See the proposal in John Broome, *Climate Matters. Ethics in a Warming World* (New York/London: W.W. Norton, 2012). For a critique of this proposal, see Gardiner, "The Threat of Intergenerational Extortion," 377–385. An optimistic approach arguing a win-win scenario as a climate contract has also been proposed in Eric A. Posner and David A. Weisbach, *Climate Change Justice* (Princeton, NJ: Princeton University Press, 2010). For a critique, see Gardiner, "The Threat of Intergenerational Extortion," 371–373.

67 See Ottmar Edenhofer, Christian Flachsland, and Steffen Brunner, "Wer besitzt die Atmosphäre?: Zur Politischen Ökonomie des Klimawandels," *Leviathan* 39, no. 2 (June 2011): 208–211, https://doi.org/10.1007/s11578-011-0115-0.

68 Ibid., 211.

69 Ibid., 212–215.

Bibliography

Akpabio, Emmanuel M. "Cultural Notions of Water and the Dilemma of Modern Management: Evidence from Nigeria." In *Water Management Options in a Globalised World. Proceedings of an International Scientific Workshop (20–23 June 2011, Bad Schönbrunn)*, edited by Martin Kowarsch, 2nd rev. ed., 156–171. Institute for Social and Developmental Studies (IGP) at the Munich School of Philosophy, 2011. www.researchgate.net/publication/308118199_Water_management_options_in_a_globalised_world.

Amery, Hussein A., and Aaron T. Wolf. *Water in the Middle East. A Geography of Peace.* Austin: University of Texas Press, 2000.

Ammer, Margit. "Klimawandel und Migration/Flucht: Welche Rechte für die Betroffenen in Europa?" In *Klimagerechtigkeit und Klimaethik*, edited by Angela Kallhoff, 81–103. Wiener Reihe. Themen der Philosophie, vol. 18. Berlin/Boston: De Gruyter, 2015.

Anderson, Kristin M., and Lisa J. Gaines. "International Water Pricing: An Overview and Historic and Modern Case Studies." In *Managing and Transforming Water Conflicts*, edited by Jerome Delli Priscoli and Aaron T. Wolf, 249–265. International Hydrology Series. Cambridge: Cambridge University Press, 2009.

Anscombe, G. E. M. *Intention.* 2nd ed. Cambridge, MA: Harvard University Press, 2000.

Arnell, Nigel W. "Climate Change and Global Water Resources." *Global Environmental Change* 9 (1999): 31–49.

Arneson, Richard J. "The Principle of Fairness and Free-Rider Problems." *Ethics* 92 (July1982): 616–633.

Attari, Shazeen U., David H. Krantz, and Elke U. Weber. "Reasons for Cooperation and Defection in Real-World Social Dilemmas." *Judgment and Decision Making* 9, no. 4 (2014): 316–334.

Baer, Paul. "Adaptation to Climate Change. Who Pays Whom?" In *Climate Ethics. Essential Readings*, edited by Stephen M. Gardiner, Simon Caney, Dale Jamieson, and Henry Shue, 247–262. Oxford/New York: Oxford University Press, 2010.

Baier, Annette C. "Doing Things with Others: The Mental Commons." In *Commonality and Particularity in Ethics*, edited by Lili Alanen, Sara Heinämaa, and Thomas Wallgren, 15–44. Swansea Studies in Philosophy. London: MacMillan, 1997.

Bakker, Karen, ed. *Eau Canada. The Future of Canada's Water.* Vancouver/Toronto: UBC Press, 2007.

Barrett, Scott. *Why Cooperate? The Incentive to Supply Global Public Goods.* Oxford/ New York: Oxford University Press, 2007.

Barry, John. *Rethinking Green Politics: Nature, Virtue and Progress.* London/Thousand Oaks, CA: SAGE, 1999. https://doi.org/10.4135/9781446279311.

Bazargan, Saba. "Complicitous Liability in War." *Philosophical Studies* 165, no. 1 (2013): 177–195.

Beaumont, Peter. "Conflict, Coexistence, and Cooperation: A Study of Water Use in the Jordan Basin." In *Water in the Middle East. A Geography of Peace*, edited by Hussein A. Amery and Aaron T. Wolf, 19–44. Austin: University of Texas Press, 2000.

Bell, Derek. "How Should We Think About Climate Justice?" *Environmental Ethics* 35, no. 2 (2013): 189–208.

Betzler, Monika. "Valuing Interpersonal Relationships and Acting Together." In *Concepts of Sharedness. Essays on Collective Intentionality*, edited by Hans Bernhard Schmid, Katinka Schulte-Ostermann, and Nikos Psarros, 253–272. Philosophische Analyse, vol. 26. Frankfurt: Ontos Verlag, 2008.

Bollier, David. *Silent Theft: The Private Plunder of Our Common Wealth.* London: Routledge, 2002.

Botzen, W. J. Wouter, Howard Kunreuther, and Erwann Michel-Kerjan. "Divergence between Individual Perceptions and Objective Indicators of Tail Risks: Evidence from Floodplain Residents in New York City." *Judgment and Decision Making* 10, no. 4 (July2015): 365–385.

Boykoff, Maxwell T. *Who Speaks for the Climate? Making Sense of Media Reporting on Climate Change.* Cambridge/New York: Cambridge University Press, 2011.

Brandstedt, Eric. "Non-Ideal Climate Justice." *Critical Review of International Social and Political Philosophy* 22, no. 2 (February 23, 2019): 221–234. https://doi.org/10. 1080/13698230.2017.1334439.

Bratman, Michael E. *Shared Agency. A Planning Theory of Acting Together.* New York: Oxford University Press, 2014.

Bratman, Michael E. "Shared Intention." *Ethics* 104, no. 1 (1993): 97–113.

Bretschneider, Wolfgang. "The Right to Water from an Economic Point of View." In *Water Management Options in a Globalised World. Proceedings of an International Scientific Workshop (20–23 June 2011, Bad Schönbrunn)*, edited by Martin Kowarsch, 2nd rev. ed., 87–94. Institute for Social and Developmental Studies (IGP) at the Munich School of Philosophy, 2011. www.researchgate.net/publication/308118199_ Water_management_options_in_a_globalised_world.

Brock, Dan W. "Forgoing Life-Sustaining Food and Water: Is It Killing?" In *By No Extraordinary Means. The Choice to Forgo Life-Sustaining Food and Water*, edited by Joanne Lynn, 117–131. Bloomington: Indiana University Press, 1989.

Broome, John. *Climate Matters. Ethics in a Warming World.* Amnesty International Global Ethics Series. New York/London: W.W. Norton, 2012.

Brown, Peter G., and Jeremy J. Schmidt, eds. *Water Ethics. Foundational Readings for Students and Professionals.* Washington, DC: Island Press, 2010.

Bruns, Bryan Randolph, and Ruth S. Meinzen-Dick. "Negotiating Water Rights: Implications for Research and Action." In *Negotiating Water Rights*, edited by Bryan Randolph Bruns and Ruth S. Meinzen-Dick, 353–380. London: ITDG Publishing, 2000.

Brzezinski, Zbigniew. *Strategic Vision: America and the Crisis of Global Power.* New York: Basic Books, 2012.

Buck, Susan J. *The Global Commons. An Introduction.* Foreword by Elinor Ostrom. Washington, DC/Covelo, CA: Island Press, 1998.

Burdack, Doreen. "The Australian Water Trade." In *Water Management Options in a Globalised World. Proceedings of an International Scientific Workshop (20–23 June 2011, Bad Schönbrunn)*, edited by Martin Kowarsch, 2nd rev. ed., 172–181. Institute for Social and Developmental Studies (IGP) at the Munich School of Philosophy, 2011. www.researchgate.net/publication/308118199_Water_management_options_in_a_globalised_world.

Cafaro, Philip, and Ronald Sandler, eds. *Virtue Ethics and the Environment.* Repr. from *Journal of Agricultural and Environmental Ethics* 23. Dordrecht: Springer, 2010.

Caney, Simon. "Climate Change and Non-Ideal Theory: Six Ways of Responding to Non-Compliance." In *Climate Justice in a Non-Ideal World*, edited by Jennifer Clare Heyward and Dominic Roser, 21–42. Oxford: Oxford University Press, 2016.

Caney, Simon. "Climate Change and the Duties of the Advantaged." *Critical Review of International Social and Political Philosophy* 13, no. 1 (March2010): 203–228. https://doi.org/10.1080/13698230903326331.

Caney, Simon. "Cosmopolitan Justice, Responsibility, and Global Climate Change." *Leiden Journal of International Law* 18 (2005): 747–775.

Caney, Simon. "Cosmopolitan Justice, Responsibility, and Global Climate Change." In *Climate Ethics. Essential Readings*, edited by Stephen M. Gardiner, Simon Caney, Dale Jamieson, and Henry Shue. Oxford/New York: Oxford University Press, 2010.

Caney, Simon. "Human Rights, Climate Change, and Discounting." *Environmental Politics* 17, no. 4 (August2008): 536–555. https://doi.org/10.1080/09644010802193401.

Caney, Simon. "Just Emissions." *Philosophy and Public Affairs* 40, no. 4 (2012): 255–300.

Caney, Simon. "Justice and the Distribution of Greenhouse Gas Emissions." *Journal of Global Ethics* 5, no. 2 (2009): 125–146.

Caney, Simon. "Two Kinds of Climate Justice: Avoiding Harm and Sharing Burdens." *Journal of Political Philosophy* 22, no. 2 (2014): 125–149.

Coady, C. A. J. *Messy Morality: The Challenge of Politics.* Uehiro Series in Practical Ethics. Oxford: Clarendon Press, 2008.

Collins, Stephanie. "Collectives' Duties and Collectivization Duties." *Australasian Journal of Philosophy* 91, no. 2 (2013): 231–248. https://doi.org/10.1080/00048402.2012.717533.

Collins, Stephanie. "Distributing States' Duties." *Journal of Political Philosophy* 24, no. 3 (2016): 344–366. https://doi.org/10.1111/jopp.12069.

Collins, Stephanie. "Duties of Group Agents and Group Members." *Journal of Social Philosophy* 48, no. 1 (2017): 38–57. https://doi.org/10.1111/josp.12181.

Collins, Stephanie. "Filling Collective Duty Gaps." *Journal of Philosophy* 114, no. 11 (2017): 573–591. https://doi.org/10.5840/jphil20171141141.

Collins, Stephanie, and Holly Lawford-Smith. "Collectives' and Individuals' Obligations: A Parity Argument." *Canadian Journal of Philosophy* 46, no. 1 (January 2, 2016): 38–58. https://doi.org/10.1080/00455091.2015.1116350.

Costanza, Robert, Ralph d'Arge, Rudolf de Groot, Stephen Farber, Monica Grasso, Bruce Hannon, Karin Limburg, et al. "The Value of the World's Ecosystem Services and Natural Capital." *Nature* 387, no. 6630 (1997): 253–260. https://doi.org/10.1038/387253a0.

Cripps, Elizabeth. *Climate Change and the Moral Agent: Individual Duties in an Interdependent World.* Oxford: Oxford University Press, 2013.

Crutzen, Paul J. "Geology of Mankind." *Nature* 415, no. 3 (January2002): 23.

Cullity, Garrett. "Asking Too Much." *The Monist* 86, no. 3 (2003): 402–418.

Daily, Gretchen C., ed. *Nature's Services: Societal Dependence on Natural Ecosystems.* Washington, DC: Island Press, 1997.

De Jonge, Jan. *Rethinking Rational Choice Theory.* New York: Palgrave MacMillan, 2012.

Delli Priscoli, Jerome, and Aaron T. Wolf. *Managing and Transforming Water Conflicts.* International Hydrology Series. Cambridge/New York: Cambridge University Press, 2009.

Dewey, John. *The Public and Its Problems.* Athens: Swallow Press, 1991.

Diamond, Jared M., and Christopher Murney. *Collapse: How Societies Choose to Fail or Succeed.* New York: Penguin Audio, 2004.

Dryzek, John S., Richard B. Norgaard, and David Schlosberg, eds. *Oxford Handbook of Climate Change and Society,* vol. 1. Oxford: Oxford University Press, 2013.

Edenhofer, Ottmar, Christian Flachsland, and Steffen Brunner. "Wer besitzt die Atmosphäre?: Zur Politischen Ökonomie des Klimawandels." *Leviathan* 39, no. 2 (June2011): 201–221. https://doi.org/10.1007/s11578-011-0115-0.

Engel, J. Ronald, Laura Westra, and Klaus Bosselmann. *Democracy, Ecological Integrity and International Law.* Newcastle upon Tyne, UK: Cambridge Scholars Publishing, 2010.

Eriksson, Lina. *Rational Choice Theory: Potential and Limits.* New York: Palgrave Mac-Millan, 2011.

Erskine, Toni. "Assigning Responsibilities to Institutional Moral Agents: The Case of States and Quasi-States." *Ethics & International Affairs* 15, no. 2 (2011): 67–85. https://doi.org/10.1111/j.1747-7093.2001.tb00359.x.

Erskine, Toni. "Coalitions of the Willing and Responsibilities to Protect: Informal Associations, Enhanced Capacities, and Shared Moral Burdens." *Ethics & International Affairs* 28, no. 1 (2014): 115–145. https://doi.org/10.1017/S0892679414000094.

Fabre, Cécile. "Guns, Food, and Liability to Attack in War." *Ethics* 120, no. 1 (2009): 36–63. https://doi.org/10.1086/649218.

Falkenmark, Malin, and Carl Folke. "The Ethics of Socio-Ecohydrological Catchment Management: Towards Hydrosolidarity." *Hydrology and Earth System Sciences* 6, no. 1 (2002): 1–9.

Fehr, Ernst, and Simon Gächter. "Cooperation and Punishment in Public Goods Experiments." *American Economic Review* 90, no. 4 (September2000): 980–994.

Feitelson, Eran. "A Hierarchy of Water Needs and Their Implications for Allocation Mechanisms." In *Global Water Ethics. Towards a Global Ethics Charter,* edited by Rafael Ziegler and David Groenfeldt, 149–166. Earthscan Studies in Water Resource Management. London/New York: Routledge, 2017.

Feldman, David Lewis. *Water Policy for Sustainable Development.* Baltimore, MD: Johns Hopkins University Press, 2007.

Fischer, Franklin M. "Water Value, Water Management, and Water Conflict: A Systematic Approach." In *Mountains: Sources of Water, Sources of Knowledge,* edited by Ellen Wiegandt, 123–148. Advances in Global Change Research, vol. 31. Dordrecht: Springer, 2008.

French, Peter A. *Collective and Corporate Responsibility.* New York: Columbia University Press, 1984.

French, Peter A. "The Corporation as a Moral Person." *American Philosophical Quarterly* 16, no. 3 (1979): 207–215.

Gajevic Sayegh, Alexandre. "Justice in a Non-Ideal World: The Case of Climate Change." *Critical Review of International Social and Political Philosophy* 21, no. 4 (July 4, 2018): 407–432. https://doi.org/10.1080/13698230.2016.1144367.

Gardiner, Stephen M. "A Call for a Global Constitutional Convention Focused on Future Generations." *Ethics & International Affairs* 28, no. 3 (2014): 299–315. https://doi.org/10.1017/S0892679414000379.

Gardiner, Stephen M. "Climate Ethics in a Dark and Dangerous Time." *Ethics* 127, no. 2 (January2017): 430–465. https://doi.org/10.1086/688746.

Gardiner, Stephen M. "A Core Precautionary Principle." *Journal of Political Philosophy* 14, no. 1 (2006): 33–60.

Gardiner, Stephen M. "Ethics and Global Climate Change." *Ethics* 114 (2004): 555–600.

Gardiner, Stephen M. *A Perfect Moral Storm. The Ethical Tragedy of Climate Change.* Oxford/New York: Oxford University Press, 2011.

Gardiner, Stephen M. "The Real Tragedy of the Commons." *Philosophy and Public Affairs* 30, no. 4 (2001): 387–416.

Gardiner, Stephen M. "The Threat of Intergenerational Extortion: On the Temptation to Become the Climate Mafia, Masquerading as an Intergenerational Robin Hood." *Canadian Journal of Philosophy* 47, no. 2–3 (2017): 368–394. https://doi.org/10.1080/00455091.2017.1302249.

Gardiner, Stephen M., and David A. Weisbach. *Debating Climate Ethics.* New York: Oxford University Press, 2016.

Gardiner, Stephen M., Simon Caney, Dale Jamieson, and Henry Shue, eds. *Climate Ethics. Essential Readings.* Oxford/New York: Oxford University Press, 2010.

Gesang, Bernward. "Gibt es politische Pflichten zum individuellen Klimaschutz?" In *Klimagerechtigkeit und Klimaethik,* edited by Angela Kallhoff, 135–142. Wiener Reihe. Themen Der Philosophie, vol. 18. Berlin/Boston: De Gruyter, 2015.

Gilbert, Margaret. "Group Membership and Political Obligation." *The Monist* 76, no. 1 (1993): 119–131.

Gilbert, Margaret. *Living Together. Rationality, Sociality, and Obligation.* New York: Rowman and Littlefield, 1996.

Gilbert, Margaret. *On Social Facts.* Princeton, NJ: Princeton University Press, 1989.

Gilbert, Margaret. "Rationality in Collective Action." *Philosophy of the Social Sciences* 36, no. 1 (March2006): 3–17.

Gilbert, Margaret. "The Structure of the Social Atom: Joint Commitment as the Foundation of Human Social Behaviour." In *Socializing Metaphysics – The Nature of Social Reality,* edited by Frederick F. Schmitt, 39–64. Lanham, MD: Rowman and Littlefield, 2003.

Gilbert, Margaret. *A Theory of Political Obligation.* Oxford: Oxford University Press, 2006.

Gilboa, Itzhak. *Rational Choice.* Cambridge, MA: MIT Press, 2010.

Gleick, Peter H. "The Human Right to Water." *Water Policy* 1 (1998): 487–503.

Goodin, Robert E. "Excused by the Unwillingness of Others?" *Analysis* 72, no. 1 (2012): 18–24. https://doi.org/10.1093/analys/anr128.

Goodin, Robert E. "Selling Environmental Indulgences." *Kyklos* 47 (1994): 573–596.

Griffiths, P. E., and R. D. Gray. "Developmental Systems and Evolutionary Explanation." *Journal of Philosophy* 91, no. 6 (June1994): 277–304.

Gyawali, Dipak. "Water and Conflict: Whose Ethics is to Prevail?" In *Water Ethics. Marcelino Botin Water Forum 2007,* edited by Manuel Ramón Llamas, Luis Martinez-Cortina, and Aditi Mukherji, 13–24. Boca Raton: CRC Press, 2009.

Hardin, Garrett. "The Tragedy of the Commons." *Science* 162, no. 3859 (1968): 1243–1248. https://doi.org/10.1126/science.162.3859.1243.

Hardisty, David J., Ben Orlove, David H. Krantz, Arthur A. Small, Kerry F. Milch, and Daniel E. Osgood. "About Time: An Integrative Approach to Effective Environmental Policy." *Global Environmental Change* 22 (2012): 684–694.

Hayward, Tim. *Constitutional Environmental Rights.* Oxford: Open University Press, 2005.

Hefny, Magdy A. "Water Management Ethics in the Framework of Environmental and General Ethics: The Case of Islamic Water Ethics." In *Water Ethics. Marcelino Botin Water Forum 2007,* edited by Manuel Ramón Llamas, Luis Martinez-Cortina, and Aditi Mukherji, 25–44. Boca Raton: CRC Press, 2009.

Held, Virginia. "Can a Random Collection of Individuals Be Morally Responsible?" *Journal of Philosophy* 67, no. 14 (1970): 471–481. https://doi.org/10.2307/2024108.

Heyward, Jennifer Clare, and Dominic Roser, eds. *Climate Justice in a Non-Ideal World.* Oxford: Oxford University Press, 2016.

Hiskes, Richard P. *The Human Right to a Green Future: Environmental Rights and Intergenerational Justice.* Cambridge: Cambridge University Press, 2008. https://doi.org/10.1017/CBO9780511575396.

Holmes, Stephen, and Cass R. Sunstein. *The Cost of Rights: Why Liberty Depends on Taxes.* New York: Norton, 1999.

Hourdequin, Marion. "Climate, Collective Action and Individual Ethical Obligation." *Environmental Values* 19, no. 4 (2010): 443–464.

Hurka, Thomas. "Liability and Just Cause." *Ethics & International Affairs* 21, no. 2 (2017): 199–218.

Hurka, Thomas. "Proportionality in the Morality of War." *Philosophy and Public Affairs* 33, no. 1 (2005): 34–66.

Hursthouse, Rosalind. "Environmental Virtue Ethics." In *Environmental Ethics,* edited by Rebecca L. Walker and P. J. Ivanhoe, 155–172. Oxford/New York: Oxford University Press, 2007.

Igneski, Violetta. "Distance, Determinacy and the Duty to Aid: A Reply to Kamm." *Law and Philosophy* 20, no. 6 (2001): 605–616.

IPCC. "5th Report," n.d. Accessed October 25, 2020. www.ipcc.ch/report/ar5/.

IPCC. "AR6 Synthesis Report," n.d. Accessed April 3, 2019. www.ipcc.ch/report/sixth-assessment-report-cycle/.

Isaacs, Tracy. "Collective Responsibility and Collective Obligation." *Midwest Studies in Philosophy* 38, no. 1 (September2014): 40–57. https://doi.org/10.1111/misp.12015.

Jamieson, Dale. "Adaptation, Mitigation, and Justice." In *Perspectives on Climate Change,* edited by Richard Howarth, 221–253. Amsterdam: Elsevier, 2005.

Jamieson, Dale. *Ethics and the Environment. An Introduction.* Cambridge: Cambridge University Press, 2008.

Jamieson, Dale. *Reason in a Dark Time. Why the Struggle Against Climate Change Failed – and What it Means for Our Future.* Oxford/New York: Oxford University Press, 2014.

Johnson, Baylor. "Ethical Obligations in a Tragedy of the Commons." *Environmental Values* 12, no. 3 (2003): 271–287.

Johnson, Baylor. "The Possibility of a Joint Communiqué: My Response to Hourdequin." *Environmental Values* 20, no. 2 (2011): 147–156.

Kallhoff, Angela. "Addressing the Commons: Normative Approaches to Common Pool Resources." In *Climate Change and Sustainable Development: Ethical Perspectives* on *Land Use and Food Production*, edited by Thomas Potthast and Simon Meisch, 63–68. Wageningen: Wageningen Academic Publishers, 2012.

Kallhoff, Angela. "Klimakooperation: Kollektives Handeln für ein Öffentliches Gut." In *Klimagerechtigkeit und Klimaethik*, edited by Angela Kallhoff, 143–167. Wiener Reihe. Themen Der Philosophie, vol. 18. Berlin/Boston: De Gruyter, 2015.

Kallhoff, Angela. "The Normative Limits of Consumer Citizenship." *Journal of Agricultural and Environmental Ethics* 29, no. 1 (2016): 23–34.

Kallhoff, Angela. "Plants in Ethics: Why Flourishing Deserves Moral Respect." *Environmental Values* 23, no. 6 (2014): 685–700.

Kallhoff, Angela. "Transcending Water Conflicts: An Ethics of Water Cooperation." In *Global Water Ethics. Towards a Global Ethics Charter*, edited by Rafael Ziegler and David Groenfeldt, 91–106. Earthscan Studies in Water Resource Management. London/New York: Routledge, 2017.

Kallhoff, Angela. "Water Ethics: Toward Ecological Cooperation." In *The Oxford Handbook of Environmental Ethics*, edited by Stephen M. Gardiner and Allen Thompson. Oxford/New York: Oxford University Press, 2017.

Kallhoff, Angela. "Water Justice: A Multilayer Term and its Role in Cooperation." *Analyse & Kritik* 36, no. 2 (2014): 367–382.

Kallhoff, Angela. *Why Democracy Needs Public Goods*. Lanham, MD: Lexington Books, 2011.

Kallhoff, Angela. "Why Societies Need Public Goods." *Critical Review of International Social and Political Philosophy* 17, no. 6 (2014): 635–651.

Kallhoff, Angela, and Michael Bruckner. "Biozentrismus." In *Handbuch Tierethik. Grundlagen-Kritik-Perspektiven*, edited by Johann S. Ach and Dagmar Borchers, 161–166. Stuttgart: Metzler, 2018.

Kallhoff, Angela, Marcello Di Paola, and Maria Schörgenhumer, eds. *Plant Ethics: Concepts and Applications*. Routledge Environmental Humanities. Abingdon/New York: Routledge, 2018.

Kamm, Frances M. "Famine Ethics: The Problem of Distance in Morality and Singer's Ethical Theory." In *Singer and His Critics*, edited by Dale Jamieson, 162–208. Philosophers and Their Critics, vol. 8. Oxford: Blackwell, 1999.

Kaul, Inge, Isabelle Grunberg, and Marc Stern, eds. *Global Public Goods: International Cooperation in the 21st Century*. New York/Oxford: Oxford University Press, 1999.

Kawall, Jason. "Reverence for Life as a Viable Environmental Virtue." *Environmental Ethics* 25, no. 4 (2003): 339–358.

Kortenkamp, Katherine V., and Colleen F. Moore. "Ecocentrism and Anthropocentrism: Moral Reasoning about Ecological Commons Dilemmas." *Journal of Environmental Psychology* 21, no. 3 (September2001): 261–272.

Kowarsch, Martin. "Beyond General Principles: Water Ethics in a Deweyan Perspective." In *Global Water Ethics. Towards a Global Ethics Charter*, edited by Rafael Ziegler and David Groenfeldt, 57–74. Earthscan Studies in Water Resource Management. London/New York: Routledge, 2017.

Kowarsch, Martin. "Ethical Targets and Questions of Water Management." In *Water Management Options in a Globalised World. Proceedings of an International Scientific Workshop (20–23 June 2011, Bad Schönbrunn)*, edited by Martin Kowarsch, 2nd rev. ed., 38–49. Institute for Social and Developmental Studies (IGP) at the Munich School of Philosophy, 2011. www.researchgate.net/publication/308118199_Water_management_options_in_a_globalised_world.

Krantz, David H. "Individual Values and Social Goals in Environmental Decision Making." In *Decision Modeling and Behavior in Complex and Uncertain Environments*, edited by T. Kugler, J. C. Smith, T. Connolly, and Y.-J. Son, 165–198. Dordrecht: Springer, 2008.

Krupp, Fred, Nathaniel Keohane, and Eric Pooley. "Less Than Zero. Can Carbon-Removal Technologies Curb Climate Change?" *Foreign Affairs* 98, no. 2 (April 2019): 142–152.

Kutz, Christopher. *Complicity: Ethics and Law for a Collective Age*. Cambridge Studies in Philosophy and Law. Cambridge: Cambridge University Press, 2000.

Lawford-Smith, Holly. "The Feasibility of Collectives' Actions." *Australasian Journal of Philosophy* 90, no. 3 (2012): 453–467. https://doi.org/10.1080/00048402.2011.594446.

Lawrence, Peter, Jeremy Meigh, and Caroline Sullivan. *The Water Poverty Index: An International Comparison*. Keele Economics Research Papers, vol. 19. Newcastle-under-Lyme, UK: Keele University, 2002.

Leist, Anton. "Klima auf Gegenseitigkeit." *Jahrbuch für Wissenschaft und Ethik* 16, no. 1 (2012): 159–178.

Leist, Anton. "Klimagerechtigkeit." *Information Philosophie* 5 (2011): 1–9.

Lichtenberg, Judith. *Distant Strangers: Ethics, Psychology, and Global Poverty*. Cambridge: Cambridge University Press, 2013.

List, Christian, and Philip Pettit. "Aggregating Sets of Judgments: An Impossibility Result." *Economics and Philosophy* 18, no. 1 (2002): 89–110.

List, Christian, and Philip Pettit . *Group Agency: The Possibility, Design, and Status of Corporate Agents*. Oxford/New York: Oxford University Press, 2011.

Llamas, Manuel Ramón, and Luis Martinez-Cortina. "Specific Aspects of Groundwater Use in Water Ethics." In *Water Ethics. Marcelino Botin Water Forum 2007*, edited by Manuel Ramón Llamas, Luis Martinez-Cortina, and Aditi Mukherji, 187–204. Boca Raton: CRC Press, 2009.

Markowitz, Ezra M., and Azim F. Shariff. "Climate Change and Moral Judgement." *Nature Climate Change* 2 (2012): 243–247. https://doi.org/10.1038/NCCLIMATE1378.

May, Larry. *The Morality of Groups: Collective Responsibility, Group-Based Harm, and Corporate Rights*. Soundings, vol. 1. Notre Dame, IN: University of Notre Dame Press, 1987.

May, Larry. *Sharing Responsibility*. Chicago: University of Chicago Press, 1996.

McGee, W. J. "Water as a Resource." *Annals of the American Academy of Political and Social Science* 33, no. 3 (May1909): 521–534.

McGinn, Colin. "Our Duties to Animals and the Poor." In *Singer and His Critics*, edited by Dale Jamieson, 150–161. Philosophers and Their Critics, vol. 8. Oxford/Malden, MA: Blackwell, 1999.

McMahan, Jeff. "The Basis of Moral Liability to Defensive Killing." *Philosophical Issues* 15, no. 1 (2005): 386–405.

Meadows, Donella H., and Club of Rome, eds. *The Limits to Growth: A Report for the Club of Rome's Project on the Predicament of Mankind*. New York: Universe Books, 1972.

Meyer, Lukas H., and Axel Gosseries, eds. *Intergenerational Justice*. Oxford/New York: Oxford University Press, 2009.

Miller, Seumas. *The Moral Foundations of Social Institutions: A Philosophical Study*. Cambridge/New York: Cambridge University Press, 2010.

Miller, Seumas. *Social Action*. Cambridge: Cambridge University Press, 2001.

Moellendorf, Darrel. *The Moral Challenge of Dangerous Climate Change: Values, Poverty, and Policy.* New York: Cambridge University Press, 2014.

Munthe, Christian. *The Price of Precaution and the Ethics of Risk.* Dordrecht: Springer, 2011.

Myers, Robert H. "Cooperating to Promote the Good." *Analyse & Kritik* 33, no. 1 (2011): 123–139.

Myles, Allen. "Liability for Climate Change. Will It Ever be Possible to Sue Anyone for Damaging the Climate?" *Nature* 421 (February 27, 2003): 891–892.

Naff, Thomas, and Joseph Dellapenna. "Can There be Confluence? A Comparative Consideration of Western and Islamic Fresh Water Law." *Water Policy* 4 (2002): 465–489.

Nitzan, Shmuel. *Collective Preference and Choice.* Cambridge: Cambridge University Press, 2010.

Nolt, John. "Future Generations in Environmental Ethics." In *The Oxford Handbook of Environmental Ethics,* edited by Stephen M. Gardiner and Allen Thompson, 344–354. New York: Oxford University Press, 2017.

Nolt, John. "How Harmful are the Average American's Greenhouse Gas Emissions?" *Ethics, Policy & Environment* 14, no. 1 (March 2011): 3–10. https://doi.org/10.1080/21550085.2011.561584.

Norton, Bryan G. "Sustainability as the Multigenerational Public Interest." In *The Oxford Handbook of Environmental Ethics,* edited by Stephen M. Gardiner and Allen Thompson, 355–366. New York: Oxford University Press, 2017.

Nussbaum, Martha C. "Aristotelian Social Democracy." In *Liberalism and the Good,* edited by R. B. Douglass, G. M. Mara, and H. S. Richardson, 203–252. New York: Routledge, 1990.

Nussbaum, Martha C. *Frontiers of Justice. Disability, Nationality, Species Membership.* Cambridge, MA: Harvard University Press, 2006.

Ochsenschlager, Edward L. *Iraq's Marsh Arabs in the Garden of Eden.* Philadelphia: University of Pennsylvania Museum of Archaeology and Anthropology, 2004.

Ostrom, Elinor. *Governing the Commons. The Evolution of Institutions for Collective Action.* Cambridge/New York: Cambridge University Press, 1990.

Ostrom, Elinor, and Roy Gardner. "Coping with Asymmetries in the Commons: Self-Governing Irrigation Systems Can Work." *Journal of Economic Perspectives* 7, no. 4 (1993): 93–112.

Oughton, Deborah. "Social and Ethical Issues in Environmental Risk Management." *Integrated Environmental Assessment and Management* 7, no. 3 (2011): 404–405.

Oughton, Deborah, Ingrid Bay-Larsen, and Gabriele Voigt. "Social, Ethical, Environmental and Economic Aspects of Remediation." *Radioactivity in the Environment* 14 (2009): 427–451.

Page, Edward. *Climate Change, Justice and Future Generations.* Cheltenham, UK/Northampton, MA: Edward Elgar, 2007.

Parfit, Derek. "Energy Policy and the Further Future. The Identity Problem." In *Climate Ethics. Essential Readings,* edited by Stephen M. Gardiner, Simon Caney, Dale Jamieson, and Henry Shue, 112–121. Oxford/New York: Oxford University Press, 2010.

Parfit, Derek. *Reasons and Persons.* Oxford: Clarendon Press, 1984.

Petersson, Björn. "Co-Responsibility and Causal Involvement." *Philosophia* 41, no. 3 (September 2013): 847–866. https://doi.org/10.1007/s11406-013-9413-x.

Pettit, Philip. "Collective Persons and Powers." *Legal Theory* 8 (2002): 443–470.

Pettit, Philip, and David Schweikard. "Joint Actions and Group Agents." *Philosophy of the Social Sciences* 36, no. 1 (March2006): 18–39. https://doi.org/10.1177/004839 3105284169.

Pogge, Thomas. *World Poverty and Human Rights. Cosmopolitan Responsibilities and Reforms.* Cambridge/Malden, MA: Polity Press, 2008.

Posner, Eric A., and Cass R. Sunstein. "Climate Change Justice." *Georgetown Law Journal* 96 (2008): 1565–1612.

Posner, Eric A., and David A. Weisbach. *Climate Change Justice.* Princeton, NJ: Princeton University Press, 2010.

Poteete, Amy R., and Marco A. Janssen. *Working Together: Collective Action, the Commons, and Multiple Methods in Practice.* Princeton, NJ: Princeton University Press, 2010.

Radkau, Joachim. *Natur und Macht. Eine Weltgeschichte der Umwelt.* München: Beck, 2002.

Radkau, Joachim. *Nature and Power. A World History of the Environment.* Cambridge: Cambridge University Press, 2008.

Raimondi, Francesca. "Joint Commitment and the Practice of Democracy." In *Group Process, Group Decision, Group Action,* edited by Robert S. Baron, Norbert L. Kerr, and Norman Miller, 285–299. Mapping Social Psychology Series. Pacific Grove, CA: Brooks/Cole, 1992.

Rawls, John. *Political Liberalism.* New York: Columbia University Press, 1996.

Rawls, John. *A Theory of Justice.* Cambridge, MA: Belknap Press, 2005.

Ripl, Wilhelm. "Water: The Bloodstream of the Biosphere." *Philosophical Transactions of The Royal Society of London, Biological Sciences* 358 (2003): 1921–1934.

Rockström, Johan, Will Steffen, Kevin Noone, Åsa Persson, F. StuartChapinIII, Eric Lambin, Timothy M. Lenton, et al. "Planetary Boundaries: Exploring the Safe Operating Space for Humanity." *Ecology and Society* 14, no. 2 (2009).

Rose, Carol M. "Energy and Efficiency in the Realignment of Common-Law Water Rights." *Journal of Legal Studies* 19, no. 2 (1990): 261–296.

Roser, Dominic. *Ethical Perspectives on Climate Policy and Climate Economics.* Zurich: University of Zurich, Faculty of Economics, Zurich Open Repository and Archive, 2010.

Sadoff, Claudia W., and David Grey. "Beyond the River: The Benefits of Cooperation on International Rivers." *Water Policy* 4 (2002): 389–403.

Sayer, Andrew. "Geography and Global Warming: Can Capitalism Be Greened?" *Area* 41, no. 3 (September2009): 350–353.

Scanlon, John, Angela Cassar, and Noémi Nemes. *Water as Human Right?*IUCN Environmental Policy and Law Paper, no. 51. Gland, Switzerland/Cambridge: IUCN – The World Conservation Union, 2004.

Schlosberg, David. *Defining Environmental Justice: Theories, Movements, and Nature.* Oxford: Oxford University Press, 2007. https://doi.org/10.1093/acprof:oso/ 9780199286294.001.0001.

Schmid, Hans Bernhard, Katinka Schulte-Ostermann, and Nikos Psarros. *Concepts of Sharedness: Essays on Collective Intentionality.* Philosophische Analyse, vol. 26. Frankfurt: Ontos Verlag, 2008.

Schwenkenbecher, Anne. "Joint Duties and Global Moral Obligations." *Ratio* 26, no. 3 (2013): 310–328. https://doi.org/10.1111/rati.12010.

Schwenkenbecher, Anne. "Joint Moral Duties." *Midwest Studies in Philosophy* 38, no. 1 (2014): 58–74. https://doi.org/10.1111/misp.12016.

Searle, John. *The Construction of Social Reality.* New York: The Free Press, 1995.

Searle, John. *Mind, Language and Society: Philosophy in the Real World.* London: Phoenix, 2000.

Seemann, Axel. "Why We Did It: An Anscombian Account of Collective Action." *International Journal of Philosophical Studies* 17, no. 5 (2009): 637–655.

Shelton, Dinah. "Human Rights, Environmental Rights, and the Right to Environment." *Stanford Journal of International Law* 28, no. 103 (1991): 103–138.

Shiva, Vandana. *Globalization's New Wars. Seed, Water & Life Forms.* New Delhi: Women Unlimited, 2005.

Shue, Henry. *Basic Rights. Subsistence, Affluence, and U.S. Foreign Policy.* 2nd ed. Princeton, NJ: Princeton University Press, 1996.

Shue, Henry. "Climate Hope: Implementing the Exit Strategy." *Chicago Journal of International Law* 13, no. 2 (2013): 381–402.

Shue, Henry. "Deadly Delays, Saving Opportunities. Creating a More Dangerous World?" In *Climate Ethics. Essential Readings*, edited by Stephen M. Gardiner, Simon Caney, Dale Jamieson, and Henry Shue, 146–162. Oxford/New York: Oxford University Press, 2010.

Shue, Henry. "Face Reality? After You! – A Call for Leadership on Climate Change." *Ethics & International Affairs* 25, no. 1 (March2011): 17–26. https://doi.org/10.1017/S0892679410000055.

Shue, Henry. "Global Environment and International Inequality." *International Affairs* 75, no. 3 (1999): 531–545.

Shue, Henry. "Subsistence Emissions and Luxury Emissions." *Law and Policy* 15, no. 1 (1993): 39–59.

Singer, Peter. "The Drowning Child and the Expanding Circle." *New Internationalist* (April1997).

Singer, Peter. *The Most Good You Can Do: How Effective Altruism is Changing Ideas about Living Ethically.* Castle Lectures in Ethics, Politics, and Economics. New Haven, CT/London: Yale University Press, 2015.

Singer, Peter. "One Atmosphere." In *One World. The Ethics of Globalization*, 14–50. The Terry Lectures. New Haven, CT: Yale University Press, 2002.

Singer, Peter. *One World: The Ethics of Globalization.* The Terry Lectures. New Haven, CT: Yale University Press, 2002.

Sinn, Hans-Werner. *Das grüne Paradoxon: Plädoyer für eine illusionsfreie Klimapolitik.* 2nd ed. Berlin: Econ Verlag, 2009.

Sinn, Hans-Werner. "Public Policies Against Global Warming." *International Tax and Public Finance* 15, no. 4 (2008): 360–394.

Sinnott-Armstrong, Walter. "It's Not My Fault: Global Warming and Individual Moral Obligations." *Advances in the Economics of Environmental Research* 5 (2005): 293–315.

Skillington, Tracey. *Climate Justice and Human Rights.* New York: Palgrave MacMillan, 2017.

Stern, Nicholas. *The Economics of Climate Change. The Stern Review.* Cambridge: Cambridge University Press, 2007.

Stiglitz, Joseph E. *Economics of the Public Sector.* 2nd ed. New York/London: W.W. Norton, 1988.

Stiglitz, Joseph E. *Whither Socialism?*Cambridge, MA: MIT Press, 1994.

Sudgen, Robert. "The Logic of Team Reasoning." *Philosophical Explorations: An International Journal for the Philosophy of Mind and Action* 6 (2003): 165–181.

Sudgen, Robert. "Thinking as a Team: Toward an Explanation of Nonselfish Behavior." *Social Philosophy and Policy* 10 (1993): 69–89.

Sunstein, Cass R. "Social Norms and Social Roles." *Columbia Law Review* 96 (1996): 903–968.

Taylor, Paul. *Respect for Nature: A Theory of Environmental Ethics.* Princeton, NJ: Princeton University Press, 1986.

Thalos, Mariam. "Precaution Has Its Reasons." In *Topics in Contemporary Philosophy 9: The Environment,* edited by W. Kabasenche, M. O'Rourke, and M. Slater, 171–184. Cambridge, MA: MIT Press, 2012.

The Potsdam Institute of Climate Impact Research, n.d. www.pik-potsdam.de/news/in-short/unep-emissions-gap-report-published.

Thompson, Allen, and Jeremy Bendik-Keymer. *Ethical Adaptation to Climate Change: Human Virtues of the Future.* Cambridge, MA: MIT Press, 2012.

Töpfer, Klaus. "Nachhaltigkeit im Anthropozän." *Nova Acta Leopoldina* 117, no. 398 (2013): 31–40.

Tuomela, Raimo. "Group Reasons." *Philosophical Issues* 22, no. 1 (2012): 402–418.

Tuomela, Raimo. *The Philosophy of Social Practices: A Collective Acceptance View.* Cambridge: Cambridge University Press, 2010.

Tuomela, Raimo. *The Philosophy of Sociality: The Shared Point of View.* Oxford/New York: Oxford University Press, 2007.

Tuomela, Raimo. *Social Ontology: Collective Intentionality and Group Agents.* New York: Oxford University Press, 2013.

Tuomela, Raimo. "The We-Mode and the I-Mode." In *Socializing Metaphysics – The Nature of Social Reality,* edited by F. Schmitt, 93–127. Lanham, MD: Rowman and Littlefield, 2003.

United Nations. "The Paris Agreement." Accessed October 4, 2018. https://unfccc.int/sites/default/files/english_paris_agreement.pdf.

United Nations. "United Nations Framework Convention on Climate Change," 1992. https://unfccc.int/resource/docs/convkp/conveng.pdf.

United Nations. "Universal Declaration of Human Rights," 1948. www.un.org/en/universal-declaration-human-rights/.

Valentini, Laura. "Ideal vs. Non-Ideal Theory: A Conceptual Map." *Philosophy Compass* 7, no. 9 (September2012): 654–664. https://doi.org/10.1111/j.1747-9991.2012.00500.x.

Valentini, Laura. "The Natural Duty of Justice in Non-Ideal Circumstances: On the Moral Demands of Institution Building and Reform." *European Journal of Political Theory,* November 28, 2017. https://doi.org/10.1177/1474885117742094.

van der Molen, Irma, and Antoinette Hildering. "Water: Cause for Conflict or Co-Operation?" *Journal on Science and World Affairs* 1, no. 2 (2005): 133–143.

van der Werf, Edwin, and C. Di Maria. "Imperfect Environmental Policy and Polluting Emissions: The Green Paradox and Beyond." *International Review of Environmental and Resource Economics* 6, no. 2 (March 30, 2012): 153–194. https://doi.org/10.1561/101.00000050.

Vanderheiden, Steven. *Atmospheric Justice: A Political Theory of Climate Change.* Oxford: Oxford University Press, 2008.

Wagner, Gernot, and Martin L. Weitzman. *Climate Shock. The Economic Consequences of a Hotter Planet.* Princeton, NJ/Oxford: Princeton University Press, 2015.

Walker, Gordon P. *Environmental Justice. Concepts, Evidence, and Politics.* London/New York: Routledge, 2012.

Walker, Rebecca L. "The Good Life for Non-Human Animals: What Virtue Requires of Humans." In *Working Virtue: Virtue Ethics and Contemporary Moral Problems*, edited by Rebecca L. Walker and Philip J. Ivanhoe. New York: Oxford University Press, 2007.

Walzer, Michael. *Arguing about War*. New Haven, CT/London: Yale University Press, 2004.

Weber, Elke U. "Experience-Based and Description-Based Perceptions of Long-Term Risk: Why Global Warming Does Not Scare Us (Yet)." *Climatic Change* 77 (2006): 103–120.

Weber, Elke U., and Paul C. Stern. "Public Understanding of Climate Change in the United States." *American Psychologist* 66, no. 4 (2011): 315–328. https://doi.org/10.1037/a0023253.

Wensveen, Louke van. *Dirty Virtues. The Emergence of Ecological Virtue Ethics*. Amherst, NY: Humanity Books, 2000.

Williams, Bernard, ed. "Must a Concern for the Environment Be Centred on Human Beings?" In *Making Sense of Humanity: And Other Philosophical Papers 1982–1993*, 233–240. Cambridge, UK: Cambridge University Press, 1995.

World Commission on Environment and Development, ed. *Our Common Future*. Oxford/New York: Oxford University Press, 1987.

Wringe, Bill. "Collective Obligations: Their Existence, Their Explanatory Power, and Their Supervenience on the Obligations of Individuals." *European Journal of Philosophy* 24, no. 2 (2016): 472–497. https://doi.org/10.1111/ejop.12076.

Wringe, Bill. "From Global Collective Obligations to Institutional Obligations." *Midwest Studies in Philosophy* 38, no. 1 (2014): 171–186. https://doi.org/10.1111/misp.12022.

Wringe, Bill. "Global Obligations and the Agency Objection." *Ratio* 23, no. 2 (2010): 217–231. https://doi.org/10.1111/j.1467-9329.2010.00462.x.

www.un.org/Waterforlifedecade/Human_right_to_water.shtml, n.d. Accessed March 14, 2019.

Ziegler, Rafael, and David Groenfeldt, eds. *Global Water Ethics: Towards a Global Ethics Charter*. Earthscan Studies in Water Resource Management. Abingdon/New York: Routledge, 2017.

Index